Solid-State Microwave Amplifier Design

Solid-State Microwave Amplifier Design

Tri T. Ha
GTE International Systems Corporation

A WILEY-INTERSCIENCE PUBLICATION
JOHN WILEY & SONS, NEW YORK • CHICHESTER • BRISBANE • TORONTO

Library of Congress Cataloging in Publication Data:

Ha, Tri T 1949–
Solid-state microwave amplifier design.

"A Wiley-Interscience publication."
Includes bibliographies and index.
1. Microwave amplifiers—Design and construction.
2. Solid state electronics. I. Title.

TK7871.2.H3 1981 621.381'325 81-21
ISBN 0-471-08971-0

Printed in the United States of America

10 9 8 7 6 5 4 3 2 1

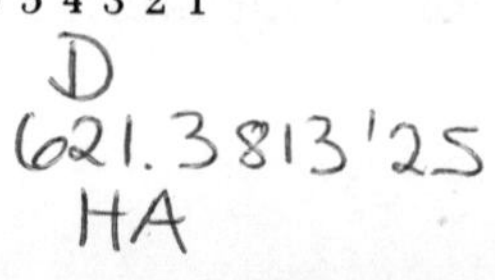

To my family and Hien

Preface

This book has grown out of the author's research in the field of solid-state microwave amplifiers. It is written primarily as a text in network theory for senior and first-year graduate students as well as a working reference for practicing engineers dealing with solid-state microwave amplifiers. To aid students in learning the material and applying it to practical design, a collection of about a hundred homework problems is included. To aid practicing engineers to quickly apply the material in the text to the design of amplifiers, numerous examples of all the aspects of design that are encountered are included for illustration.

To a large extent the text is self-contained. For the basic material, the results are derived in a clear and precise manner; for the less important material the reader is referred to appropriate, easily accessible references. In order to understand the text, the reader is assumed to know basic network analysis and synthesis and to have had an introduction to matrix and complex variable theory.

Chapter 1 begins with a review of various matrix characterizations of networks and some of their basic properties that form the foundation of broadband matching theory. Important network matrices such as the impedance, admittance, chain, scattering, and chain scattering matrices are reviewed. The scattering matrix is the most important network characterization and is discussed in the fullest detail in Chapter 2. The chain matrix is very convenient in microwave amplifier analysis on a digital computer; hence it is also discussed in detail.

The discussion in Chapter 2 focuses on the generalized scattering matrix normalized to complex impedances. The scattering matrix normalized to real impedances used in practical measurements of microwave devices is a special case of this. The generalized scattering matrix is indispensable in the derivation of various forms of power gain for two-port amplifiers, and forms the foundation of broadband matching theory. In Chapter 2 an analysis of the stability of two-port networks is also discussed; the results are concisely presented in Appendices A1 and A2. To prepare the reader for the material on the design of low noise amplifiers and receivers in Chapters 4, 5, and 6, I elect to discuss various aspects of noise in two-ports such as the noise figure, noise temperature, and design formula in terms of the minimum noise figure and the optimum noise source reflection coefficient. The noise figure of a cascaded chain of two-ports is also derived for later use.

In Chapter 3 an introduction to various active devices of microwave integrated circuits such as the GaAs FET, the silicon bipolar transistor, and the IMPATT diode is presented. Also discussed are the bias circuits of the transistor, including their configuration and operation. For passive components, a brief discussion of the directional coupler, circulator, and isolator, which are used widely in balanced and single-ended amplifiers, is presented. Formulas for lumped capacitors and inductors at microwave frequencies are also given, together with those of microstrip lines. The design tables for commonly used microstrip substrates based upon the formulas in this chapter are included in Appendix A3.

Chapter 4 deals explicitly with the design of narrowband transistor amplifiers and negative resistance reflection amplifiers with less than 10% fractional bandwidth. The design of transistor amplifiers covers three types: maximum gain, low noise, and power amplification, including the use of potentially unstable transistors. The concepts of constant power gain, constant available power gain, and constant noise figure are introduced and their design formulas are derived in detail. For negative resistance reflection amplifiers, the transducer power gain is derived and applied to the design of small- and large-signal amplifiers. The purpose of this chapter is to present simple techniques for designing narrowband amplifiers without using a digital computer.

For wider bandwidth amplifiers, these techniques usually lead to unacceptable results; hence broadband matching techniques must be employed. This is the main purpose of Chapter 5. Here a general broadband matching theory for the lumped models of unconditionally stable transistors is presented. These models are verified by data of various microwave transistors that are currently used in industry. The theory is applied to the design of transistor amplifiers that can compensate the inherent gain roll-off of the transistor. The distributed approximation of lumped matching networks is also discussed, along with the Kuroda transformation for ease of real circuit fabrication. A general theory for the design of transistor amplifiers directly from the transistor's real frequency scattering parameters using semi-infinite slope lines is also presented. Since it is impossible to design a broadband amplifier for optimal performance without optimization of the initial design, I elect to discuss, to a certain extent, various optimization techniques that are used widely in microwave network optimization programs. As in the narrowband case, the design of broadband microwave amplifiers using potentially unstable transistors is presented. The chapter ends with a discussion of broadband negative resistance reflection amplifiers.

Chapter 6 deals with two main topics, namely, the characterization of signal distortion in microwave systems and the techniques of power combining at microwave frequencies. For memoryless nonlinear systems, weak distortion is analyzed by the power series expansion method. When the nonlinear system possesses memory, the distortion characterization is analyzed by the Volterra functional series. Various aspects of distortion representation such as the intercept point, 1 dB gain compression point, cross modulation, AM-to-PM conversion, and dynamic range are discussed. Methods for computing Volterra transfer functions of various orders using multifrequency inputs are also presented. These techniques are applied to compute the intermodulation distortion products for both GaAs FET and IMPATT diode amplifiers. Also discussed in detail in this chapter are microwave power combining techniques using directional coupler and split-T combiner/divider as a building block for an n-way power combining/dividing structure that is used in an n-way amplifier whose output power is n times that of the individual amplifiers in the system. The result is then applied to analyze the performance of the balanced amplifier, which is a special two-way amplifier that can provide good input and output match. Graceful degradation of n-way amplifiers is also discussed and illustrated by numerical examples. Truly graceful degradation of variable n-way amplifiers using variable combiner/dividers is presented in detail with comparison to fixed n-way amplifiers. The chapter ends with an analysis of low noise circulator-coupled amplifiers that can provide a "fail-soft" operation for low-cost receivers.

The reader may find the material in Appendices A1–A8 useful in aiding comprehension of the text.

In summary, the material in the text is selected to focus on the fundamental principles that can be applied with ease to the design of microwave amplifiers and related solid-state circuits. A typical one-semester course would cover Chapters 1, 2, 4, and 5, Sections 6.1 to 6.4 and 6.10 in detail, with Chapter 3 used as supplementary reading material. The rest of the material in Chapter 6 can be used for advanced seminars.

Standard notation is used throughout the text. A boldface letter denotes a matrix or vector; $\mathbf{0}_n$ and $\mathbf{1}_n$ represent the $n \times n$ zero and identity matrices, respectively. The superscript T denotes matrix transposition and the superior asterisk, *, the complex conjugate. The symbols $\geqslant$ and $\leqslant$ are also used for positive and negative semidefinite matrices, respectively. The real part, the imaginary part, the magnitude, and the phase of a complex number Z will be denoted as Re Z, Im Z, |Z| and ∠Z, respectively.

TRI T. HA

Waltham, Massachusetts
March 1981

Contents

Solid-State Microwave Amplifier Design

Introduction

Over the last ten years, microwave semiconductor device technology has progressed very rapidly. Gallium arsenide field-effect transistors (GaAs FET), silicon bipolar transistors, IMPATT, Gunn, and tunnel diodes are now widely used in solid-state amplifiers of communication systems. From uhf to 2 GHz, silicon bipolar transistors are dominant in performance in low noise, high gain, and high power amplification. The current revolution in the amplifier area started with the GaAs FET, which is capable of providing low noise, high gain, and power amplification from 2 to 20 GHz. Octave bandwidth GaAs FET amplifiers within this frequency range have been built, and research in the millimeter wave region will probably provide FET's with performance up to 40 GHz. The IMPATT diodes provide higher power amplification, up to 100 GHz, especially in pulse power amplifiers and oscillators.

To see how solid-state amplifiers are used in communication systems, consider a typical rf part of an earth terminal shown in Fig. I. The low noise amplifier receives the signal from the satellite within the 3.7–4.2 GHz band; typically it is a GaAs FET amplifier with 1.5 dB noise figure at 20°C and 50 dB gain. The input signal is first down-converted to 1042 MHz with 36 MHz bandwidth and is amplified by the first IF amplifier, which is a bipolar transistor amplifier. This intermediate signal is then down-converted to a signal at 70 MHz with 36 MHz bandwidth and is amplified again by a bipolar transistor amplifier. These two IF amplifiers usually have a low noise figure so that the overall noise figure of the down-converter can be kept within an acceptable range. In the transmitter, the first IF amplifier amplifies the 70 MHz signal with 36 MHz bandwidth before it is up-converted to a signal within the 36 MHz bandwidth centered at 1182 MHz. As in the receiver, the two IF amplifiers are bipolar transistor amplifiers. The 1182 MHz signal is then up-converted to a signal within the 5.9–6.4 GHz band and is highly amplified by the power amplifier to transmit to the satellite. For a light route earth station that handles up to five SCPC signals (single channel per carrier), the power amplifier can be made of power GaAs FETs and can provide a signal level up to 10 W or 40 dBm.

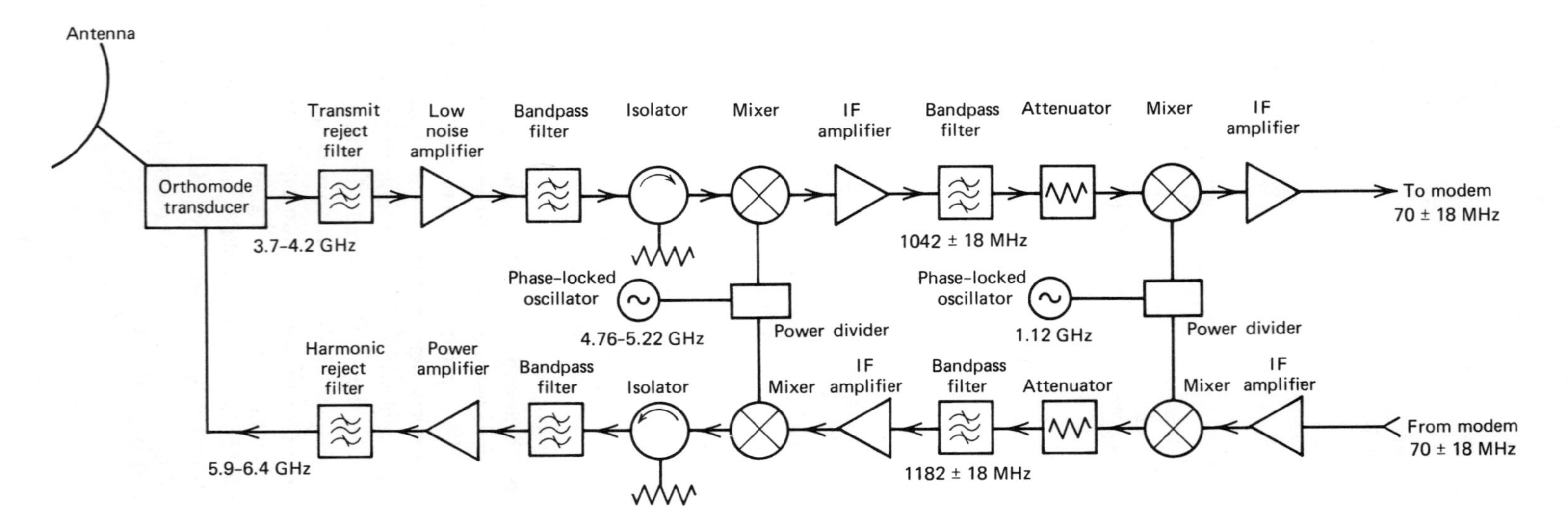

Fig. I. Block diagram of a C-band earth terminal.

Since solid-state microwave amplifiers play an important role in communication and electronic countermeasure systems, it is necessary to develop a sound design theory for them based on available solid-state devices such as the GaAs FET, the bipolar transistor, and the IMPATT diode. This is the purpose of the text.

In order to cover the general aspects of solid-state amplifier design, both narrowband and broadband matching techniques are discussed. The narrowband matching technique deserves its own detailed treatment since it can be used to design amplifiers with less than 10% fractional bandwidth with reasonable time and effort without a computer. The broadband matching technique is used to design amplifiers with an octave bandwidth or more. To date there is no unique technique for broadband matching solid-state devices; each has its own advantage and disadvantage, which can greatly influence the time, effort, and cost of an amplifier design. Some techniques require a designer with a broad background in synthesis or computer optimization; some can be used by nonspecialists. This text presents effective broadband matching techniques that can be used by specialists or nonspecialists with the least time, cost, and effort.

Because of different applications of solid-state amplifiers in telecommunications, different categories of design will be treated, including low noise, high gain or high power amplification. Since the cost per decibel of microwave gain or noise figure is very high, the designer cannot afford to sacrifice rf performance by improper dc bias circuit design. Microwave transistors have extremely high gain at low frequencies, and the matching and bias circuits have a tendency to present a very high or a very low impedance at the out-of-band frequency region. This can make the amplifiers fall into the unstable region at low frequencies and cause destructive oscillations; consequently, techniques for proper biasing and stabilization are also discussed. An analysis of unconditional stability and region of stability in the case of potentially unstable devices is given in detail and forms the cornerstone of the amplifier design. Microwave amplifiers are not perfectly linear, and their nonlinearity generates some undesirable false responses; hence important measures of amplifier distortions such as cross modulation and intermodulation are also presented, together with other aspects such as AM-to-PM conversion, intercept point, dynamic range, and so forth.

Many applications require microwave solid-state power amplifiers with an output power level that exceeds the capability of a single semiconductor device. Therefore techniques using n-way combiners and dividers to combine the output power of several devices or amplifiers will also be discussed. The failure of one amplifier or device in a combined system might not result in a complete shutdown of the overall amplifier, which makes n-way amplifier systems more attractive than a single amplifier.

In low-cost earth stations or "receive only" terminals, where the degradation of the noise figure to a certain margin is acceptable for economic reasons, circulator-coupled amplifiers can provide the necessary fail-soft operation at much lower cost than redundant amplifiers, while providing a typical 2 to 3 dB degradation in noise figure at the C-band.

While solid-state microwave amplifier design is a dynamic and rapidly growing field, its principles are well formulated. Many aspects of amplifier design stem from tube amplifiers. My purpose is to refine the techniques so that they can be applied to the new solid-state devices and at the same time to use these techniques in conjunction with computerized optimization schemes to yield a satisfactory amplifier design.

1

Matrix Characterization of Networks

An electrical network is an interconnection of electrical circuit elements according to some scheme. Associated with the network are terminals. When a voltage is applied across two terminals so that the current entering one terminal is equal to the current leaving the other terminal, the pair of terminals is called a port. If a network possesses n number of ports, it is called an n-port network or an n-port, as shown schematically in Fig. 1.1, where the voltage across port j and the current entering port j as a function of time t are denoted as $v_j(t)$ and $i_j(t)$, respectively. Since all the networks with which we are dealing in the text are time-invariant we will use the Laplace transforms of the voltage and the current, that is,

$$V_j(s) = \mathcal{L}\{v_j(t)\} \qquad \text{and} \qquad I_j(s) = \mathcal{L}\{i_j(t)\}$$

in the n-port characterization.

1.1 THE IMPEDANCE, ADMITTANCE, AND CHAIN MATRICES

For a linear time-invariant n-port network, the impedance matrix is an $n \times n$ matrix $\mathbf{Z}(s) = [Z_{ij}(s)]$ that maps the $n \times 1$ current vector $\mathbf{I}(s)$ into the $n \times 1$ voltage vector $\mathbf{V}(s)$ through

$$\mathbf{V}(s) = \mathbf{Z}(s)\mathbf{I}(s) \tag{1.1}$$

The impedance matrix $\mathbf{Z}(s)$ can only be defined if it is possible to connect arbitrary current sources at each of the ports and obtain well-defined corresponding voltage responses. Conversely, if it is possible to obtain a well-defined set of current responses for an arbitrary selection of port voltages there is an $n \times n$ admittance matrix $\mathbf{Y}(s)$ relating $\mathbf{V}(s)$ and $\mathbf{I}(s)$ as follows:

$$\mathbf{I}(s) = \mathbf{Y}(s)\mathbf{V}(s) \tag{1.2}$$

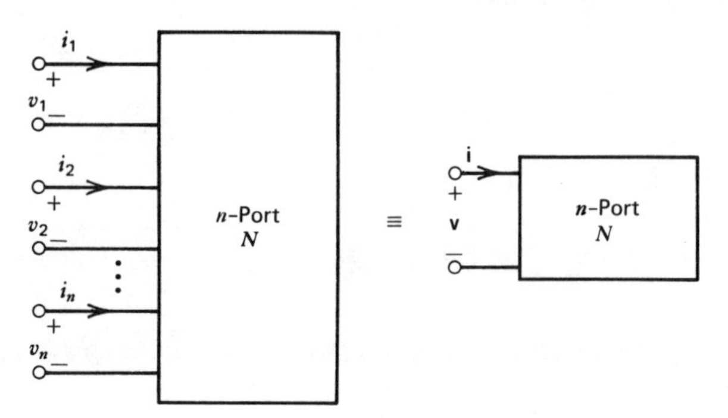

Fig. 1.1. Representation of an n-port network.

If an n-port network possesses both $\mathbf{Z}(s)$ and $\mathbf{Y}(s)$, then from (1.1) and (1.2) it is clear that

$$\mathbf{Z}(s)=\mathbf{Y}^{-1}(s) \tag{1.3}$$

Example 1.1

Consider the small-signal model of a microwave transistor in Fig. 1.2, its 2×2 admittance matrix $\mathbf{Y}(s)=[Y_{ij}(s)]$, $i=1,2$; $j=1,2$ can be computed as

$$Y_{11}(s)=\left.\frac{I_1(s)}{V_1(s)}\right|_{V_2(s)=0}=\frac{sC_i}{1+sC_i(R_i+sL_i)}$$

$$Y_{22}(s)=\left.\frac{I_2(s)}{V_2(s)}\right|_{V_1(s)=0}=\frac{R_o^{-1}+sC_o}{1+sL_o(R_o^{-1}+sC_o)}$$

$$Y_{21}(s)=\left.\frac{I_2(s)}{V_1(s)}\right|_{V_2(s)=0}=\frac{g_m}{[1+sL_o(R_o^{-1}+sC_o)][1+sC_i(R_i+sL_i)]}$$

$$Y_{12}(s)=\left.\frac{I_1(s)}{V_2(s)}\right|_{V_1(s)=0}=0$$

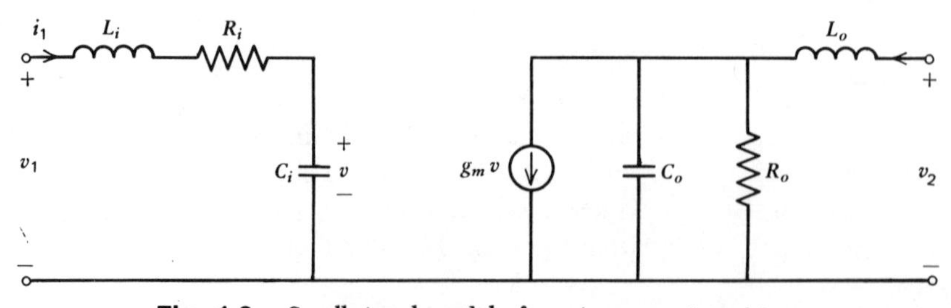

Fig. 1.2. Small-signal model of a microwave transistor.

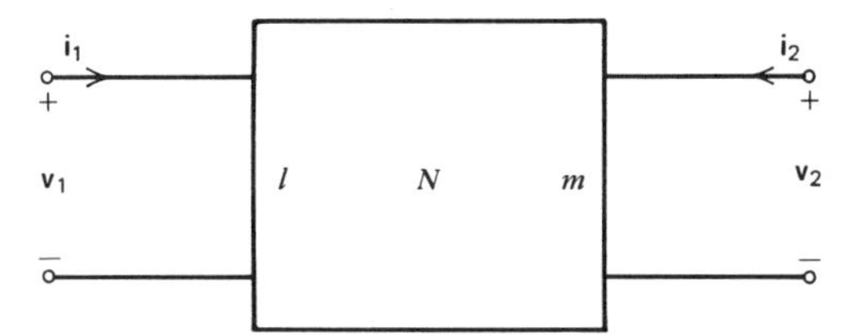

Fig. 1.3. The partitioned $(l+m)$-port network.

Therefore, $\mathbf{Y}(s)=[Y_{ij}(s)]$ is

$$\mathbf{Y}(s)=\begin{bmatrix} \dfrac{sC_i}{1+sC_i(R_i+sL_i)} & 0 \\ \dfrac{g_m}{\left[1+sL_o\left(R_o^{-1}+sC_o\right)\right]\left[1+sC_i(R_i+sL_i)\right]} & \dfrac{R_o^{-1}+sC_o}{1+sL_o\left(R_o^{-1}+sC_o\right)} \end{bmatrix}$$

When the transfer property of a network is of interest, such as in a cascade connection, the chain matrix $\mathcal{C}(s)=[\mathcal{C}_{ij}(s)]$ is often considered. In most cases consider the $(l+m)$-port network in Fig. 1.3 where l and m are the number of input ports and output ports, respectively. Then the chain matrix $\mathcal{C}(s)$ is given as

$$\begin{bmatrix} \mathbf{V}_1(s) \\ \mathbf{I}_1(s) \end{bmatrix}=\begin{bmatrix} \mathcal{C}_{11}(s) & \mathcal{C}_{12}(s) \\ \mathcal{C}_{21}(s) & \mathcal{C}_{22}(s) \end{bmatrix}\begin{bmatrix} \mathbf{V}_2(s) \\ -\mathbf{I}_2(s) \end{bmatrix} \tag{1.4a}$$

$$=\mathcal{C}(s)\begin{bmatrix} \mathbf{V}_2(s) \\ -\mathbf{I}_2(s) \end{bmatrix} \tag{1.4b}$$

where $\mathcal{C}_{11}$, $\mathcal{C}_{21}$, $\mathcal{C}_{12}$, and $\mathcal{C}_{22}$ are the $l\times m$ matrices. The chain matrix is often referred to as the ABCD matrix in the literature. As a rule, we will drop the complex variable $s=\sigma+j\omega$ in the matrix or vector notations since we are working explicitly with the Laplace transform.

Example 1.2

Consider the network in Fig. 1.2. The chain matrix is given as

$$\mathcal{C}=\begin{bmatrix} -g_m^{-1}\left(R_o^{-1}+sC_o\right)\left[1+sC_i(R_i+sL_i)\right] & -g_m^{-1}\left[1+L_o\left(R_o^{-1}+sC_o\right)\right]\left[1+sC_i(R_i+sL_i)\right] \\ -sC_ig_m^{-1}\left(R_o^{-1}+sC_o\right) & -sC_ig_m^{-1}\left[1+sL_o\left(R_o^{-1}+sC_o\right)\right] \end{bmatrix}$$

Using (1.1), (1.2), and (1.4) the relations between $\mathcal{C}$, **Z**, and **Y** can be derived in the case $l=m=n/2$

$$\mathbf{Z}=\begin{bmatrix}\mathbf{Z}_{11} & \mathbf{Z}_{12}\\ \mathbf{Z}_{21} & \mathbf{Z}_{22}\end{bmatrix}=\begin{bmatrix}\mathcal{C}_{11}\mathcal{C}_{21}^{-1} & \mathcal{C}_{11}\mathcal{C}_{21}^{-1}\mathcal{C}_{22}-\mathcal{C}_{12}\\ \mathcal{C}_{21}^{-1} & \mathcal{C}_{21}^{-1}\mathcal{C}_{22}\end{bmatrix} \tag{1.5a}$$

$$\mathbf{Y}=\begin{bmatrix}\mathbf{Y}_{11} & \mathbf{Y}_{12}\\ \mathbf{Y}_{21} & \mathbf{Y}_{22}\end{bmatrix}=\begin{bmatrix}\mathcal{C}_{22}\mathcal{C}_{12}^{-1} & \mathcal{C}_{21}-\mathcal{C}_{22}\mathcal{C}_{12}^{-1}\mathcal{C}_{11}\\ -\mathcal{C}_{12}^{-1} & \mathcal{C}_{12}^{-1}\mathcal{C}_{11}\end{bmatrix} \tag{1.5b}$$

$$\mathcal{C}=\begin{bmatrix}\mathbf{Z}_{11}\mathbf{Z}_{21}^{-1} & \mathbf{Z}_{11}\mathbf{Z}_{21}^{-1}\mathbf{Z}_{22}-\mathbf{Z}_{12}\\ \mathbf{Z}_{21}^{-1} & \mathbf{Z}_{21}^{-1}\mathbf{Z}_{22}\end{bmatrix}=\begin{bmatrix}-\mathbf{Y}_{21}^{-1}\mathbf{Y}_{22} & -\mathbf{Y}_{21}^{-1}\\ \mathbf{Y}_{12}-\mathbf{Y}_{11}\mathbf{Y}_{21}^{-1}\mathbf{Y}_{22} & -\mathbf{Y}_{11}\mathbf{Y}_{21}^{-1}\end{bmatrix} \tag{1.5c}$$

where $\mathbf{Z}_{ij}$ and $\mathbf{Y}_{ij}$ $(i=1,2;\ j=1,2)$ are $\frac{1}{2}n\times\frac{1}{2}n$ matrices.

1.2 THE SCATTERING AND CHAIN SCATTERING MATRICES

A very important description of networks, especially in the microwave frequency range, is the scattering matrix $\mathbf{S}(s)$. Consider the n-port network N in Fig. 1.1, and define the incident voltage wave $\mathbf{V}^i$ and the reflected voltage wave $\mathbf{V}^r$ as

$$\mathbf{V}^i=\frac{\mathbf{V}+\mathbf{I}}{2} \qquad \mathbf{V}^r=\frac{\mathbf{V}-\mathbf{I}}{2} \tag{1.6a}$$

Then the $n\times n$ scattering matrix **S** is given by the relation

$$\mathbf{V}^r=\mathbf{S}\mathbf{V}^i \tag{1.6b}$$

S is called the scattering matrix normalized to unit resistors at each of the ports of N. A more detailed study of **S** is given in Chapter 2. Using (1.1), (1.2), and (1.6) a relationship between **S**, **Z**, and **Y** can be established as

$$\mathbf{S}=(\mathbf{Z}+\mathbf{1}_n)^{-1}(\mathbf{Z}-\mathbf{1}_n) \tag{1.7a}$$

$$\mathbf{S}=(\mathbf{1}_n+\mathbf{Y})^{-1}(\mathbf{1}_n-\mathbf{Y}) \tag{1.7b}$$

$$\mathbf{Z}=(\mathbf{1}_n+\mathbf{S})(\mathbf{1}_n-\mathbf{S})^{-1} \tag{1.7c}$$

$$\mathbf{Y}=(\mathbf{1}_n-\mathbf{S})(\mathbf{1}_n+\mathbf{S})^{-1} \tag{1.7d}$$

In many cases where cascade connection is of interest, the chain scattering matrix is probably an appropriate description if the excitations and responses are taken as incident and reflected voltages. Consider the $(l+m)$-port network N in Fig. 1.3 and assume that each port is normalized to unit resistor; then the chain scattering matrix $\mathbf{\Phi}$ is defined as

$$\begin{bmatrix} \mathbf{V}_1^i \\ \mathbf{V}_1^r \end{bmatrix} = \begin{bmatrix} \mathbf{\Phi}_{11} & \mathbf{\Phi}_{12} \\ \mathbf{\Phi}_{21} & \mathbf{\Phi}_{22} \end{bmatrix} \begin{bmatrix} \mathbf{V}_2^r \\ \mathbf{V}_2^i \end{bmatrix} = \mathbf{\Phi} \begin{bmatrix} \mathbf{V}_2^r \\ \mathbf{V}_2^i \end{bmatrix} \tag{1.8}$$

where $\mathbf{\Phi}_{ij}$ $(i=1,2;\ j=1,2)$ are $l \times m$ matrices, and $\mathbf{V}^i$ and $\mathbf{V}^r$ are defined in (1.6a).

In the case $l=m$, a relationship between $\mathbf{\Phi}$ and $\mathbf{S}$ can be established as

$$\mathbf{\Phi} = \begin{bmatrix} \mathbf{S}_{21}^{-1} & -\mathbf{S}_{21}^{-1}\mathbf{S}_{22} \\ \mathbf{S}_{11}\mathbf{S}_{21}^{-1} & \mathbf{S}_{12} - \mathbf{S}_{11}\mathbf{S}_{21}^{-1}\mathbf{S}_{22} \end{bmatrix} \tag{1.9}$$

$$\mathbf{S} = \begin{bmatrix} \mathbf{\Phi}_{21}\mathbf{\Phi}_{11}^{-1} & \mathbf{\Phi}_{22} - \mathbf{\Phi}_{21}\mathbf{\Phi}_{11}^{-1}\mathbf{\Phi}_{12} \\ \mathbf{\Phi}_{11}^{-1} & -\mathbf{\Phi}_{11}^{-1}\mathbf{\Phi}_{12} \end{bmatrix} \tag{1.10}$$

1.3 PASSIVITY AND LOSSLESSNESS

The concept of passivity and losslessness is very important in network synthesis. It can be best defined in terms of energy. For an n-port network N to be passive it is necessary that

$$\mathcal{E}(t_0) = \int_{-\infty}^{t_0} \mathbf{v}^T(t)\mathbf{i}(t)\,dt \geqslant 0 \tag{1.11}$$

for every finite time t_0. Relation (1.11) implies that for a passive network the input energy is always nonnegative. Furthermore, if

$$\int_{-\infty}^{\infty} \mathbf{v}^T(t)\mathbf{v}(t)\,dt < \infty \qquad \text{and} \qquad \int_{-\infty}^{\infty} \mathbf{i}^T(t)\mathbf{i}(t)\,dt < \infty$$

then N is said to be lossless if and only if it is passive and

$$\mathcal{E}(\infty) = \int_{-\infty}^{\infty} \mathbf{v}^T(t)\mathbf{i}(t)\,dt = 0 \tag{1.12}$$

If N is not passive it is called active. In the frequency domain, the passivity condition in (1.11) can be expressed as follows [1, Chapter 4]:

$$\mathrm{Re}\left[\mathbf{V}^{T*}(s)\mathbf{I}(s)\right]\geqslant 0 \tag{1.13}$$

for all s in $\mathrm{Re}\, s>0$. Losslessness includes (1.13) and the following equality

$$\int_{-\infty}^{\infty}\mathrm{Re}\left[\mathbf{V}^{T*}(j\omega)\mathbf{I}(j\omega)\right]d\omega=0 \tag{1.14}$$

for almost all real ω.

1.4 BOUNDED-REAL, *J*-CONTRACTIVE REAL, AND POSITIVE-REAL CONCEPTS

This section concerns various properties of an $n\times n$ real matrix of the complex variable s and their relationships to various descriptions of passive networks.

Definition 1.1

An $n\times n$ matrix $\mathbf{S}(s)$ is termed bounded-real if and only if [1, Chapter 4]:

(a) $\mathbf{S}(s)$ is analytic in $\mathrm{Re}\, s>0$.
(b) $\mathbf{S}(s)$ is real for real positive s.
(c) $\mathbf{1}_n-\mathbf{S}^{T*}(s)\mathbf{S}(s)\geqslant\mathbf{0}$ in $\mathrm{Re}\, s>0$.

Furthermore if $\mathbf{S}(s)$ is rational in s, it is a rational bounded real matrix (**BR**). The following theorem is useful for testing **BR** property.

BR Theorem

An $n\times n$ matrix $\mathbf{S}(s)$ is **BR** if and only if:

(a) $\mathbf{S}(s)$ is rational and satisfies conditions a and b of Definition 1.1,
(b) $\mathbf{1}_n-\mathbf{S}^{T*}(j\omega)\mathbf{S}(j\omega)\geqslant\mathbf{0}$ for all real ω.

$\mathbf{S}(s)$ is said to be rational lossless bounded-real (**LBR**) if it is **BR** and

$$\mathbf{1}_n-\mathbf{S}^{T}(-s)\mathbf{S}(s)=\mathbf{0}$$

for all s.

As seen from (1.13), an n-port is passive if $\operatorname{Re}\mathbf{V}^{T*}I \geqslant 0$ for all s in $\operatorname{Re} s > 0$. Using (1.6) we get

$$\operatorname{Re}\mathbf{V}^{T*}\mathbf{I} = \tfrac{1}{2}(\mathbf{V}^{T*}\mathbf{I} + \mathbf{I}^{T*}\mathbf{V}) = \mathbf{V}^{iT*}\mathbf{V}^{i} - \mathbf{V}^{rT*}\mathbf{V}^{r}$$

$$= \mathbf{V}^{iT*}(\mathbf{1}_n - \mathbf{S}^{T*}\mathbf{S})\mathbf{V}^{i} \geqslant 0 \qquad \text{in } \operatorname{Re} s > 0 \tag{1.15}$$

Since (1.15) can be taken to hold for all complex constant n-vectors $\mathbf{V}^i$, we conclude that

$$\mathbf{1}_n - \mathbf{S}^{T*}\mathbf{S} \geqslant \mathbf{0} \qquad \text{in } \operatorname{Re} s > 0$$

This condition shows that $\mathbf{S}(s)$ must be analytic in $\operatorname{Re} s > 0$, since if $\mathbf{S}(s)$ were to have a pole in $\operatorname{Re} s > 0$, a constant $\mathbf{V}^i$ could be found to make $(\mathbf{S}\mathbf{V}^i)^{T*}(\mathbf{S}\mathbf{V}^i)$ unbounded for s near the pole. As $\mathbf{V}^{iT*}\mathbf{V}^i$ is independent of s, the right side of (1.15) could be made as negative as desired, which contradicts the passivity of the n-port. This also implies that $\mathbf{1}_n - \mathbf{S}^{T*}(j\omega)\mathbf{S}(j\omega) \geqslant \mathbf{0}$ for all real ω, since $\mathbf{S}(s)$ cannot have poles on the $j\omega$-axis. For a finite lumped network, $\mathbf{S}(s)$ is real and rational; hence we can conclude that the scattering matrix of a linear lumped finite time-invariant network is **BR**.

Example 1.3

We want to establish the **BR** property of the following real and rational matrix:

$$\mathbf{S}(s) = \frac{1}{s+2}\begin{bmatrix} 1 & 1 \\ 1 & -s-1 \end{bmatrix}$$

It is seen that $\mathbf{S}(s)$ is real for real s and analytic in $\operatorname{Re} s > 0$, since the only pole is at $s = -2$. It remains to examine $\mathbf{1}_2 - \mathbf{S}^{T*}(j\omega)\mathbf{S}(j\omega)$. We have

$$\mathbf{1}_2 - \mathbf{S}^{T*}(j\omega)\mathbf{S}(j\omega) = \begin{bmatrix} 1 & 0 \\ 0 & 1 \end{bmatrix} - \frac{1}{\omega^2+4}\begin{bmatrix} 1 & 1 \\ 1 & j\omega - 1 \end{bmatrix}\begin{bmatrix} 1 & 1 \\ 1 & -j\omega - 1 \end{bmatrix}$$

$$= \frac{1}{\omega^2+4}\begin{bmatrix} \omega^2+2 & j\omega \\ -j\omega & 2 \end{bmatrix}$$

which is positive-definite since the $(1,1)$ and the $(2,2)$ terms are positive, as is the determinant, which is equal to one. Therefore, according to the **BR** theorem, $\mathbf{S}(s)$ is rational and bounded-real.

Definition 1.2

An $n \times n$ matrix $\mathbf{\Sigma}(s)$ is said to be *J*-contractive real with a given real symmetric matrix $\mathbf{J}$, such that $\mathbf{J}^2 = \mathbf{1}_n$, if and only if

(a) $\mathbf{\Sigma}(s)$ is real for real positive s.
(b) $\mathbf{J} - \mathbf{\Sigma}^{T*}(s)\mathbf{J}\mathbf{\Sigma}(s) \geqslant \mathbf{0}$ in $\operatorname{Re} s > 0$.

Also if $\mathbf{\Sigma}(s)$ is rational in s, it is said to be *J*-contractive real rational (**JR**), and if in addition $\mathbf{J} - \mathbf{\Sigma}^{T}(-s)\mathbf{J}\mathbf{\Sigma}(s) = \mathbf{0}$ for all s, it is called *J*-lossless real rational (**LJR**).

Consider the $2n$-port network with the set of ports partitioned as in Fig. 1.3 ($l = m$) where the chain matrix $\mathcal{C}$ is defined as in (1.4). Consequently we have

$$2\operatorname{Re} \mathbf{V}^{T*}I = \begin{bmatrix} \mathbf{V}_2^{T*} & -\mathbf{I}_2^{T*} \end{bmatrix} \mathbf{J} \begin{bmatrix} \mathbf{V}_2 \\ -\mathbf{I}_2 \end{bmatrix} - \begin{bmatrix} \mathbf{V}_1^{T*} & \mathbf{I}_1^{T*} \end{bmatrix} \mathbf{J} \begin{bmatrix} \mathbf{V}_1 \\ \mathbf{I}_1 \end{bmatrix}$$

$$= \begin{bmatrix} \mathbf{V}_2^{T*} & -\mathbf{I}_2^{T*} \end{bmatrix} (\mathbf{J} - \mathcal{C}^{T*}\mathbf{J}\mathcal{C}) \begin{bmatrix} \mathbf{V}_2 \\ -\mathbf{I}_2 \end{bmatrix} \tag{1.16}$$

where

$$\mathbf{J} = \begin{bmatrix} \mathbf{0}_n & -\mathbf{1}_n \\ -\mathbf{1}_n & \mathbf{0}_n \end{bmatrix}$$

Passivity requires that the right side of (1.16) be nonnegative definite Hermitian for all s in $\operatorname{Re} s > 0$. Since (1.16) can be taken to hold for all complex constant $2n$-vector $[\mathbf{V}_2 \ \ -\mathbf{I}_2]^T$, we conclude that $\mathbf{J} - \mathcal{C}^{T*}\mathbf{J}\mathcal{C} \geqslant \mathbf{0}$ in $\operatorname{Re} s > 0$. Similarly, for the chain scattering matrix $\mathbf{\Phi}$ defined in (1.8), we have

$$\operatorname{Re} \mathbf{V}^{T*}\mathbf{I} = \begin{bmatrix} \mathbf{V}_2^{rT*} & \mathbf{V}_2^{iT*} \end{bmatrix} (\mathbf{J} - \mathbf{\Phi}^{T*}\mathbf{J}\mathbf{\Phi}) \begin{bmatrix} \mathbf{V}_2^r \\ \mathbf{V}_2^i \end{bmatrix} \tag{1.17}$$

where

$$\mathbf{J} = \begin{bmatrix} -\mathbf{1}_n & \mathbf{0}_n \\ \mathbf{0}_n & \mathbf{1}_n \end{bmatrix}$$

Passivity requires that $\mathbf{J} - \mathbf{\Phi}^{T*}\mathbf{J}\mathbf{\Phi} \geqslant \mathbf{0}$ in $\operatorname{Re} s > 0$. Hence, the chain and chain scattering matrices of a linear, finite, lumped, time-invariant passsive $2n$-port is **JR**.

Definition 1.3

An $n \times n$ matrix $\mathbf{A}(s)$ is said to be positive-real if and only if:

(a) $\mathbf{A}(s)$ is analytic in $\operatorname{Re} s>0$.
(b) $\mathbf{A}(s)$ is real for real positive s.
(c) $\mathbf{A}(s)+\mathbf{A}^{T*}(s)\geqslant\mathbf{0}$ in $\operatorname{Re} s>0$.

If $\mathbf{A}(s)$ is also rational, it is said to be positive-real rational (**PR**). The following theorem is of importance in testing **PR** property [1, Chapter 5].

PR Theorem

An $n\times n$ matrix $\mathbf{A}(s)$ is **PR** if and only if

(a) $\mathbf{A}(s)$ is rational and satisfies conditions a and b of Definition 1.3.
(b) $\mathbf{A}^{T*}(j\omega)+\mathbf{A}(j\omega)\geqslant\mathbf{0}$ for all real ω, with $j\omega$ not a pole of any element of $\mathbf{A}(j\omega)$. If $j\omega_0$ is a pole of any element of $\mathbf{A}(j\omega)$, it must be a simple pole, and $\lim(s-j\omega_0)\mathbf{A}(s)$, $s\to j\omega_0$, in the case $\omega_0<\infty$ and $\lim \mathbf{A}(j\omega)/j\omega$, $\omega\to\infty$, in the case $\omega_0=\infty$, is nonnegative definite Hermitian.

In addition, if $\mathbf{A}^T(-s)+\mathbf{A}(s)=\mathbf{0}$ for all s, such that s is not a pole of any element of $\mathbf{A}(s)$, then $\mathbf{A}(s)$ is said to be lossless positive real rational (**LPR**).

In the case the n-port possesses an impedance or an admittance matrix, (1.7) shows that $\mathbf{Z}$ and $\mathbf{Y}$ exist if $\mathbf{1}_n-\mathbf{S}$ and $\mathbf{1}_n+\mathbf{S}$ are nonsingular in $\operatorname{Re} s>0$; this implies that $\mathbf{Z}$ and $\mathbf{Y}$ must be analytic in $\operatorname{Re} s>0$ for a passive network. Furthermore, $\mathbf{S}^*(s)=\mathbf{S}(s^*)$ (reality of $\mathbf{S}$) also yields $\mathbf{Z}^*(s)=\mathbf{Z}(s^*)$ and $\mathbf{Y}^*(s)=\mathbf{Y}(s^*)$, and if $\mathbf{S}$ is rational, so are $\mathbf{Z}$ and $\mathbf{Y}$. The energy constraint $\operatorname{Re}\mathbf{V}^{T*}\mathbf{I}\geqslant 0$ in $\operatorname{Re} s>0$ can be written as

$$\begin{aligned}\operatorname{Re}\mathbf{V}^{T*}\mathbf{I}&=\tfrac{1}{2}(\mathbf{V}^{T*}\mathbf{I}+\mathbf{I}^{T*}\mathbf{V})=\tfrac{1}{2}(\mathbf{V}^{T*}\mathbf{Y}\mathbf{V}+\mathbf{V}^{T*}\mathbf{Y}^{T*}\mathbf{V})\\&=\tfrac{1}{2}\mathbf{V}^{T*}(\mathbf{Y}+\mathbf{Y}^{T*})\mathbf{V}\geqslant 0 \qquad \text{in } \operatorname{Re} s>0 \qquad (1.18)\end{aligned}$$

We note that (1.18) holds for all complex constant n-vector $\mathbf{V}$, hence $\mathbf{Y}^{T*}+\mathbf{Y}\geqslant\mathbf{0}$ in $\operatorname{Re} s>0$. Similarly, $\mathbf{Z}^{T*}+\mathbf{Z}>\mathbf{0}$ in $\operatorname{Re} s>0$ for a passive network. Thus, the admittance matrix and impedance matrix of a linear finite time-invariant lumped passive network are necessarily **PR**.

Example 1.4

We wish to verify that

$$Z(s)=\frac{s(s^2+2)}{s^2+1}$$

is **PR**. Obviously, $Z(s)$ is real and rational. Its poles are ∞, $+j1$ and $-j1$, so that

$Z(s)$ is analytic in $\text{Re}\, s > 0$. Furthermore,

$$Z(s) = s + \frac{\frac{1}{2}}{s-j} + \frac{\frac{1}{2}}{s+j}$$

Thus the residues at its poles are all positive. We also have $Z^{T*}(j\omega)+Z(j\omega)=0$. Therefore, according to the **PR** theorem, $Z(s)$ is **PR**. Since $Z^{T*}(-s)+Z(s)=0$, it is also **LPR**.

1.5 INTERCONNECTIONS OF NETWORKS

In this section we consider some basic interconnections of networks: the series, the parallel, the cascade-load, and the cascade connections.

(a) *Series and parallel connections.* Consider the two n-port networks N_1 and N_2 connected in series as shown in Fig. 1.4a, where port j of N_1 is connected in series with port j of N_2 such that the current entering port j of

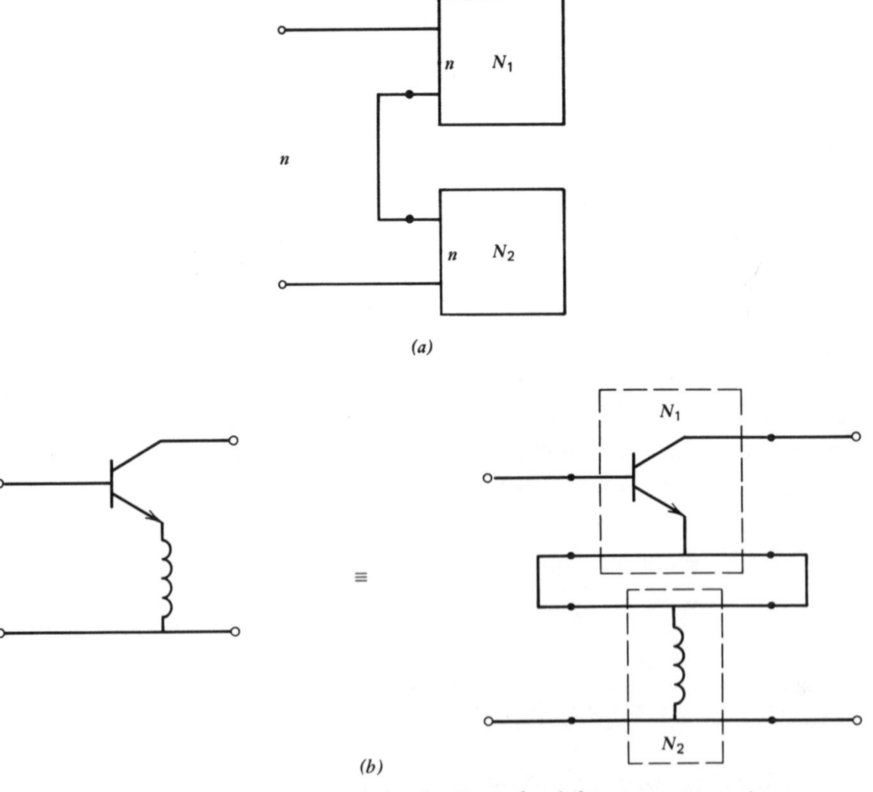

Fig. 1.4. (a) Series connection. (b) Example of the series connection.

N_1 must be equal to the current leaving port j of N_2. Since the port currents must be equal and the port voltages add, the series connection has an impedance matrix

$$\mathbf{Z}=\mathbf{Z}_1+\mathbf{Z}_2 \tag{1.19}$$

where $\mathbf{Z}_1$ and $\mathbf{Z}_2$ are the impedance matrices of the n-port N_1 and N_2, respectively. An example of a two-port series connection is shown in Fig 1.4b. The parallel connection of N_1 and N_2 is shown in Fig. 1.5a where port j of N_1 is connected in parallel with port j of N_2. Since the port voltages are identical and the port currents add, the parallel connection has an admittance matrix

$$\mathbf{Y}=\mathbf{Y}_1+\mathbf{Y}_2 \tag{1.20}$$

where $\mathbf{Y}_1$ and $\mathbf{Y}_2$ are the admittance matrices of N_1 and N_2, respectively. An example of a two-port parallel connection is shown in Fig. 1.5b.

(b) *Cascade-load connection.* Consider the arrangement of Fig. 1.6a where an m-port N_L cascade-loads an $(n+m)$-port N_c to produce an n-port N. Let $\mathbf{S}_L$ be the scattering matrix of N_L and

$$\mathbf{S}_c=\begin{bmatrix} \mathbf{S}_{11} & \mathbf{S}_{12} \\ \mathbf{S}_{21} & \mathbf{S}_{22} \end{bmatrix}$$

be the scattering matrix of N_c partitioned according to ports. Then the

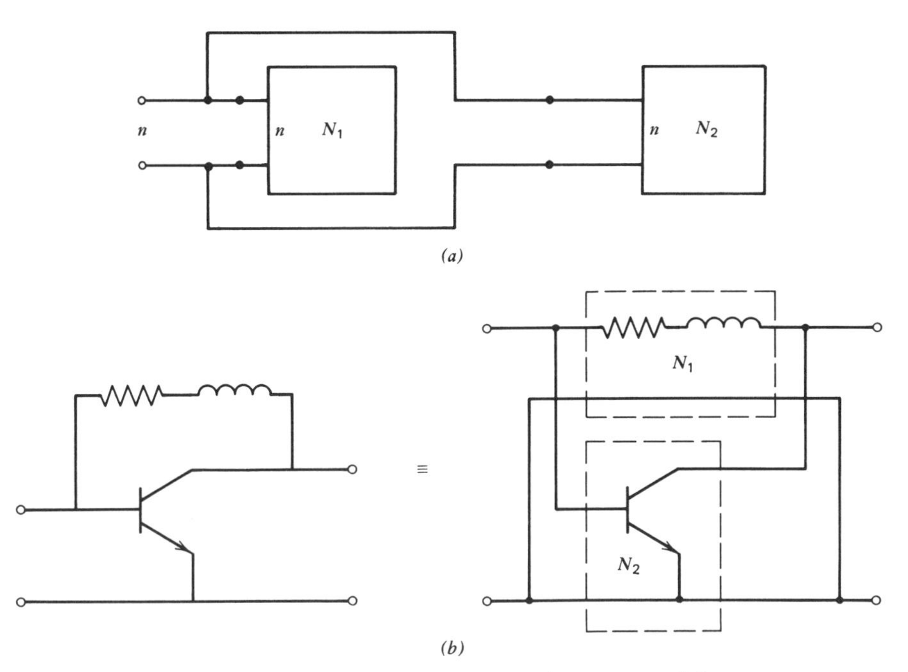

Fig. 1.5. (a) Parallel connection. (b) Example of the parallel connection.

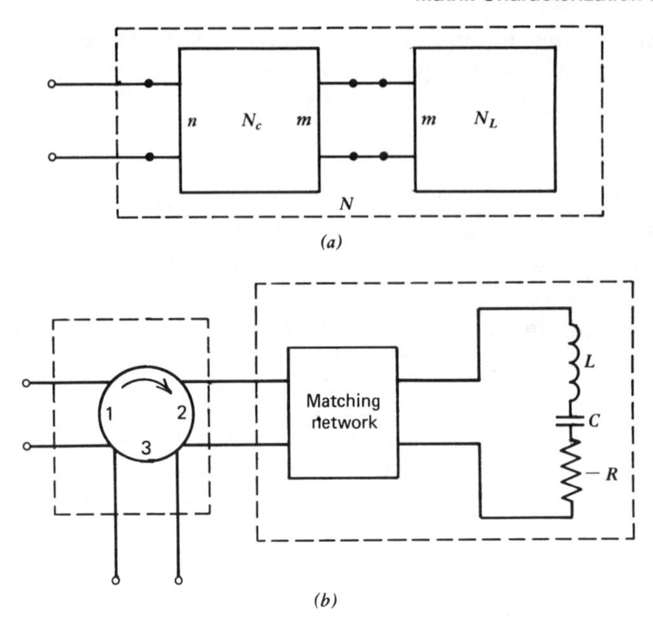

Fig. 1.6. (*a*) Cascade-load connection. (*b*) Example of the cascade-load connection.

scattering matrix $\mathbf{S}$ of the cascade-load network N is given by

$$\mathbf{S}=\mathbf{S}_{11}+\mathbf{S}_{12}\mathbf{S}_L(\mathbf{1}_m-\mathbf{S}_{22}\mathbf{S}_L)^{-1}\mathbf{S}_{21} \tag{1.21a}$$

$$=\mathbf{S}_{11}+\mathbf{S}_{12}(\mathbf{1}_m-\mathbf{S}_L\mathbf{S}_{22})^{-1}\mathbf{S}_L\mathbf{S}_{21} \tag{1.21b}$$

with the assumption that the inverses exist. Equation (1.21a) results from substituting $\mathbf{V}_2^i=\mathbf{S}_L\mathbf{V}_2^r$ into $\mathbf{V}_2^r=\mathbf{S}_{21}\mathbf{V}_1^i+\mathbf{S}_{22}\mathbf{V}_2^i$ and solving for $\mathbf{V}_2^r$, which in turn is substituted into the first of these two equations, using $\mathbf{V}_1^r=\mathbf{S}_{11}\mathbf{V}_1^i+\mathbf{S}_{12}\mathbf{V}_2^i$. Equation (1.21b) is obtained in a similar manner, except that the equation for $\mathbf{V}_2^r$ is first multiplied by $\mathbf{S}_L$ and the result then substituted into the equation for $\mathbf{V}_1^r$. An example of the cascade-load connection of an IMPATT diode (modeled by a $-RLC$ circuit) at port 2 of a circulator is shown in Fig. 1.6*b* where the input is at port 1 and the output is at port 3.

(c) *Cascade connection.* The cascade connection of N_1 and N_2 is shown in Fig. 1.7*a*, where it is seen that $\mathbf{V}_2$ and $-\mathbf{I}_2$ of N_1 are now $\mathbf{V}_1$ and $\mathbf{I}_1$ of N_2; therefore from (1.4), the chain matrix $\mathcal{C}$ of the cascade connection N is given as

$$\mathcal{C}=\mathcal{C}_1\mathcal{C}_2 \tag{1.22}$$

where $\mathcal{C}_1$ and $\mathcal{C}_2$ are the chain matrices of N_1 and N_2, respectively. Similarly since $\mathbf{V}_2^r$ and $\mathbf{V}_2^i$ of N_1 are now $\mathbf{V}_1^i$ and $\mathbf{V}_1^r$ of N_2, the chain scattering matrix $\mathbf{\Phi}$ of the cascade connection N can be related to the chain scattering matrices $\mathbf{\Phi}_1$ and $\mathbf{\Phi}_2$ of N_1 and N_2 as

$$\mathbf{\Phi}=\mathbf{\Phi}_1\mathbf{\Phi}_2 \tag{1.23}$$

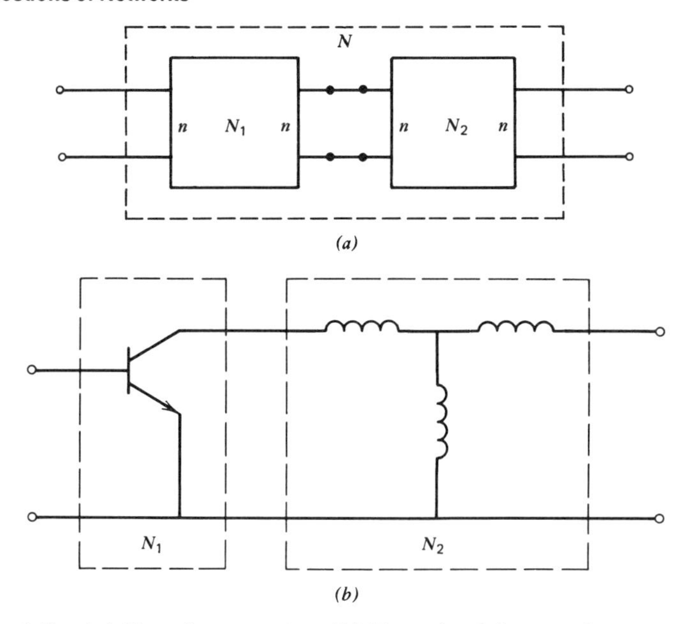

Fig. 1.7. (*a*) Cascade connection. (*b*) Example of the cascade connection.

An example of the cascade connection is shown in Fig. 1.7*b*.

These network matrices can be used efficiently in the computer-aided analysis of n-port networks, especially for microwave two-port amplifiers. Figure 1.8 shows a typical microwave transistor circuit and its method of analysis. The circuit contains, from left to right, a shunt open stub (open-circuited transmission line in parallel), a series transmission line, a transistor, a series transmission line and a shunt short stub (short-circuited transmission line in parallel). Both source and load are 50 Ω. As shown, all five elements are connected in cascade; hence it is natural to use the chain matrix or chain scattering matrix in the analysis. Since all microwave transistor data are given in scattering parameters, a conversion to chain or chain scattering parameters is necessary. The two-port scattering matrix **S** can be converted to the two-port chain matrix $\mathcal{C}$ by the following relations:

$$\mathcal{C}_{11}=\frac{-\Delta+S_{11}-S_{22}+1}{2S_{21}} \tag{1.24a}$$

$$\mathcal{C}_{12}=Z_0\frac{\Delta+S_{11}+S_{22}+1}{2S_{21}} \tag{1.24b}$$

$$\mathcal{C}_{21}=Z_0^{-1}\frac{\Delta-S_{11}-S_{22}+1}{2S_{21}} \tag{1.24c}$$

$$\mathcal{C}_{22}=\frac{\Delta-S_{11}-S_{22}+1}{2S_{21}} \tag{1.24d}$$

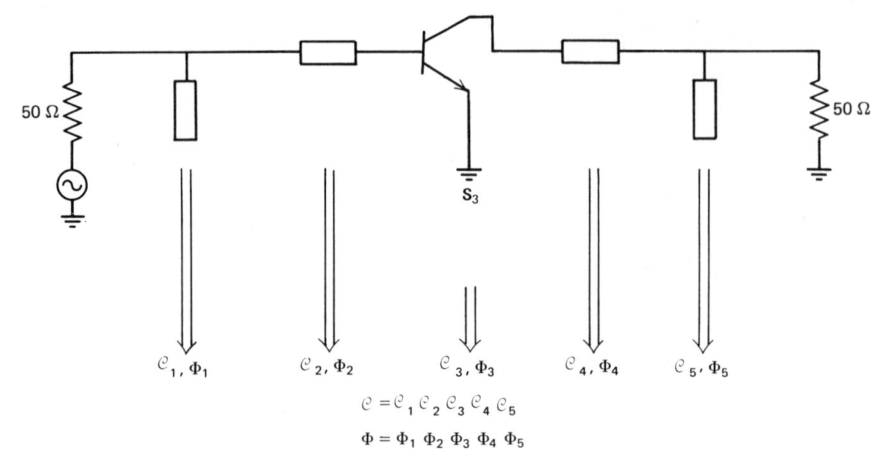

Fig. 1.8. A microwave transistor amplifier and its analysis by the chain matrix and chain scattering matrix methods.

where

$$\Delta = S_{11}S_{22} - S_{12}S_{21} \tag{1.24e}$$

and Z_0 is the normalized impedance of the scattering matrix **S**. (Detailed discussion of scattering matrix normalization will be postponed to Chapter 2.) The conversion from scattering matrix to chain scattering matrix can be done using (1.9). And by comparing (1.9) and (1.24) it is apparent that the transformations from **S** to $\mathcal{C}$ are more complex than those required from **S** to $\mathbf{\Phi}$. Therefore for greater calculation efficiency, the chain scattering matrix should be used for cascading involving many transistors.

For each frequency, the chain or chain scattering matrix of each element is calculated and multiplied from left to right to obtain the total chain or chain scattering matrix. From this, all the figures of merit such as input and output reflection coefficients, VSWR (voltage standing wave ratio) can be computed using (1.21). Power gain can also be computed from various formulas given in Chapter 2.

The interesting feature of this type of analysis is the possibility of storing the matrix representation of individual network elements in the computer memory by appropriate code. This can then be called upon to perform the computation.

Example 1.5

The chain matrices $\mathcal{C}$ representing series and shunt elements with impedances Z_s and Z_p shown in Figs. 1.9*a* and *b* are given respectively by

$$\mathcal{C} = \begin{bmatrix} 1 & Z_s \\ 0 & 1 \end{bmatrix} \tag{1.25a}$$

$$\mathcal{C} = \begin{bmatrix} 1 & 0 \\ Z_p^{-1} & 1 \end{bmatrix} \tag{1.25b}$$

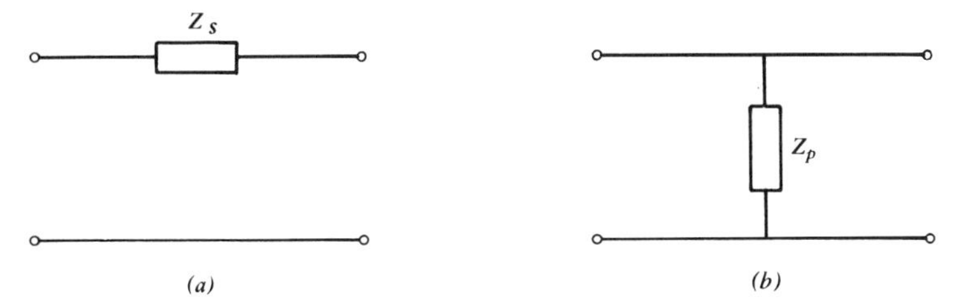

Fig. 1.9. (a) A series element. (b) A parallel element.

The corresponding scattering matrices **S** are given respectively by

$$\mathbf{S}=\frac{1}{1+2(Z_0/Z_s)}\begin{bmatrix} 1 & 2Z_0/Z_s \\ 2Z_0/Z_s & 1 \end{bmatrix} \tag{1.26a}$$

$$\mathbf{S}=\frac{1}{1+2(Z_p/Z_0)}\begin{bmatrix} 1 & 2Z_p/Z_0 \\ 2Z_p/Z_0 & 1 \end{bmatrix} \tag{1.26b}$$

The corresponding chain scattering matrices **Φ** are given respectively by

$$\mathbf{\Phi}=\frac{1}{2Z_0/Z_s}\begin{bmatrix} 1+2(Z_0/Z_s) & -1 \\ 1 & 2(Z_0/Z_s)-1 \end{bmatrix} \tag{1.27a}$$

$$\mathbf{\Phi}=\frac{1}{2Z_p/Z_0}\begin{bmatrix} 1+2(Z_p/Z_0) & -1 \\ 1 & 2(Z_p/Z_0)-1 \end{bmatrix} \tag{1.27b}$$

Example 1.6

For a transmission line of length l and characteristic impedance Z_0, the voltages and currents at each port shown in Fig. 1.10 are related by the equations

$$V_1=V_2\cosh(\gamma l)+I_2Z_0\sinh(\gamma l) \tag{1.28a}$$

$$I_1=V_2Z_0^{-1}\sinh(\gamma l)+I_2\cosh(\gamma l) \tag{1.28b}$$

where $\gamma=\alpha+j\beta$ with α=loss/unit length, $\beta=\omega/v_p$ is the phase constant, and v_p is the phase velocity of the travelling wave. If the line is lossless, that is, $\alpha=0$, then $\gamma=j\beta$ ($j^2=-1$) and $\sinh(\gamma l)=j\sin(\beta l)$ and $\cosh(\gamma l)=\cos(\beta l)$. The chain matrix of the series transmission line is given as

$$\mathcal{C}=\begin{bmatrix} \cosh(\gamma l) & Z_0\sinh(\gamma l) \\ Z_0^{-1}\sinh(\gamma l) & \cosh(\gamma l) \end{bmatrix} \tag{1.29}$$

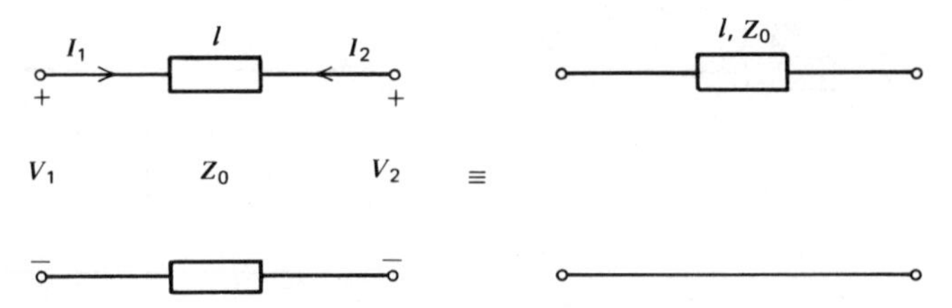

Fig. 1.10. A series transmission line.

If the line is open-circuited at one end, the input impedance is given as

$$Z = Z_0/\tan(\gamma l) \tag{1.30a}$$

If the line is short-circuited at one end, the input impedance is given as

$$Z = Z_0/\cot(\gamma l) \tag{1.30b}$$

From (1.25) the chain matrices $\mathcal{C}$ of a shunt open stub and of a shunt short stub shown in Figs. 1.11*a* and *b* are given respectively by

$$\mathcal{C} = \begin{bmatrix} 1 & 0 \\ Z_0^{-1}\tan(\gamma l) & 1 \end{bmatrix} \tag{1.31a}$$

$$\mathcal{C} = \begin{bmatrix} 1 & 0 \\ Z_0^{-1}\cot(\gamma l) & 1 \end{bmatrix} \tag{1.31b}$$

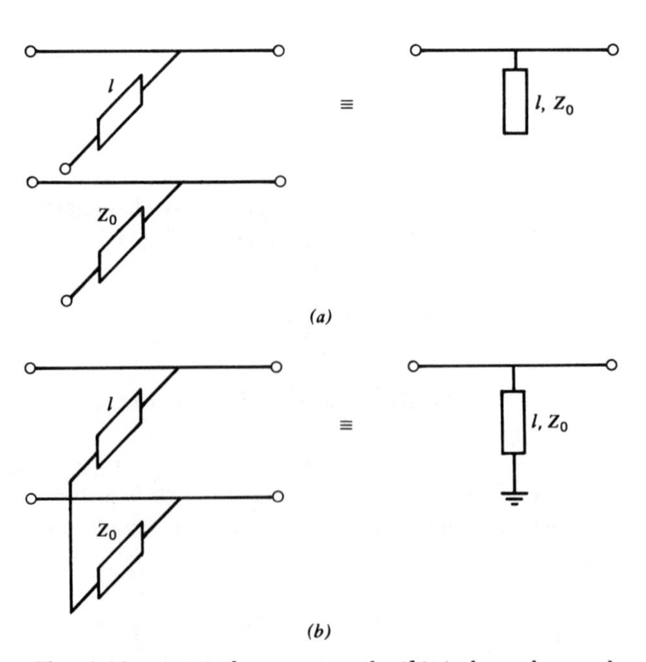

Fig. 1.11. (*a*) A shunt open stub. (*b*) A shunt short stub.

1.6 CONCLUDING REMARKS

This chapter presented several important network matrices and summarized their main properties so they can be applied in subsequent chapters. Knowing how to use these network matrices can simplify mathematical manipulations considerably. The concept of **BR** and **PR** properties is very important in the design of lossless equalizers. A detailed discussion of these two properties and their applications to passive network synthesis can be found in [1–6]. For readers who are interested in state-variables **BR** and **PR** properties, the material in [7] is excellent. The **JR** property is not as widely used as the **BR** and **PR** properties in network synthesis since it is useful only in cascade synthesis; the reader can find a more detailed discussion in [8, 9].

PROBLEMS

1.1. Find the scattering matrix of 2 two-ports in direct cascade in terms of the individual scattering matrices.

1.2. Express the scattering matrix $\mathbf{S}$ of a two-port in terms of its chain matrix $\mathcal{C}$. Generalize the result for n-ports.

1.3. A time-variable capacitor $C(t)$ constrains its voltage and current by the relation $i(t)=(d/dt)\{C(t)v(t)\}$. Find the necessary and sufficient conditions on $C(t)$ for the capacitor to be passive and lossless.

1.4. Find the scattering matrix of the resistor, inductor, capacitor, transformer, and gyrator.

1.5. Verify Equations (1.21a) and (1.21b).

1.6. Find the impedance matrix $\mathbf{Z}$ in terms of the impedance matrices $\mathbf{Z}_c$ and $\mathbf{Z}_L$ of the coupling network N_c and the load network N_L as in the case of the scattering matrix.

1.7. From (1.21) show that if the last m ports of N_c are terminated in 1 Ω resistors, then $\mathbf{S}=\mathbf{S}_{11}$.

1.8. Show that

$$\mathbf{S}(s)=\begin{bmatrix} 0 & 1 \\ \dfrac{s-1}{s+1} & 0 \end{bmatrix}$$

is **LBR**.

1.9. Show that if $\mathbf{S}_1$ and $\mathbf{S}_2$ are **LBR**, then $\mathbf{S}_1\mathbf{S}_2$ is **LBR**.

1.10. Test the following function to see if it is **BR**:

$$S(s)=\frac{s^2+4s-1}{2s^3+5s^2+6s+3}$$

1.11. Test the following function to see if it is **PR**:

$$A(s)=\frac{2s^4+7s^3+11s^2+12s+4}{s^4+5s^3+9s^2+11s+6}$$

1.12. Verify that

$$A(s)=\frac{s^3+3s^2+5s+1}{s^3+2s^2+s+2}$$

is **PR**.

1.13. Show that if $\mathbf{\Sigma}_1$ and $\mathbf{\Sigma}_2$ are **JR(LJR)**, then $\mathbf{\Sigma}_1\mathbf{\Sigma}_2$ is **JR(LJR)**.

1.14. Let $(\mathbf{A},\mathbf{B},\mathbf{C},\mathbf{D})$ be a minimal state-space realization of $\mathbf{\Sigma}(s)$ [7, Chapter 3], $\mathbf{\Sigma}(s)=\mathbf{D}+\mathbf{C}(s\mathbf{1}_r-\mathbf{A})^{-1}\mathbf{B}$, where r is the degree of $\mathbf{\Sigma}(s)$. Show that $\mathbf{\Sigma}(s)$ is **LJR** if and only if there exists a real symmetric positive-definite matrix **P** such that

$$\mathbf{PA}+\mathbf{A}^T\mathbf{P}+\mathbf{C}^T\mathbf{JC}=\mathbf{0}$$

$$\mathbf{PB}+\mathbf{C}^T\mathbf{JC}=\mathbf{0}$$

$$\mathbf{J}-\mathbf{D}^T\mathbf{JD}=\mathbf{0}$$

1.15. Show that if $\mathbf{A}_1(s)$ and $\mathbf{A}_2(s)$ is **PR**, then $\mathbf{A}_1(s)+\mathbf{A}_2(s)$ is **PR**.

1.16. Write a simple computer program to calculate the chain matrix, the chain scattering matrix, and the scattering matrix of a high frequency transistor amplifier shown in Fig. P1.16. The scattering parameters of the transistor are given at 750 MHz as $S_{11}=0.277\angle-59°$, $S_{12}=0.078\angle 93°$, $S_{11}=1.92\angle 64°$, $S_{22}=0.848\angle-31°$.

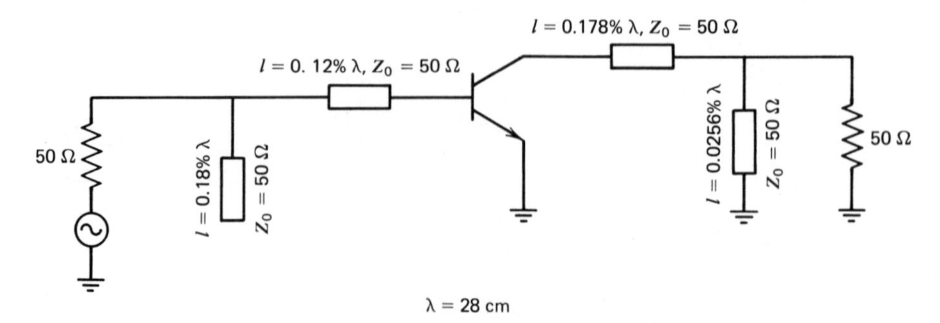

Fig. P.1.16. A 750 MHz transistor amplifier.

REFERENCES

1. R. W. Newcomb, *Linear Multiport Synthesis*. New York: McGraw-Hill, 1966.
2. G. C. Temes and J. W. LaPatra, *Introduction to Circuits Synthesis and Design*. New York: McGraw-Hill, 1977.
3. M. E. Van Valkenburg, *Introduction to Modern Network Synthesis*. New York: Wiley, 1960.
4. E. A. Guillemin, *Synthesis of Passive Networks*. New York: Wiley, 1957.
5. L. Weinberg, *Network Analysis and Synthesis*. New York: McGraw-Hill, 1962.
6. V. Belevitch, *Classical Network Theory*. San Francisco: Holden-Day, 1968.
7. B. D. O. Anderson and S. Vonpanitlerd, *Network Analysis and Synthesis: A Modern Systems Theory Approach*. Englewood Cliffs, NJ: Prentice-Hall, 1973.
8. T. T. Ha, "State-space properties of real rational *J*-lossless matrices and their applications to network synthesis," *IEE J. Electron. Circuits and Syst.*, **1**, no. 6, 198–206, 1977.
9. T. T. Ha, "Computer-aided cascade synthesis of passive networks," *IEE J. Electron. Circuits Syst.*, **2**, no. 5, 155–158, 1978.

2

Scattering Matrix

In Chapter 1 various matrix characterizations of networks and their relationships to each other were discussed. Each of these network matrices has certain advantages and disadvantages, as shown in Section 1.5, and if used appropriately can simplify the mathematical manipulations a great deal. However, at microwave frequencies (above 1 GHz $=10^9$ Hz) there are some problems, including the following:

(a) Equipment to measure total voltage and total current at the ports of the network is not readily available.

(b) Short and open circuits are difficult to achieve over a broad band of frequencies because of lead inductance and capacitance. Furthermore, these measurements typically require tuning stubs separately adjusted at each frequency to reflect short or open circuits to the device terminals, and this makes the process inconvenient and tedious.

(c) Active devices such as transistors and negative resistance diodes are very often not short- or open-circuit stable.

To overcome these problems, traveling waves rather than voltages and currents are used at microwave frequencies. Examples of waves as variables were seen in Section 1.2, where the incident and reflected voltage waves were used in the description of scattering matrices of n-port networks with unit resistor normalization at each port. In this case, the network is completely embedded between resistive load and source; hence there is little chance for oscillations to occur. Another important advantage of scattering matrices is that incident and reflected waves, unlike terminal voltages and currents, do not vary in magnitude at points along lossless transmission lines. This means that the scattering matrix can be measured on a device located at some distance from the measurement transducers, if they are connected by low-loss transmission lines.

The scattering parameters actually originated from the theory of waves traveling in transmission lines, and since uniform transmission lines play an important role in the design of microwave frequency matching networks, we briefly review them in the following section.

2.1 THE DISTRIBUTED ELEMENT: UNIFORM TRANSMISSION LINE

At low frequencies all circuit elements are lumped, since their physical dimensions are negligible compared with the wavelength of the electromagnetic signal inside the components. In such a network, since Kirchhoff's voltage and current relationships are applicable, the current within any branch is the same at all points of the branch between its terminating nodes, and the voltage along any perfect conductor is constant and changes only at the components.

At high frequencies, if the signal wave length λ is comparable to the physical dimension of the components, then the spatial variations of voltages and currents along the wires and in the components must be taken into consideration. In such circuits, called distributed networks, Kirchhoff's laws are no longer valid and Maxwell's laws must be applied. Consider a signal of frequency $f=10^{10}$ Hz $=10$ GHz traveling with velocity $v=3\times 10^{8}$ m/sec, then the wavelength λ is given by

$$\lambda=\frac{v}{f}=\frac{3\times 10^{8}\ \text{m/sec}}{10^{10}\ \text{Hz}}=3\times 10^{-2}\ \text{m}=3\ \text{cm} \tag{2.1}$$

As seen from (2.1), it is not practical to make circuits with lumped discrete components that have dimensions negligibly small compared with 3 cm. In practice, for frequencies above 1 GHz, most microwave circuits are designed with distributed elements that can be manufactured inexpensively.

The basic building block in distributed circuits is the uniform transmission line. This is a two-port element whose symbol is shown in Fig. 2.1*a*. Some actual lines are shown in Fig. 2.1*b*–*d*. For an elementary section of such a line, Δx meters long, the equivalent circuit is shown in Fig. 2.2, where L and R represent the inductance and resistance, respectively, of each conductor of the line per meter, and C and G represent the capacitance and the leakage conductance, respectively, of the insulation between the conductors per meter. If L, R, C, and G do not depend on the location x of the segment along the line, the line is a uniform transmission line. Assume that the line parameters satisfy $R\ll\omega L$ and $G\ll\omega C$ at all frequencies of interest. Then the approximation $R=0$ and $G=0$ can be made and the line is called a uniform lossless transmission line and abbreviated as UL. A UL is completely characterized by its characteristic impedance Z_0 and its electrical line length θ given by

$$Z_0=(L/C)^{1/2} \tag{2.2a}$$

$$\theta=\omega l/v=\omega l(LC)^{1/2}=2\pi l/\lambda \tag{2.2b}$$

where L and C are defined above, l is the physical length of the line, and v is the propagation velocity of the signal in the line. At microwave frequencies, most ULs have a 50 Ω characteristic impedance. Other lines of 75, 90, and 300 Ω impedance are also used.

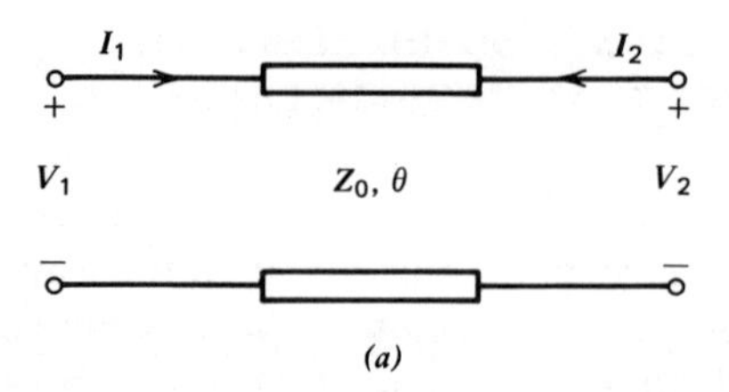

(a)

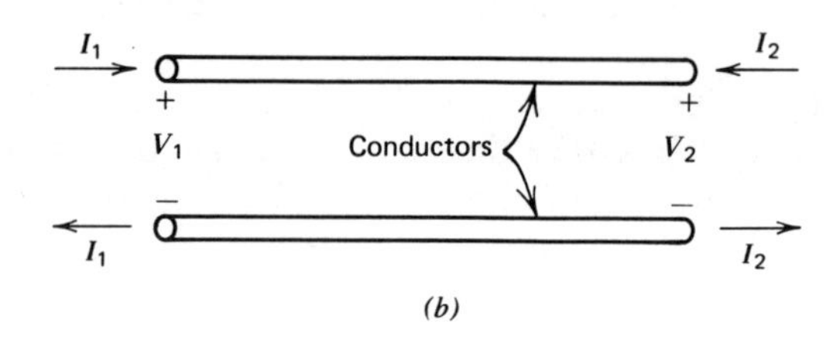

(b)

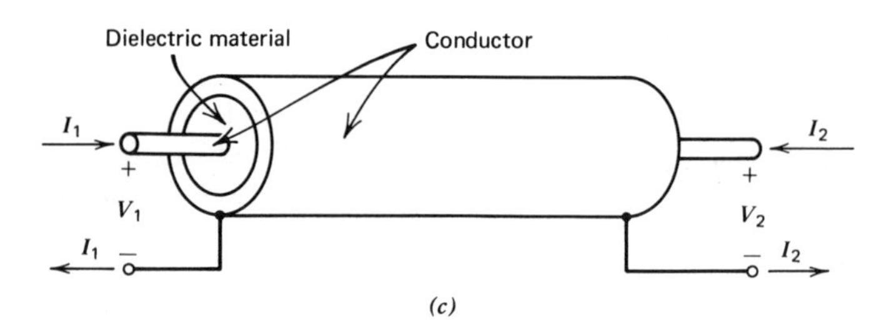

(c)

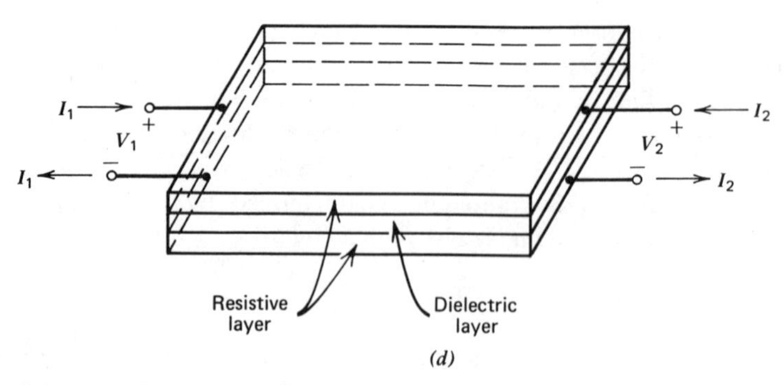

(d)

Fig. 2.1. (*a*) Transmission line symbol. (*b*) Parallel-wire line. (*c*) Coaxial line. (*d*) Microstrip line.

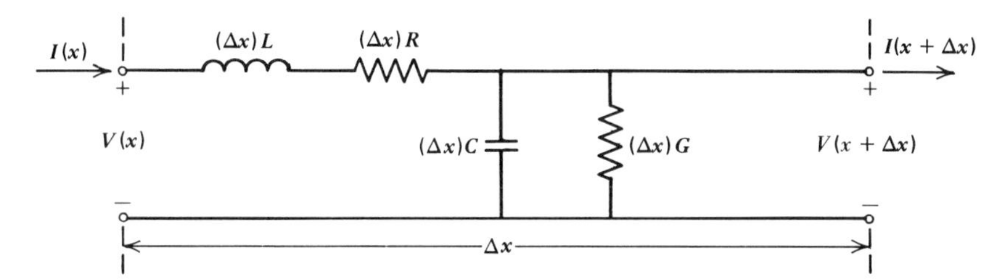

Fig. 2.2. Equivalent circuit of a segment of transmission line.

Now consider a UL with a characteristic impedance Z_0 connected between a load Z_L and a source E_s with internal impedance Z_s as in Fig. 2.3. Voltage, current, and power can be considered to be waves traveling in both directions along the line. A portion of the waves incident on the load will be reflected. It then becomes incident on the source and in turn rereflects from the source (if $Z_s \neq Z_0$), resulting in a standing wave on the line. According to the theory of transmission lines, the voltage at a given point along the line is the sum of the incident and reflected voltage waves at that point, and the current on the line is the difference between the incident and reflected voltage waves divided by Z_0, that is,

$$V = V^i + V^r \tag{2.3a}$$

$$I = \frac{V^i - V^r}{Z_0} \tag{2.3b}$$

From (2.3) we can derive

$$V^i = \tfrac{1}{2}(V + Z_0 I) \tag{2.4a}$$

$$V^r = \tfrac{1}{2}(V - Z_0 I) \tag{2.4b}$$

Note that (2.4) is identical to (1.6) if $Z_0 = 1\ \Omega$. As in (1.6c) we define the reflection coefficient Γ for the load Z_L as

$$V^r = \Gamma V^i \tag{2.5}$$

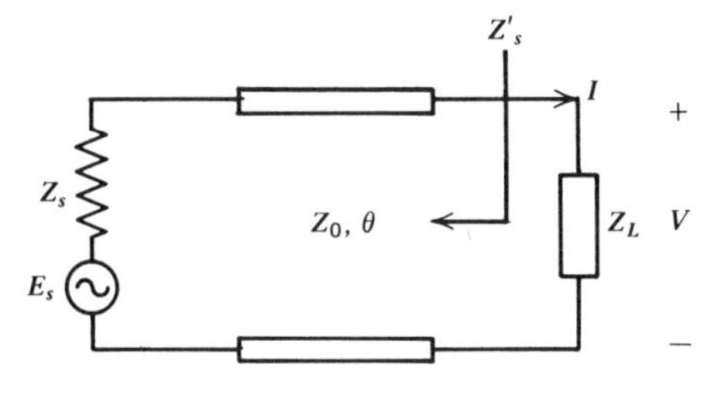

Fig. 2.3. A uniform lossless transmission line embedded between a load and a source.

From (2.4) and (2.5) we get

$$\Gamma=\frac{V-Z_0 I}{V+Z_0 I}=\frac{(V/I)-Z_0}{(V/I)+Z_0} \tag{2.6}$$

At the load we have $V/I=Z_L$. Hence the load reflection coefficient is

$$\Gamma=\frac{Z_L-Z_0}{Z_L+Z_0}=\frac{Y_0-Y_L}{Y_0+Y_L} \tag{2.7}$$

where $Y_L=1/Z_L$ is the load admittance. Now instead of using V^i and V^r as variables, we define new variables

$$a=V^i/Z_0^{1/2}=\tfrac{1}{2}Z_0^{-1/2}(V+Z_0 I) \tag{2.8a}$$

$$b=V^r/Z_0^{1/2}=\tfrac{1}{2}Z_0^{-1/2}(V-Z_0 I) \tag{2.8b}$$

From (2.5) and (2.8) we remark that Γ remains unchanged, that is, $\Gamma=V^r/V^i=b/a$. Furthermore the square of the magnitude of these new variables has the dimension of power. Thus, $|a|^2$ can be thought of as the incident power on Z_L and $|b|^2$ as power reflected from Z_L. We term a and b the incident and reflected scattering variables, respectively, and Γ the load reflection coefficient normalized to the real characteristic impedance. In the following section we will generalize the above concept to multiport networks instead of the one-port network, normalized to arbitrary complex impedances at each port.

2.2 THE GENERALIZED SCATTERING MATRIX

Consider the network in Fig. 2.3, and assume that the generator impedance Z_s has a positive real part, that is, $\operatorname{Re} Z_s(j\omega)>0$† for all ω in the frequency band of interest B. Let Z_s' be Thevenin's impedance looking toward the generator from the load. Then it can be shown that [1, 2]

$$Z_s'=Z_0\frac{Z_s+jZ_0\tan\theta}{Z_0+jZ_s\tan\theta} \tag{2.9}$$

where Z_0 and θ are the characteristic impedance and the electrical length of the transmission line. It is clear that $\operatorname{Re} Z_s'(j\omega)>0$ for all ω in the frequency band B. The incident and reflected power wave a and b to and from the load are defined by

$$a=\tfrac{1}{2}(\operatorname{Re} Z_s')^{-1/2}(V+Z_s' I) \tag{2.10a}$$

$$b=\tfrac{1}{2}(\operatorname{Re} Z_s')^{-1/2}(V-Z_s'^{*} I) \tag{2.10b}$$

†All the network matrices and their corresponding variables in this chapter are functions of the real frequency $j\omega$, and not the complex frequency $s=\sigma+j\omega$ unless otherwise specified.

We term a and b the generalized incident and reflected scattering variables, respectively. From (2.10) we defined the generalized reflection coefficient Γ of the load Z_L as

$$\Gamma=\frac{b}{a}=\frac{V-Z_s'^*I}{V+Z_s'I}=\frac{(V/I)-Z_s'^*}{(V/I)+Z_s'} \tag{2.11}$$

At the load we get $V/I=Z_L$, hence

$$\Gamma=\frac{Z_L-Z_s'^*}{Z_L+Z_s'} \tag{2.12}$$

The physical meaning of this concept should be explained. We observe that the power P_L into the load Z_L is given by $\operatorname{Re} Z_L|I|^2$ where I is the current into the load. Since $I=E_s/(Z_s'+Z_L)$, we get

$$P_L=\operatorname{Re} Z_L\left|\frac{E_s}{Z_s'+Z_L}\right|^2=\frac{R_L|E_s|^2}{(R_L+R_s')^2+(X_L+X_s')^2} \tag{2.13a}$$

$$=\frac{|E_s|^2}{4R_s'+R_L^{-1}\left[(R_L-R_s')^2+(X_L+X_s')^2\right]} \tag{2.13b}$$

where R_L' and R_s' and the real parts of Z_L and Z_s', respectively, and X_L and X_s' are the imaginary parts. By assumption, $R_s'>0$, hence we see from (2.13b) that P_L is maximum when $R_L=R_s'$ and $X_L=-X_s'$, that is,

$$P_{L,\max}=P_a=\frac{|E_s|^2}{4R_s'} \tag{2.14}$$

The power in (2.14) is called the available power from the generator. We note that for a given Z_s', if V and I are given, a and b are readily calculated from (2.10). On the other hand, if a and b are given, V and I are obtained as

$$V=(\operatorname{Re} Z_s')^{-1/2}(Z_s'^*a+Z_s'b) \tag{2.15a}$$

$$I=(\operatorname{Re} Z_s')^{-1/2}(a-b) \tag{2.15b}$$

Referring to Fig. 2.3 and using its Thevenin equivalent circuit, we get the relationship between the voltage V across the load Z_L and the current I into Z_L

$$V=E_s-Z_s'I \tag{2.16}$$

Substituting (2.16) into (2.10a) we obtain

$$|a|^2=\frac{|E_s|^2}{4R_s'}=P_a \tag{2.17}$$

From (2.10) and (2.16) we get by direct substitution

$$|a|^2-|b|^2=\tfrac{1}{4}(R_s')^{-1}(V+Z_s'I)(V^*+Z_s'^*I^*)-(V-Z_s'^*I)(V^*-Z_s'I^*) \tag{2.18a}$$

$$=\tfrac{1}{4}(R_s')^{-1}(Z_s'+Z_s'^*)(VI^*+V^*I)=\mathrm{Re}(V^*I) \tag{2.18b}$$

$$=P_L \tag{2.18c}$$

Equations (2.17) and (2.18c) can be interpreted as follows. The generator E_s sends the power $|a|^2$ toward the load, regardless of the load impedance Z_L. However, when the load is not matched, that is, when the equations $R_L=R_s'$ and $X_L=-X_s'$ are not satisfied, part of the incident power $|a|^2$ is reflected back to the generator. This reflected power is given by $|b|^2$; hence the net power absorbed in the load is $|a|^2-|b|^2$.

Next, to determine the generalized scattering matrix of an n-port network N, consider the network in Fig. 2.4, where E_i and Z_i at port i are the Thevenin equivalent voltage and impedance looking toward the generator. We assume that $Z_i(j\omega)$, $i=1,2,\ldots,n$ have positive real parts for all ω in B. Let **a**, **b**, **V** and **I** be the n-vectors whose ith component a_i, b_i, V_i and I_i are the generalized incident and reflected scattering variables and the voltage across and the current into port i, respectively. Then **a** and **b** can be written in terms of V and I as follows

$$\mathbf{a}=\mathbf{R}(\mathbf{V}+\mathbf{Z}\mathbf{I}) \tag{2.19a}$$

$$\mathbf{b}=\mathbf{R}(\mathbf{V}-\mathbf{Z}^{T*}\mathbf{I}) \tag{2.19b}$$

where **R** and **Z** are diagonal matrices whose ith components are given by $\tfrac{1}{2}(\mathrm{Re}\,Z_i)^{-1/2}$ and Z_i, respectively. The $n\times n$ generalized scattering matrix **S** of N, normalized with respect to the n impedances $Z_1, Z_2,\ldots, Z_n$ is defined by

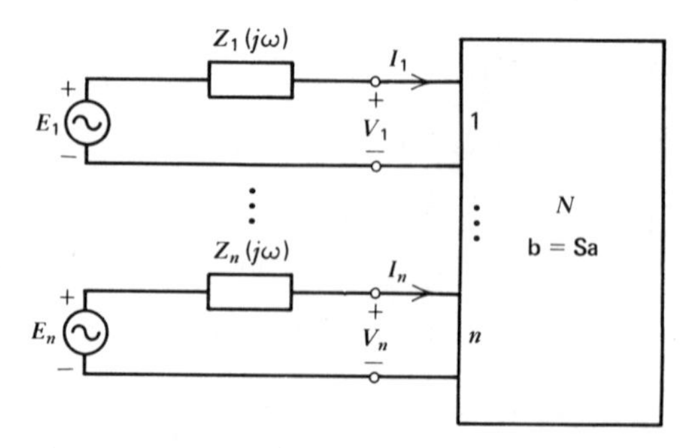

Fig. 2.4. The n-port N excited by generators with positive real part internal impedance.

means of the linear matrix equation

$$\mathbf{b}=\mathbf{S}\mathbf{a} \tag{2.20}$$

Let $\mathbf{Z}_N$ be the impedance matrix of N, that is, $\mathbf{V}=\mathbf{Z}_N\mathbf{I}$. Then from (2.19) and (2.20) we can get S in terms of $\mathbf{Z}_N$ as

$$\mathbf{S}=\mathbf{R}(\mathbf{Z}_N-\mathbf{Z}^{T*})(\mathbf{Z}_N+\mathbf{Z})^{-1}\mathbf{R}^{-1} \tag{2.21}$$

Similarly

$$\mathbf{Z}_N=\mathbf{R}^{-1}(\mathbf{1}_n-\mathbf{S})^{-1}(\mathbf{S}\mathbf{Z}+\mathbf{Z}^{T*})\mathbf{R} \tag{2.22}$$

Some physical interpretations of $\mathbf{S}$ in terms of power are discussed below. Let P_i be the average power entering N at port i, that is, $P_i=\mathrm{Re}(V_i^*I_i)$. Then as in the one-port case we get

$$P_i=|a_i|^2-|b_i|^2, \qquad i=1,2,\ldots,n \tag{2.23}$$

Therefore the total average power absorbed by N is given by

$$P=\sum_{i=1}^{n}P_i=\sum_{i=1}^{n}|a_i|^2-\sum_{i=1}^{n}|b_i|^2 \tag{2.24a}$$

$$=\mathbf{a}^{T*}\mathbf{a}-\mathbf{b}^{T*}\mathbf{b} \tag{2.24b}$$

which is a generalization of (2.18) for the one-port network. Substituting (2.20) into (2.24b) yields

$$P=\mathbf{a}^{T*}\left[\mathbf{1}_n-\mathbf{S}^{T*}\mathbf{S}\right]\mathbf{a} \tag{2.25}$$

We remark that if N is passive then $P\geqslant 0$ for all real ω and $a(j\omega)$ hence, $\mathbf{1}_n-\mathbf{S}^{T*}\mathbf{S}\geqslant 0$ for all real ω. Furthermore if N is lossless, then $P=0$ and consequently $\mathbf{1}_n-\mathbf{S}^{T*}\mathbf{S}=0$. Thus the complex normalization has succeeded in preserving the important properties possessed by a scattering matrix normalized with respect to 1 Ω impedance as in Chapter 1.

Now let all ports of N except port i (excited by E_i) be terminated in their respective normalizing impedances as shown in Fig. 2.5. Then

$$V_k=-Z_kI_k, \qquad k\neq i \tag{2.26}$$

Substituting (2.26) into the kth equation in (2.19a) yields $a_k=0$, $k\neq i$. Therefore, from (2.20) we get, using $E_i=V_i+Z_iI_i$

$$b_k=S_{ki}a_i=\tfrac{1}{2}(\mathrm{Re}\,Z_i)^{-1/2}S_{ki}E_i, \qquad i=1,2,\ldots,n \tag{2.27}$$

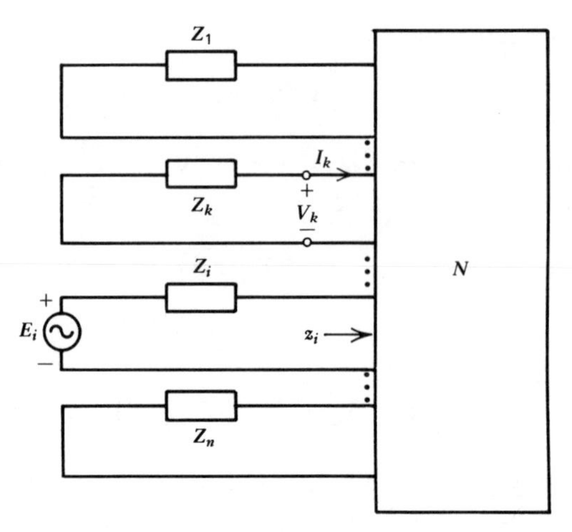

Fig. 2.5. Matched terminations to determine the transducer power gain.

From (2.26) and the kth equation of (2.19b) we have

$$b_k = -(\operatorname{Re} Z_k)^{1/2} I_k, \qquad k \neq i \tag{2.28}$$

Therefore from (2.27) and (2.28) we have

$$|S_{ki}|^2 = \frac{(\operatorname{Re} Z_k)|I_k|^2}{|E_i|^2/4(\operatorname{Re} Z_i)} = \frac{|b_k|^2}{|a_i|^2} = \frac{P_k}{(P_a)_i} \tag{2.29}$$

We note that $(P_a)_i = |E_i|^2/4(\operatorname{Re} Z_i)$ is the maximum available power to N at port i and $P_k = |b_k|^2 = (\operatorname{Re} Z_k)|I_k|^2$ is the power delivered to the termination Z_k at port k. By definition, the ratio $P_k/(P_a)_i = |S_{ki}|^2$ is the transducer power gain from port i to port k. To determine S_{ii} of S, let z_i be the impedance looking into port i of N with other ports terminated in their respective normalized impedances as in Fig. 2.5. Then $V_i = z_i I_i$ and from the ith equations in (2.19) and (2.26) we have

$$S_{ii} = \frac{b_i}{a_i} = \frac{V_i - Z_i^* I_i}{V_i + Z_i I_i} = \frac{z_i - Z_i^*}{z_i + Z_i} \tag{2.30}$$

In other words, S_{ii} is the input reflection coefficient at port i with all other ports terminated in their respective normalized impedances (i.e., with all other ports matched). In summary, the elements of **S** in (2.20) can be evaluated as

$$S_{ii} = \left.\frac{b_i}{a_i}\right|_{a_k=0,\, k\neq i} \tag{2.31a}$$

$$S_{ki} = \left.\frac{b_k}{a_i}\right|_{a_k=0,\, k\neq i} \tag{2.31b}$$

The condition $a_k=0$, $k \neq i$ means that all ports, except port i, are terminated in their respective normalized impedances, such that the waves incident on these terminations are totally absorbed. We observed that in the special case when $Z_k = Z_0$, then S_{kk} in (2.30) is exactly (2.7) for the one-port case. Also if $Z_0 = 1\ \Omega$ for each port, then (2.21) reduces to (1.7a). Besides the commonly used transducer power gain G, there are two other forms of power gain: the available power gain G_a and the power gain G_p (Problem 2.16).

The generalized scattering parameters are defined in terms of specific load and generator impedances and are very useful in analyzing the stability and complex conjugate match of two-port networks, as will be seen in subsequent discussions. They also form the foundation for the theory of broadband matching discussed in Chapter 5. Since in practice the scattering parameters are usually normalized to 50 Ω to make broadband measurements. Then, to proceed with the design or to utilize the measured parameters, we must have an expression for the generalized scattering parameters in terms of the measured parameters and arbitrary generator and load impedances. Suppose the normalized impedance Z_i at each port i of N in Fig. 2.4 is changed to Z_i' $(i=1,2,\ldots,n)$. Then the incident and reflected scattering variables have to be redefined accordingly. The generalized scattering matrix $\mathbf{S}'$ normalized to Z_i' is, of course, different from $\mathbf{S}$. From (2.21) we get

$$\mathbf{S}' = \mathbf{R}'(\mathbf{Z}_N - \mathbf{Z}'^{T*})(\mathbf{Z}_N + \mathbf{Z}')^{-1}\mathbf{R}'^{-1} \tag{2.32}$$

where $\mathbf{R}'$ and $\mathbf{Z}'$ represent $\mathbf{R}$ and $\mathbf{Z}$, respectively, when Z_i is replaced by Z_i' for $i=1,2,\ldots,n$. Defining

$$\mathbf{\Gamma} = (\mathbf{Z}' - \mathbf{Z})(\mathbf{Z}' + \mathbf{Z}^{T*})^{-1} \tag{2.33}$$

and substituting (2.22) into (2.32) we get

$$\mathbf{S}' = \mathbf{R}'\mathbf{R}^{-1}(\mathbf{1}_n - \mathbf{S})^{-1}(\mathbf{S} - \mathbf{\Gamma}^{T*})(\mathbf{1}_n - \mathbf{\Gamma}^{T*})^{-1}(\mathbf{1}_n - \mathbf{\Gamma})(\mathbf{1}_n - \mathbf{S}\mathbf{\Gamma})^{-1}(\mathbf{1}_n - \mathbf{S})\mathbf{R}\mathbf{R}'^{-1} \tag{2.34}$$

Since

$$(\mathbf{1}_n - \mathbf{S})^{-1}(\mathbf{S} - \mathbf{\Gamma}^{T*})(\mathbf{1}_n - \mathbf{\Gamma}^{T*})^{-1} = (\mathbf{1}_n - \mathbf{\Gamma}^{T*})^{-1}(\mathbf{S} - \mathbf{\Gamma}^{T*})(\mathbf{1}_n - \mathbf{S})^{-1}$$

and

$$(\mathbf{1}_n - \mathbf{\Gamma})(\mathbf{1}_n - \mathbf{S}\mathbf{\Gamma})^{-1}(\mathbf{1}_n - \mathbf{S}) = (\mathbf{1}_n - \mathbf{S})(\mathbf{1}_n - \mathbf{\Gamma}\mathbf{S})^{-1}(\mathbf{1}_n - \mathbf{\Gamma})$$

then $\mathbf{S}'$ becomes

$$\mathbf{S}' = \mathbf{A}^{-1}(\mathbf{S} - \mathbf{\Gamma}^{T*})(\mathbf{1}_n - \mathbf{\Gamma}\mathbf{S})^{-1}\mathbf{A}^{T*} \tag{2.35}$$

where **A** is the diagonal matrix defined by

$$\mathbf{A}=\mathbf{R}'^{-1}\mathbf{R}(\mathbf{1}_n-\mathbf{\Gamma}^{T*}) \tag{2.36}$$

The ith diagonal element of **A** is given by

$$A_{ii}=|1-\Gamma_i|^{-1}(1-\Gamma_i^*)(1-|\Gamma_i|^2)^{1/2} \tag{2.37}$$

where

$$\Gamma_i=\frac{Z_i'-Z_i}{Z_i'+Z_i^*} \tag{2.38}$$

2.3 ANALYSIS OF TWO-PORT NETWORKS

This section is devoted to the analysis of two-ports, since they play an important role in the design of two-port microwave amplifiers. Throughout the analysis we let $\mathbf{S}=[S_{ij}]$ be the scattering matrix of the two-port N in Fig. 2.6 normalized to the real characteristic impedance Z_0 at the generator and at the load sides. We also let $\mathbf{S}'=[S_{ij}']$ be the generalized scattering matrix of N normalized to the generator impedance Z_s and the load impedance Z_L with the assumption that Z_s and Z_L have positive-real parts for all ω in the frequency band of interest B. In the design of microwave amplifiers, the most useful measure of the power flow through the amplifier is the transducer power gain G. It is defined as the ratio of the average power P_L delivered to the load Z_L to the maximum power available from the generator P_a, that is,

$$G=\frac{P_L}{P_a}=\frac{(\operatorname{Re} Z_L)|-I_2|^2}{|E_s|^2/4(\operatorname{Re} Z_s)} \tag{2.39}$$

From (2.29) it is clear that

$$G=|S_{21}'|^2 \tag{2.40}$$

We note that $\mathbf{S}'$ and $\mathbf{S}$ can be related by the equation (2.35), that is,

$$\mathbf{S}'=\mathbf{A}^{-1}(\mathbf{S}-\mathbf{\Gamma}^{T*})(\mathbf{1}_2-\mathbf{\Gamma}\mathbf{S})^{-1}\mathbf{A}^{T*} \tag{2.41a}$$

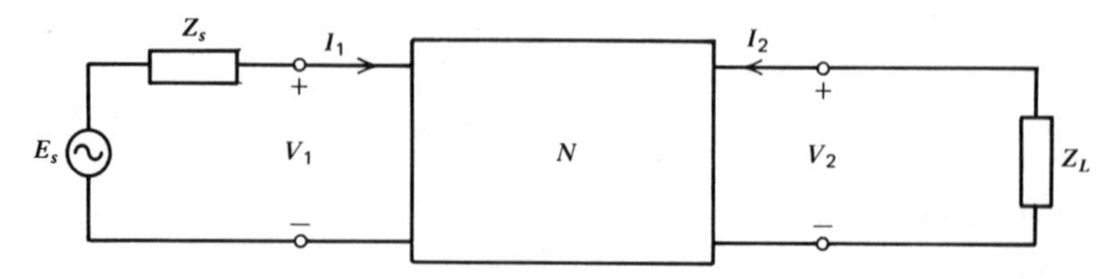

Fig. 2.6. The two-port network N embedded between the source and the load.

where

$$\mathbf{A}=\tfrac{1}{2}Z_0^{-1/2}\mathbf{R}'^{-1}(\mathbf{1}_2-\mathbf{\Gamma}^{T*})=\begin{bmatrix} A_1 & 0 \\ 0 & A_2 \end{bmatrix} \tag{2.41b}$$

$$\mathbf{\Gamma}=(\mathbf{Z}'-\mathbf{Z})(\mathbf{Z}'+\mathbf{Z}^{T*})^{-1}=\begin{bmatrix} \Gamma_s & 0 \\ 0 & \Gamma_L \end{bmatrix} \tag{2.41c}$$

$$\mathbf{Z}'=\begin{bmatrix} Z_s & 0 \\ 0 & Z_L \end{bmatrix}, \qquad \mathbf{Z}=Z_0\mathbf{1}_2 \tag{2.41d}$$

From (2.41) we get

$$S'_{21}=\frac{A_1^*}{A_2}\frac{S_{21}(1-|\Gamma_L|^2)}{(1-\Gamma_s S_{11})(1-\Gamma_L S_{22})-\Gamma_s\Gamma_L S_{12}S_{21}} \tag{2.42}$$

$$G=|S'_{21}|^2=|S_{21}|^2\frac{(1-|\Gamma_s|^2)(1-|\Gamma_L|^2)}{|(1-\Gamma_s S_{11})(1-\Gamma_L S_{22})-\Gamma_s\Gamma_L S_{12}S_{21}|^2} \tag{2.43}$$

If we now assume the network to be unilateral, that is, $S_{12}=0$, then (2.43) becomes

$$G_u=|S_{21}|^2\frac{1-|\Gamma_s|^2}{|1-\Gamma_s S_{11}|^2}\cdot\frac{1-|\Gamma_L|^2}{|1-\Gamma_L S_{22}|^2} \tag{2.44}$$

G_u is called the unilateral transducer power gain. The first term $|S_{21}|^2$ in (2.44) is a parameter of the two-port N and is fixed. For example, if N is a two-port transistor, then $|S_{21}|^2$ is determined and invariant once the bias conditions are established. The other two terms, however, are not only related to the remaining S-parameters of N, S_{11} and S_{22}, but also to the source and load reflection coefficients $\Gamma_s=(Z_s-Z_0)/(Z_s+Z_0)$ and $\Gamma_L=(Z_L-Z_0)/(Z_L+Z_0)$. It is these two quantities that we are able to control in the design of the two-port amplifier. From (2.44) we see that by choosing $\Gamma_s=S_{11}^*$ and $\Gamma_L=S_{22}^*$, G_u achieves its maximum, that is,

$$G_{u,\max}=|S_{21}|^2\frac{1}{1-|S_{11}|^2}\cdot\frac{1}{1-|S_{22}|^2} \tag{2.45}$$

Equation (2.45) shows that the maximum unilateral gain of N is determined solely by its S-parameters and is independent of the load or generator impedances.

In the case $S_{12}\neq 0$, the condition under which both input and output ports can be matched simultaneously to achieve a maximum G in (2.43) is much more complicated. From section 2.2, we know that the generator impedance $Z_s=R_s$

$+jX_s$ can be matched if and only if for a given load Z_L, the input impedance at port 1 of N in Fig. 2.6 is equal to Z_s^*, or equivalently, the reflection coefficient at port 1, Γ_i, is equal to Γ_s^*. From (1.21a) this implies that

$$\Gamma_i = \Gamma_s^* = S_{11} + \frac{S_{12}S_{21}\Gamma_L}{1-\Gamma_L S_{22}} \tag{2.46a}$$

Similarly, the load impedance Z_L can be matched if and only if for a given generator impedance Z_s, the output reflection coefficient Γ_o at port 2 of N in Fig. 2.6 is equal to Γ_L^*, that is,

$$\Gamma_o = \Gamma_L^* = S_{22} + \frac{S_{12}S_{21}\Gamma_s}{1-\Gamma_s S_{11}} \tag{2.46b}$$

where Γ_s and Γ_L are given in (2.41c). It is worth noting that (2.46a) and (2.46b) can be obtained by setting $S'_{11}=0$ and $S'_{22}=0$ (matched condition means no reflection). For simultaneous matching, (2.46a) and (2.46b) have to be satisfied at the same time. This yields the matched source and load reflection coefficients Γ_{sM} and Γ_{LM} as

$$\Gamma_{sM} = C_1^*\left[B_1 \pm \left(B_1^2 - 4|C_1|^2\right)^{1/2}\right] / \left(2|C_1|^2\right) \tag{2.47a}$$

$$\Gamma_{LM} = C_2^*\left[B_2 \pm \left(B_2^2 - 4|C_2|^2\right)^{1/2}\right] / \left(2|C_2|^2\right) \tag{2.47b}$$

where

$$B_1 = 1 + |S_{11}|^2 - |S_{22}|^2 - |\Delta|^2 \tag{2.48a}$$

$$B_2 = 1 - |S_{11}|^2 + |S_{22}|^2 - |\Delta|^2 \tag{2.48b}$$

$$C_1 = S_{11} - \Delta S_{22}^* \tag{2.48c}$$

$$C_2 = S_{22} - \Delta S_{11}^* \tag{2.48d}$$

$$\Delta = S_{11}S_{22} - S_{12}S_{21} \tag{2.48e}$$

The minus sign is used when $B_j > 0$ and the plus sign when $B_j < 0$ $(j=1,2)$.

We note that $\Gamma_{sM} = (Z_{sM} - Z_0)/(Z_{sM} + Z_0)$ and $\Gamma_{LM} = (Z_{LM} - Z_0)/(Z_{LM} + Z_0)$ where Z_{sM} and Z_{LM} are the matched source and load impedances. To ensure that the real parts of Z_{sM} and Z_{LM} are always positive, it is necessary that $|\Gamma_{sM}| < 1$ and $|\Gamma_{LM}| < 1$. A derivation from (2.47) yields the following necessary and sufficient conditions for simultaneous matching of a two-port [3]:

$$K = \frac{1 - |S_{11}|^2 - |S_{22}|^2 + |S_{11}S_{22} - S_{12}S_{21}|^2}{2|S_{12}S_{21}|} > 1 \tag{2.49a}$$

if $|S_{12}S_{21}| \neq 0$, or

$$|S_{11}| < 1, \qquad |S_{22}| < 1 \tag{2.49b}$$

if $|S_{12}S_{21}| = 0$.

The transducer gain under the simultaneously matched condition is obtained upon the substituting (2.47)–(2.49a) into (2.43)

$$G_{\max} = |S'_{21}|^2 = \left|\frac{S_{21}}{S_{12}}\right|\left(K + \sqrt{K^2 - 1}\right) \tag{2.50a}$$

if $|S_{12}S_{21}| \neq 0$ and $B_1 < 0$ and

$$G_{\max} = |S'_{21}|^2 = \left|\frac{S_{21}}{S_{12}}\right|\left(K - \sqrt{K^2 - 1}\right) \tag{2.50b}$$

if $|S_{12}S_{21}| \neq 0$ and $B_1 > 0$. When $|S_{12}S_{21}| = 0$, $G_{\max}$ in (2.50) reduce to $G_{u,\max}$ in (2.45).

In the design of amplifiers, stability is very important. A two-port amplifier is said to be unconditionally stable if the real parts of its input and output impedances remain positive for all passive load and source impedances. We note that the input reflection coefficient for a given load Z_L is given by the right side of (2.46a) and the output reflection coefficient is given by the right side of (2.46b) for a given source Z_s. Unconditional stability is satisfied if and only if

$$\left|S_{11} + \frac{\Gamma_L S_{12}S_{21}}{1 - \Gamma_L S_{22}}\right| < 1, \qquad |\Gamma_L| < 1 \tag{2.51a}$$

and

$$\left|S_{22} + \frac{\Gamma_s S_{12}S_{21}}{1 - \Gamma_s S_{11}}\right| < 1, \qquad |\Gamma_s| < 1 \tag{2.51b}$$

Note that (2.51a) and (2.51b) are equivalent to $|S'_{11}| < 1$ and $|S'_{22}| < 1$ for all $|\Gamma_s| < 1$ and $|\Gamma_L| < 1$. From (2.51) the necessary and sufficient conditions for unconditional stability is given by (Appendix A1):

$$K = \frac{1 - |S_{11}|^2 - |S_{22}|^2 + |S_{11}S_{22} - S_{12}S_{21}|^2}{2|S_{12}S_{21}|} > 1 \tag{2.51c}$$

and

$$|\Delta| = |S_{11}S_{22} - S_{12}S_{21}| < 1 \tag{2.51d}$$

We observe that (2.51c) is identical with (2.49a), that is, the simultaneous

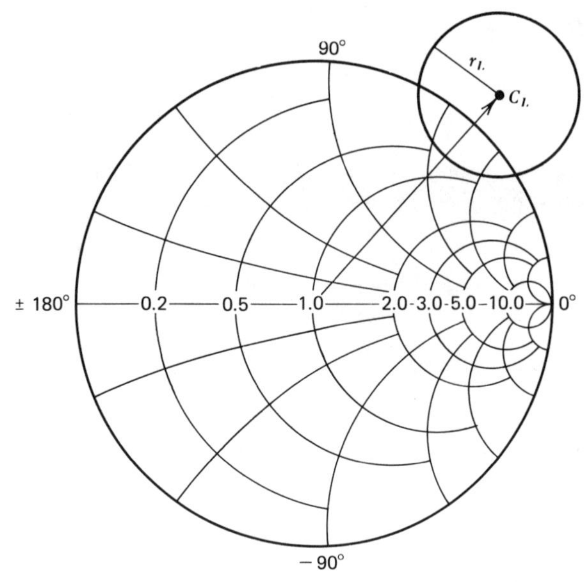

Fig. 2.7. Stability circle plotted on the Smith chart.

matching of two-ports is possible if they are unconditionally stable, but not vice versa.

If the two-port network is not unconditionally stable, it is termed potentially unstable, that is, the real part of its input and output impedances can be negative for some source and load impedances. Since many transistors are potentially unstable over a certain range of required frequency, it is necessary to know which load and source impedances provide stable operations. First, the load Z_L that causes the input impedance to have a negative real part is examined, that is, $|\Gamma_i|>1$ where Γ_i is given by (2.46a). The stability circle on the load plane has center C_L and radius r_L given as (Appendix A2)

$$C_L=\frac{S_{11}\Delta^*-S_{22}^*}{|\Delta|^2-|S_{22}|^2} \tag{2.52a}$$

$$r_L=\left|\frac{S_{12}S_{21}}{|\Delta|^2-|S_{22}|^2}\right| \tag{2.52b}$$

where $\Delta=S_{11}S_{22}-S_{12}S_{21}$, as shown in Fig. 2.7.

If we plot this load stability circle on the Smith chart,* we can determine all values of Γ_L that make $|\Gamma_i|=1$. The circle represents the boundary. To determine which area represents the stable region, let $Z_L=Z_0$ where Z_0 is the

*Explanation and use of the Smith chart can be found in electromagnetic theory texts such as references 1, 2, and 4.

normalized real characteristic impedance, then $\Gamma_L = 0$, and this represents the point at the center of the Smith chart. In this case $\Gamma_i = S_{11}$. If $|S_{11}| < 1$ then $|\Gamma_i| < 1$ and the whole region outside the load stability circle represents the stable region for Γ_L. If $|S_{11}| > 1$ then $|\Gamma_i| > 1$ and the whole region inside the load stability circle represents the stable region for Γ_L. If only passive loads are considered, then the portion of the Smith chart that lies outside the load stable circle represents the stable region of Γ_L if $|S_{11}| < 1$ and the one that lies inside the load stable circle if $|S_{11}| > 1$.

Similarly, the stability circle on the source plane has center C_s and radius r_s given as

$$C_s = \frac{S_{22}\Delta^* - S_{11}^*}{|\Delta|^2 - |S_{11}|^2} \tag{2.53a}$$

$$r_s = \left| \frac{S_{12}S_{21}}{|\Delta|^2 - |S_{11}|^2} \right| \tag{2.53b}$$

Example 2.1

A microwave transistor has the following scattering parameters at 6 GHz:

$$S_{11} = 0.614\angle -167.4^\circ \qquad S_{21} = 2.187\angle 32.4^\circ$$

$$S_{12} = 0.046\angle 65^\circ \qquad S_{22} = 0.716\angle -83^\circ$$

To check the stability of the transistor, we refer to (2.51) and get $|\Delta| = |0.342\angle 113.16^\circ| = 0.342 < 1$ and $K = 1.1296 > 1$. Thus the transistor is unconditionally stable. From (2.50b) we have $B_1 = 0.747 > 0$, therefore $G_{max} = 28.7$ or 14.58 dB. The maximum unilateral gain $G_{u,max} = 12$ dB as computed from (2.45). From (2.47) and (2.48) we have $B_2 = 1.02 > 0$ and hence $\Gamma_{sM} = 0.868\angle 169.75^\circ$ and $\Gamma_{LM} = 0.9\angle 84.5^\circ$.

Example 2.2

A microwave transistor has the following scattering parameters at 2 GHz:

$$S_{11} = 0.894\angle -60.6^\circ \qquad S_{21} = 3.122\angle 123.6^\circ$$

$$S_{12} = 0.020\angle 62.4^\circ \qquad S_{22} = 0.781\angle -27.6^\circ$$

Using (2.51) we get $|\Delta| = |0.697\angle -83^\circ| = 0.697 < 1$ and $K = 0.618 < 1$. Thus the transistor is potentially unstable. The load stability circle with center C_L and

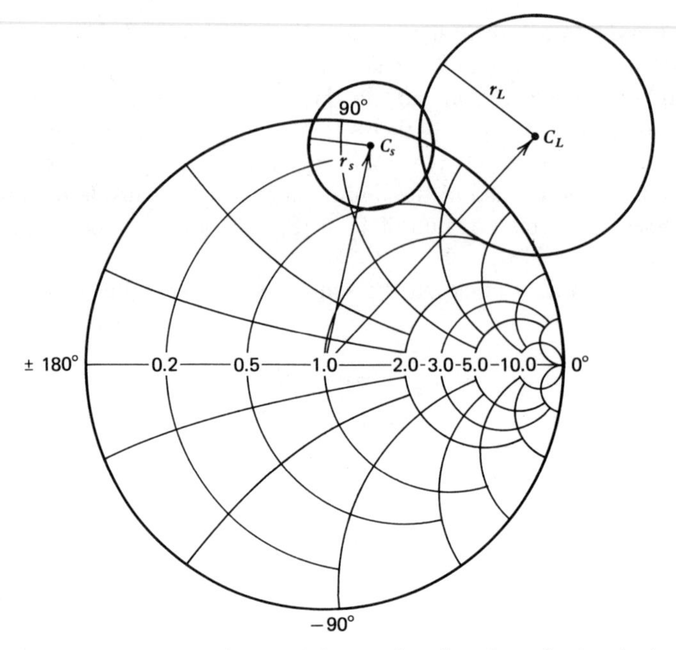

Fig. 2.8. Source and load stability circles plotted on the Smith chart.

radius r_L and the source stability circle with center C_s and radius r_s are given by (2.52) and (2.53) as

$$C_L = 1.367 \angle 46.9° \qquad r_L = 0.5$$

$$C_s = 1.136 \angle 68.4° \qquad r_s = 0.198$$

These circles are plotted on the Smith chart as shown in Fig. 2.8. The region of the Smith chart that lies outside these two circles is the stable load and source region.

2.4 CHANGE IN REFERENCE PLANE

In the measurement of the scattering parameters of two-port transistors or other active devices, it is impractical to attach the rf connectors to the actual device terminals because of their small size. Therefore transmission lines are usually added at the input and output port for ease of measurement, as shown in Fig. 2.9. The scattering matrix **S** of the device, represented by the two-port N, can be related to the scattering matrix **S**′ measured at the two new ports and the characteristic impedance Z_0 and electrical lengths θ_1 and θ_2 of the transmission lines. In order to find a relationship for **S** in terms of **S**′, Z_0, θ_1 and θ_2, consider the two-port transmission line (UL) shown in Fig. 2.1a. It can be

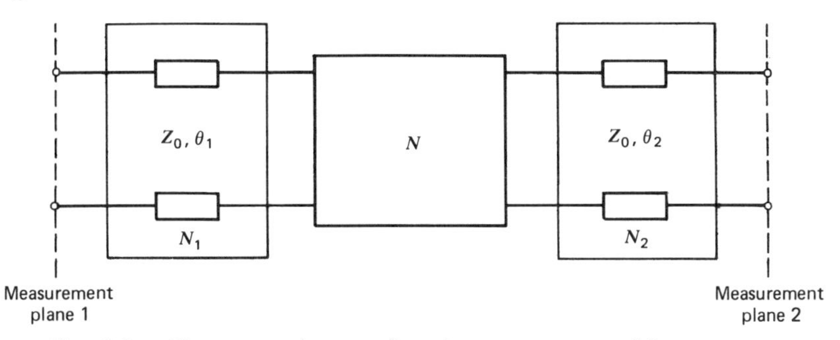

Fig. 2.9. Change in reference plane for measurement of S-parameters.

shown that its impedance matrix **Z** is given by [5, p. 328]

$$\mathbf{Z} = -jZ_0 \begin{bmatrix} \cot\theta & \csc\theta \\ \csc\theta & \cot\theta \end{bmatrix} \tag{2.54a}$$

Substituting the normalized impedance matrix $Z_0^{-1}\mathbf{Z}$ into (1.7a) yields the scattering matrix $\mathbf{\Sigma}$ normalized to Z_0 of the UL as

$$\mathbf{\Sigma} = \begin{bmatrix} 0 & e^{-j\theta} \\ e^{-j\theta} & 0 \end{bmatrix} \tag{2.54b}$$

From (1.9) the corresponding chain scattering matrix $\mathbf{\Phi}$ is given as

$$\mathbf{\Phi} = \begin{bmatrix} e^{j\theta} & 0 \\ 0 & e^{-j\theta} \end{bmatrix} \tag{2.54c}$$

Now let $\mathbf{\Phi}_1$, $\mathbf{\Phi}_2$, and $\mathbf{\Phi}_3$ be the chain scattering matrices of the two-ports N_1, N, and N_2 respectively. Then the overall chain scattering matrix $\mathbf{\Phi}'$ of the cascade chain is given by (1.23) as

$$\begin{aligned} \mathbf{\Phi}' &= \mathbf{\Phi}_1 \quad \mathbf{\Phi}_2 \quad \mathbf{\Phi}_3 \\ &= \begin{bmatrix} e^{j\theta_1} & 0 \\ 0 & e^{-j\theta_1} \end{bmatrix} \begin{bmatrix} S_{21}^{-1} & -S_{21}^{-1}S_{22} \\ S_{11}S_{21}^{-1} & S_{12} - S_{11}S_{21}^{-1}S_{22} \end{bmatrix} \begin{bmatrix} e^{j\theta_2} & 0 \\ 0 & e^{-j\theta_2} \end{bmatrix} \\ &= \begin{bmatrix} e^{j(\theta_1+\theta_2)}S_{21}^{-1} & -e^{j(\theta_1-\theta_2)}S_{21}^{-1}S_{22} \\ e^{-j(\theta_1-\theta_2)}S_{11}S_{21}^{-1} & e^{-j(\theta_1+\theta_2)}\left(S_{12} - S_{11}S_{21}^{-1}S_{22}\right) \end{bmatrix} \end{aligned} \tag{2.55}$$

The corresponding scattering matrix $\mathbf{S}'$ of $\mathbf{\Phi}'$ can be evaluated using (1.10) as

$$\mathbf{S}' = \begin{bmatrix} S_{11}e^{-j2\theta_1} & S_{12}e^{-j(\theta_1+\theta_2)} \\ S_{21}e^{-j(\theta_1+\theta_2)} & S_{22}e^{-j2\theta_2} \end{bmatrix} = \begin{bmatrix} S'_{11} & S'_{12} \\ S'_{21} & S'_{22} \end{bmatrix} \tag{2.56a}$$

We observe that $|S'_{ij}| = |S_{ij}|$ $(i, j = 1, 2)$, that is, the added ULs change the phase of the scattering parameters only and leave the magnitude invariant. From (2.56a), $\mathbf{S}$ can be directly calculated as

$$\mathbf{S} = \begin{bmatrix} S'_{11}e^{j2\theta_1} & S'_{12}e^{j(\theta_1+\theta_2)} \\ S'_{21}e^{j(\theta_1+\theta_2)} & S'_{22}e^{j2\theta_2} \end{bmatrix} \tag{2.56b}$$

This process is called the de-embedding of N.

2.5 NOISE IN LINEAR TWO-PORTS

In detection systems, weak signals are always disturbed by accompanying spurious signals called noise. In addition, any detection system adds additional noise in the process of both detection and amplification. The noise contribution of the detection system is usually the larger of these two, and since there is no control over the level of the input noise, the designer has little choice but to study, measure, and design a receiver such that its noise contribution is minimized. In the following discussion we assume the receiving system is linear because noise usually enters the system where the signal is low. Hence the receiver will always operate in its small-signal mode where linearity exists in the sense that the principle of superposition applies, that is, the effect of signals and noise is additive. Furthermore, the receiving systems considered here are assumed to be two-ports, where the noise entering the input port is processed to the output port in the same manner as the input signal. The noise generated by internal noise sources of the two-port is considered to be independent (uncorrelated) of the input signals and noise. The noise performance of the receiver is usually rated by its noise figure F defined by [6, 7]

$$F = \frac{N_o}{N_a G} = \frac{N_o}{kT_0 BG} > 1 \tag{2.57}$$

when the noise temperature of the source is standard (290°K). G is the two-port transducer power gain, N_o is the noise-output power, and N_a is the available noise-input power, which is equal to $kT_0 B$, where k is Boltzmann's constant (1.374×10^{-23} joule/°K), B is the noise bandwidth and $T_0 = 290$°K.

Of the total noise-output power $N_o = kT_0BGF$, a portion is the result of the amplified noise-input power kT_0BG; hence the amount of noise power added by the two-port is $(F-1)kT_0BG$. Another way to determine the noise performance of a two-port is the effective input noise temperature T_e defined as follows

$$T_e = T_0(F-1) \tag{2.58}$$

which is the temperature of the source when it is connected to a noise-free equivalent two-port that would result in the same noise-output power as that of the actual two-port connected to a noise-free source.

From the definition, the noise figure F depends on the structure of the two-port and the internal impedance of the source and can be expressed as (Appendix A8)

$$F = F_m + \frac{R_n}{G_s}\left[(G_s - G_m)^2 + (B_s - B_m)^2\right] \tag{2.59a}$$

where $Y_s = G_s + jB_s$ is the input admittance of the source, and F_m is called the optimum (minimum) noise figure of the two-port. This is the lowest noise figure that can be obtained through the adjustment of the source admittance Y_s. The optimum source admittance $Y_m = G_m + jB_m$ is that particular value of Y_s for which F_m is achieved. R_n is a positive number and has the dimension of resistance. Let $\Gamma_s = (Y_0 - Y_s)/(Y_0 + Y_s)$ and $\Gamma_m = (Y_0 - Y_m)/(Y_0 + Y_m)$ be the reflection coefficient of the source admittance and the optimum source admittance, normalized to the real characteristic admittance $Y_0 = 1/Z_0$. Then the noise figure F in (2.59a) can also be expressed as follows:

$$F = F_m + 4\frac{R_n}{Z_0} \cdot \frac{|\Gamma_s - \Gamma_m|^2}{|1+\Gamma_m|^2(1-|\Gamma_s|^2)} \tag{2.59b}$$

From (2.59a) we see that F is completely specified by four parameters: F_m, R_n, G_m and B_m or F_m, R_n, $|\Gamma_m|$ and $\angle\Gamma_m$ as in (2.59b).

In making a noise figure measurement of a two-port, a broadband noise source is used at the input. By ranging its equivalent absolute temperature, two levels of noise-output power are obtained at the output of a two-port as shown schematically in Fig. 2.10. When the noise source operates as an input termination at T_0(290°K), the maximum available noise power is kT_0B; hence, according to (2.57), the noise power added by the two-port is $(F-1)kT_0BG$ where G is the transducer power gain of the two-port. The noise-output power N_o in this case is $N_o = FkT_0BG$. When the noise source operates as an input termination at $T_1 \gg T_0$, the maximum available noise power is kT_1B and the noise-power output N_1 is $N_1 = kT_1BG + (F-1)kT_0BG$. Taking the ratio N_1/N_o we have

$$\frac{N_1}{N_o} = \frac{kT_1BG + (F-1)kT_0BG}{FkT_0BG} = \frac{T_1 + (F-1)T_0}{FT_0} \tag{2.60}$$

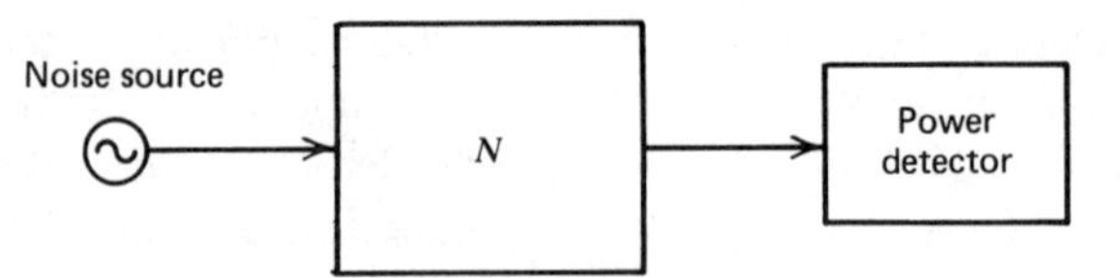

Fig. 2.10. Measurement of the noise figure of a two-port network.

Solving for F we get

$$F = \frac{T_1 - T_0}{T_0}\left(\frac{N_1}{N_o} - 1\right)^{-1} \tag{2.61a}$$

or

$$F\,(\mathrm{dB}) = 10\log\frac{T_1 - T_0}{T_0} - 10\log\left(\frac{N_1}{N_o} - 1\right) \tag{2.61b}$$

In (2.61), the ratio $(T_1 - T_0)/T_0$ is a measure of the relative excess noise power available from the noise source and is specified by the manufacturer of that source as a function of frequency. Therefore, in order to determine F in (2.61), only N_1/N_o needs to be measured. In practice N_1/N_o can be measured conveniently by available automatic noise figure measurement equipment.

In communication systems, the two-port subsystems are often connected in cascade. For example, a low noise amplifier in a receiver can consist of many stages in cascade. Hence, in order to design a system with a low noise figure, it is necessary to know the noise contribution of individual stages in the system. Consider a chain of n two-port networks $N_1, N_2, \ldots, N_n$ in direct cascade whose transducer power-gains over the noise bandwidth B are $G_1, G_2, \ldots, G_n$, respectively. The source supplies a noise-input power kT_0B which is amplified by n two-ports and hence appears at the output as $kT_0B\prod_{i=1}^{n}G_i$. The noise power added by N_1 is $(F_1 - 1)kT_0BG_1$. This is amplified by $N_2, N_3, \ldots, N_n$ and appears at the output as $(F_1 - 1)kT_0B\prod_{i=1}^{n}G_i$. Similarly, the output noise contributed by N_2 is $(F_2 - 1)kT_0B\prod_{i=2}^{n}G_i$ and of N_n is $(F_n - 1)kT_0BG_n$. Therefore, the noise figure F of the cascade chain is

$$F = \frac{kT_0B\left[\prod_{i=1}^{n} G_i + (F_1 - 1)\prod_{i=1}^{n} G_i + (F_2 - 1)\prod_{i=2}^{n} G_i + \cdots + (F_n - 1)G_n\right]}{kT_0B\prod_{i=1}^{n} G_i} \tag{2.62a}$$

$$= F_1 + \frac{F_2 - 1}{G_1} + \frac{F_3 - 1}{G_1G_2} + \cdots + \frac{F_n - 1}{\prod_{i=1}^{n-1} G_i} \tag{2.62b}$$

From (2.62b) we see that the overall noise figure of a cascade of two-ports depends strongly upon the noise figure of the first stage if $G_1 \gg 1$. The noise contribution of subsequent stages is reduced by the gain of the chain up to that point.

In the design of low noise amplifiers, the designer is faced with the task of determining the parameters F_m, G_m, B_m, and R_n in (2.59a) or F_m, $|\Gamma_m|$, $\angle\Gamma_m$ and R_n in (2.59b) to characterize the noise behavior of the two-port. A general procedure for obtaining these parameters is as follows:

(a) To determine F_m, low loss tuners are placed at the input and output of the two-port with the purpose of adjusting for the minimum noise figure of the two-port. This gives F_m.

(b) To determine Γ_m, the two-port is removed and the source reflection coefficient is measured (looking toward the noise source).

(c) To determine R_n, the input tuner is removed. In this case the source impedance is just Z_0, the real characteristic impedance of the noise source, and hence $\Gamma_s = 0$. The noise figure F' is measured when $\Gamma_s = 0$.

(d) R_n is then given by (2.59b)

$$R_n = Z_0(F' - F_m)\frac{|1+\Gamma_m|^2}{4|\Gamma_m|^2} \tag{2.63}$$

Often many (Γ_s, F) measurement pairs are obtained, and then F_m, Γ_m, and R_n are determined by a best fit algorithm.

2.6 CONCLUDING REMARKS

This chapter presented a detailed discussion of generalized reflection coefficients of one-port networks and generalized scattering matrices of n-ports. The results were applied to the analysis of linear two-ports. The concept of transducer power gain, maximum available power gain, and power gain (Problem 2.16) can be easily understood with complex source and load normalization. Since the scattering matrices of most active devices are measured with respect to real normalization Z_0 (50 Ω usually), a relationship between complex and real normalization scattering matrices was derived and used to obtain various forms of power gains in terms of the load and source reflection coefficients and the real normalized scattering matrix of the two-port network.

In the design of two-port amplifiers, one of the most important aspects is stability; a careless design can turn an amplifier into an oscillator, and in many cases the oscillation is destructive for devices such as GaAs FETs. Therefore a study of unconditional stability and potential unstability was discussed; the analysis was based entirely on the cascade-load expression of scattering matrices developed in Chapter 1, since it is simple and applies directly to the definition of stability.

The cascade-load formula is also useful in the complex conjugate match of two-ports; this provides the maximum transducer power gain. If the two-port is unconditionally stable then it can always be conjugately matched but not vice versa. If the two-port is potentially unstable ($K > 1$ can happen in this case with $|\Delta| > 1$) the use of a complex conjugate match must be tested by the load and source stability circles. In such cases it is more appropriate to use the power gain G_p (Problem 2.16) together with the constant power gain concept to be discussed in detail in Chapter 4.

It is worth noting that the concepts of stability and complex conjugate match can also be interpreted by the use of the generalized scattering matrix (Problems 2.20 and 2.21). Since this derivation is lengthy and complicated, especially in the stability analysis, we undertook the cascade-load approach. Information on unconditional stability and potential unstability is given in the references, but some are misleading and even wrong, so the reader must be cautious. A good discussion of unconditional stability is given in [8].

In practical situations, scattering parameters of active devices such as bipolar transistors, FETs, or diodes are measured on test fixtures with microstrip lines connecting their terminals to measurement test sets. In order to relate the measured parameters with the device parameters, a change in reference is necessary. Since most microstrip lines on test fixtures are very low loss (0.1 dB per wavelength) because of short length (less than a quarter wavelength) an approximation using uniform lossless transmission lines is good enough for many practical cases.

In space or satellite communications, where the received signal is very weak (below -130 dBm) the design of low noise receivers is of utmost importance. In order to understand the role of noise in two-ports, the last section of this chapter was devoted to this. Expressions for the noise figure in terms of the source admittance or reflection coefficient were given. This enables us to design narrow and broadband low noise amplifiers in Chapters 4 and 5. The cascade noise expression was derived to measure the contribution of subsequent cascaded stages in a chain of two-ports; this plays an important role in the design of low noise receivers.

PROBLEMS

2.1. The impedance matrix of a UL is given as

$$\mathbf{Z} = -jZ_0 \begin{bmatrix} \cot\theta & \csc\theta \\ \csc\theta & \cot\theta \end{bmatrix}$$

Verify Equations (2.9).

2.1. Derive Equations (2.47a) and (2.47b).

2.2. Derive Equations (2.50a) and (2.50b).

2.3. Show that the voltage gain A_v of a two-port with arbitrary source and load reflection coefficients Γ_s and Γ_L is given as

$$A_v = S_{21}(1+\Gamma_L)(1-S_{22}\Gamma_L)^{-1}(1+S'_{11})^{-1}$$

where

$$S'_{11} = S_{11} + S_{12}S_{21}\Gamma_L(1-S_{22}\Gamma_L)^{-1}$$

2.4. Define the unilateral figure of merit of a two-port as

$$u = |S_{11}S_{22}S_{12}S_{21}|\left|1-|S_{11}|^2\right|^{-1}\left|1-|S_{22}|^2\right|^{-1}$$

Show that

$$\frac{1}{1+u^2} < \frac{G}{G_u} < \frac{1}{1-u^2}$$

2.5. The scattering parameters of a transistor are measured to be $S_{11} = 0.277\angle -59°$, $S_{12} = 0.078\angle 93°$, $S_{21} = 1.92\angle 64°$, $S_{22} = 0.848\angle -31°$ at $f = 750$ MHz. Show that the transistor is stable at this frequency. Find the matched source and load reflection coefficients and the transducer power gain.

2.6. Verify Equation (2.53).

2.7. Another definition of two-port unconditional stability is that all port currents are zero under all passive source and load impedances. Let $\mathbf{Z}(j\omega) = \mathbf{Z}^T(j\omega)$ be the 2×2 impedance matrix of a reciprocal two-ports N. Prove that N is strictly passive, that is, $\mathrm{Re}\,\mathbf{Z}(j\omega) > 0$, if and only if it is unconditionally stable. Generalize the result to n-ports.

2.8. Let $\mathbf{Z}_i(j\omega)$ be the impedance matrices of 2 two-ports N_1 and N_2. Prove that if $\mathbf{Z}_1$ and $\mathbf{Z}_2$ possess identical principal minors and N_1 is unconditionally stable, then so is N_2. Generalize the result to n-ports.

2.9. Show that the necessary and sufficient conditions for a two-port with impedance matrix $\mathbf{Z}$ to be unconditionally stable are

$$\mathrm{Re}\, Z_{11} > 0, \qquad \mathrm{Re}\, Z_{22} > 0$$

and

$$2(\mathrm{Re}\, Z_{11})(\mathrm{Re}\, Z_{22}) > \mathrm{Re}(Z_{12}Z_{21}) + |Z_{12}Z_{21}|$$

2.10. Use the result of Problem 2.8 to show that if the two-port N with impedance matrix $\mathbf{Z}$ is unconditionally stable then so is the two-port N' with impedance matrix

$$\mathbf{Z}' = \begin{bmatrix} Z_{11} & (Z_{12}Z_{21})^{1/2} \\ (Z_{12}Z_{21})^{1/2} & Z_{22} \end{bmatrix}$$

2.11. Show that the two-port N with scattering matrix $\mathbf{S}$ is unconditionally stable if and only if the reciprocal two-port defined by

$$\mathbf{S}' = \begin{bmatrix} S_{11} & (S_{12}S_{21})^{1/2} \\ (S_{12}S_{21})^{1/2} & S_{22} \end{bmatrix}$$

is unconditionally stable.

2.12. Use the fact that a reciprocal two-port is unconditionally stable if and only if it is strictly passive to show (2.51c) and (2.51d).

2.13. Prove that the noise figure of a cascade of infinite numbers of two-ports with identical gain G and noise figure F is given by $1+[(F-1)/(1-G^{-1})]$ (in this case it is called noise measure).

2.14. The noise parameters of a GaAs FET are given at $f=4$ GHz as $R_n=23.14\ \Omega$, $\Gamma_m=0.618\angle 98°$ and $F_m=1.6$ dB with $Z_0=50\ \Omega$. Find the noise figure F if $\Gamma_s=0$.

2.15. The scattering parameters of a GaAs FET are measured to be $S_{11}=0.398\angle -81°$, $S_{12}=0.1175\angle 78°$, $S_{21}=1.38\angle 84°$, $S_{22}=0.75\angle -35°$ at $f=4$ GHz. Show that the FET is stable at this frequency. Show that the matched source and load reflection coefficients are $\Gamma_{sM}=0.651\angle 113.2°$, $\Gamma_{LM}=0.857\angle 40.7°$ with the maximum transducer power gain $G_{\max}=8.2$ dB.

2.16. In addition to the transducer power gain G, there are two other forms of power gain: the power gain G_p and the available power gain G_a defined as

$$G_p = \frac{P_L}{P_i} = \frac{\text{power delivered to load}}{\text{input power}}$$

$$G_a = \frac{(P_a)_L}{(P_a)_s} = \frac{\text{maximum available power at output}}{\text{maximum available power at input}}$$

Let $\mathbf{S}'$ be the scattering matrix of the two-port N normalized to the source and load impedances Z_s and Z_L. Show that $G_p=|S'_{21}|^2/(1-|S'_{11}|^2)$ and $G_a=|S'_{21}|^2/(1-|S'_{22}|^2)$. Now let $\mathbf{S}$ be the scattering matrix of N normalized to Z_0 at both input and output ports. Show that

$$G_p = \frac{|S_{21}|^2(1-|\Gamma_L|^2)}{(1-|S_{11}|^2)+|\Gamma_L|^2(|S_{22}|^2-|\Delta|^2)-2\,\mathrm{Re}(\Gamma_L C_2)}$$

and

$$G_a = \frac{|S_{21}|^2(1-|\Gamma_s|^2)}{(1-|S_{22}|^2)+|\Gamma_s|^2(|S_{11}|^2-|\Delta|^2)-2\,\mathrm{Re}(\Gamma_s C_1)}$$

where C_1 and C_2 are given in (2.48c) and (2.48d), respectively.

2.17. Show that G_p in Problem 2.16 is maximum when the load reflection coefficient

$$\Gamma_L = \left(S_{22} + \frac{S_{12}S_{21}}{1+S_{11}} \right)^*$$

and $G_{p,\max}$ is given by $G_{p,\max} = \frac{1}{2}K^{-1}|S_{21}/S_{12}|$ where K is in (2.51c) and $K<0.5$. For $K>1$ show $G_{p,\max} = G_{\max}$ in (2.50).

2.18. Show that for $K<1$, then

$$\Gamma_L = \left(S_{22} + \frac{S_{12}S_{21}}{1+S_{11}} \right)^*$$

and

$$\Gamma_s = \left(S_{11} + \frac{S_{12}S_{21}\Gamma_L}{1-S_{22}\Gamma_L} \right)^*$$

are stable load and source reflection coefficients if and only if $K>0$.

2.19. Show that $G_{a,\max} = G_{\max}$ in (2.50) if $K>1$ and $G_{a,\max}$ and $G_{\max}$ are infinite when $K<1$.

2.20. Derive the stability results of (2.51), (2.52), and (2.53) using the generalized scattering matrix of (2.41a). Hint: Note that

$$S'_{11} = \frac{A_1^*}{A_1} \frac{(1-\Gamma_L S_{22})(S_{11}-\Gamma_s^*) + \Gamma_L S_{12}S_{21}}{(1-\Gamma_s S_{11})(1-\Gamma_L S_{22}) - \Gamma_s \Gamma_L S_{12}S_{21}}$$

and $|S'_{11}|<1$ does not depend on the particular source or load reflection coefficients. Hence, let $\Gamma_s = 0$ and examine $|S'_{11}|=1$ in terms of Γ_L, and in a similar way let $\Gamma_L = 0$ and examine $|S'_{22}|=1$ in terms of Γ_s.

2.21. Derive (2.47) using the generalized scattering matrix. Hint: If a two-port is conjugately matched then $|S'_{11}|=0$ and $|S'_{22}|=0$. That is,

$$(1-\Gamma_L S_{22})(S_{11}-\Gamma_s^*) + \Gamma_L S_{12}S_{21} = 0$$

$$(1-\Gamma_s S_{11})(S_{22}-\Gamma_L^*) + \Gamma_s S_{12}S_{21} = 0$$

Eliminating Γ_L and Γ_s yields $\Gamma_L^2 - (B_2/C_2)\Gamma_L + C_2^*/C_2 = 0$ and $\Gamma_s^2 - (B_1/C_1)\Gamma_s + C_1^*/C_1 = 0$ where B_i and C_i $(i=1,2)$ are given in (2.48).

REFERENCES

1. G. J. Wheeler, *Introduction to Microwaves*. Englewood Cliffs, NJ: Prentice-Hall, 1963.
2. R. E. Collins, *Foundations for Microwave Engineering*. New York: McGraw-Hill, 1966.

3. K. Kurokawa, "Power waves and the scattering matrix," *IEEE Trans. Microwave Theory Tech.*, **MTT-13**, no. 3, 194–202, 1965.
4. D. T. Paris and F. K. Hurd, *Basic Electromagnetic Theory*. New York: McGraw-Hill, 1969.
5. G. C. Temes and J. W. LaPatra, *Introduction to Circuit Synthesis and Design*. New York: McGraw-Hill, 1977.
6. "IRE standards on methods of measuring noise in linear two-ports," *Proc. IRE*, **48**, no. 1, 60–68, 1960.
7. "Representation of noise in linear two-ports," *Proc. IRE*, **48**, no. 1, 69–74, 1960.
8. D. Wood, "Reappraisal of unconditional stability criteria for active two-port networks in terms of S-parameters," *IEEE Trans. Circuits Syst.*, **CAS-23**, no. 2, 73–81, 1976.
9. K. Kurokawa, *An Introduction to the Theory of Microwave Circuits*. New York: Academic Press, 1969.
10. D. C. Youla, "On scattering matrices normalized to complex port numbers," *Proc. IRE*, **49**, no. 7, 122, 1961.
11. D. C. Youla, "A stability characterization of the reciprocal linear passive *n*-port," *Proc. IRE*, **47**, no. 6, 1150–1151, 1959.
11. D. C. Youla, "A note on the stability of linear nonreciprocal *n*-ports," *Proc. IRE*, **48**, no. 1, 121–122, 1960.
12. W. H. Ku, "A simple derivation for the stability criterion of linear active two-ports," *Proc. IEEE*, **53**, no. 3, 310–311, 1965.
13. W. H. Ku, "Unilateral gain and stability criterion of active two-ports in terms of scattering parameters," *Proc. IEEE*, **54**, no. 11, 1617–1618, 1966.
14. K. Kurokawa, "Actual noise measure of linear amplifiers," *Proc. IRE*, **49**, no. 9, 1391–1397, 1961.
15. G. Caruso and M. Sanino, "Computer-aided determination of microwave two-port noise parameters," *IEEE Trans. Microwave Theory Tech.*, **MTT-26**, no. 9, 639–642, 1978.
16. R. Q. Lane, "The determination of device noise parameters," *Proc. IEEE*, **57**, no. 8, 1461–1462, 1969.
17. R. P. Meys, "A wave approach to the noise properties of linear microwave devices," *IEEE Trans. Microwave Theory Tech.*, **MTT-26**, no. 1, 34–37, 1978.
18. "Description of noise performance of amplifiers and receiving systems," *Proc. IEEE*, **51**, no. 3, 436–442, 1963.
19. S-Parameters...circuit analysis and design. Hewlett-Packard application note 95, Sept. 1968.
20. S-Parameter design. Hewlett-Packard application note 154, April 1972.

3

Introduction to Passive and Active Microwave Integrated Circuit Components

Microwave circuits can be integrated in hybrid or monolithic forms. Since the technology for the monolithic form is not yet well developed, most microwave integrated circuits (MICs) are in hybrid form in which the circuit interconnections are made by metal lines on insulating substrates. The active devices can be made of entirely different materials than the substrates and are attached to the substrates by soldering, bonding, and so forth. The passive components can be either distributed or lumped elements and may be fabricated separately and then attached to the substrates in the same way as the active devices. The active devices can be in either package or chip form. The package devices are commonly used at a lower frequency range than the chip devices since they exhibit more parasitic circuit elements that limit their performance at higher frequency. The advantage of packaging is to protect the devices during mounting in the circuit.

The purpose of this chapter is to introduce passive and active hybrid MICs suitable for the design of small-signal, low noise, and medium power solid-state amplifiers. The passive circuit elements are the microstrip lines and the lumped elements (inductors, capacitors, and resistors) as well as other passive elements such as circulators, isolators, and directional couplers. Approximations of lumped elements such as inductors and capacitors by distributed lines are also discussed. The main active devices to be studied are bipolar transistors, metal semiconductor field-effect transistors (MESFETs) and IMPATT diodes. Design models of these active devices will be given, and the bias circuits for amplifiers will be analyzed.

3.1 MICROSTRIP LINES

A section of a microstrip line is shown in Fig. 3.1. The metal conductor strip with width W and thickness t is deposited by thin film or thick film technology on a dielectric substrate of relative dielectric constant ϵ_r. Typical

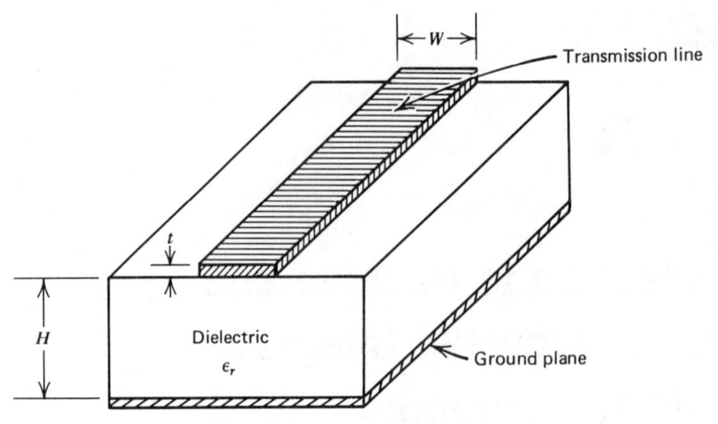

Fig. 3.1. A microstrip line.

substrate materials are alumina, Duroid®, sapphire, quartz and fused silica with ϵ_r varying from 2 to 12. Above the top of the substrate is air with $\epsilon_r = 1$. Microstrip lines are conveniently used for hybrid MICs since they can be fabricated cheaply. Passive lumped elements can be fabricated on the same substrate and chip or beam-leaded active devices can be bonded directly to the metal strips. The characteristic impedance of the microstrip line with width W, height H and negligible thickness t ($t/H \leq 0.005$) is given as [1] follows. For $W/H \leq 1$

$$Z_0 = \frac{60}{\sqrt{\epsilon_r'}} \ln\left(\frac{8H}{W} + \frac{W}{4H}\right) \tag{3.1a}$$

where the effective dielectric constant ϵ_r' is given as

$$\epsilon_r' = \frac{\epsilon_r + 1}{2} + \frac{\epsilon_r - 1}{2}\left[\left(1 + \frac{12H}{W}\right)^{-1/2} + 0.04\left(1 - \frac{W}{H}\right)^2\right] \tag{3.1b}$$

For $W/H \geq 1$

$$Z_0 = \frac{120\pi/\sqrt{\epsilon_r'}}{W/H + 1.393 + 0.667 \ln(W/H + 1.444)} \tag{3.1c}$$

where

$$\epsilon_r' = \frac{\epsilon_r + 1}{2} + \frac{\epsilon_r - 1}{2}\left(1 + 12\frac{H}{W}\right)^{-1/2} \tag{3.1d}$$

The expressions for W/H in terms of Z_0 and ϵ_r are as follows. For $W/H \leq 2$

$$\frac{W}{H} = \frac{8 \exp(A)}{\exp(2A) - 2} \tag{3.2a}$$

For $W/H \geq 2$

$$\frac{W}{H}=\frac{2}{\pi}\left[B-1-\ln(2B-1)+\frac{\epsilon_r-1}{2\epsilon_r}\left\{\ln(B-1)+0.39-\frac{0.61}{\epsilon_r}\right\}\right] \quad (3.2b)$$

where

$$A=\frac{Z_0}{60}\left(\frac{\epsilon_r+1}{2}\right)^{1/2}+\frac{\epsilon_r-1}{\epsilon_r+1}\left(0.23+\frac{0.11}{\epsilon_r}\right) \quad (3.2c)$$

$$B=\frac{377\pi}{2Z_0\epsilon_r^{1/2}} \quad (3.2d)$$

The zero-thickness ($t=0$) formulas given above can be modified to consider the thickness of the strip when W is replaced by an effective strip width W_e as follows ($t<H$ and $t<W/2$). For $W/H \geq \frac{1}{2}\pi$

$$W_e=W+\frac{t}{\pi}\left(1+\ln\frac{2H}{t}\right) \quad (3.3a)$$

For $W/H \leq \frac{1}{2}\pi$

$$W_e=W+\frac{t}{\pi}\left(1+\ln\frac{4\pi W}{t}\right) \quad (3.3b)$$

At higher frequencies, the effective dielectric constant ϵ_r' and Z_0 are both functions of the frequency. The dispersion in ϵ_r' is given as

$$\epsilon_r'(f)=\epsilon_r-\frac{\epsilon_r-\epsilon_r'}{1+(f/f_p)^2G} \quad (3.4a)$$

where

$$f_p=Z_0/8\pi H \quad (3.4b)$$

$$G=0.6+0.009Z_0 \quad (3.4c)$$

Here, f is in GHz and H is in centimeters. For $f_p \gg f$, $\epsilon_r'(f)=\epsilon_r'$. This means that high impedance lines on thin substrates are less dispersive. The dispersion in Z_0 is

$$Z_0(f)=\frac{377H}{W_e(f)\sqrt{\epsilon_r'(f)}} \quad (3.5a)$$

where

$$W_e(f) = W + \frac{W_e(0) - W}{1 + (f/f_p)^2} \tag{3.5b}$$

$$W_e(0) = \frac{377H}{Z_0(0)\sqrt{\epsilon_r'(0)}} \tag{3.5c}$$

In microstrip lines, the two sources of dissipative loss are conductor loss and dielectric loss. The conductor loss α_c is given as follows. For $W/H \leqslant 1/2\pi$

$$\alpha_c = \frac{8.68 R_s P}{2\pi Z_0 H}\left[1 + \frac{H}{W_e} + \frac{H}{\pi W_e}\left(\ln\frac{4\pi W}{t} + \frac{t}{W}\right)\right] \tag{3.6a}$$

For $1/2\pi < W/H \leqslant 2$

$$\alpha_c = \frac{8.68 R_s P Q}{2\pi Z_0 H} \tag{3.6b}$$

For $W/H \geqslant 2$

$$\alpha_c = \frac{8.68 R_s Q}{Z_0 H}\left[\frac{W_e}{H} + \frac{2}{\pi}\ln\left\{2\pi\left(\frac{W_e}{2H} + 0.94\right)\right\}\right]^{-2} \cdot \left[\frac{W_e}{H} + \frac{W_e/\pi H}{W_e/2H + 0.94}\right] \tag{3.6c}$$

where

$$P = 1 - (W_e/4H)^2 \tag{3.6d}$$

$$Q = 1 + \frac{H}{W_e} + \frac{H}{\pi W_e}\left(\ln\frac{2H}{t} - \frac{t}{\pi}\right) \tag{3.6e}$$

and the surface resistivity R_s is given in terms of the free space permeability μ_0 and the conductivity σ of the strip metal as

$$R_s = (\pi f \mu_0/\sigma)^{1/2} \tag{3.6f}$$

The dielectric loss with loss tangent, $\tan\delta$, is given by

$$\alpha_d = 27.3\frac{\epsilon_r(\epsilon_r' - 1)\tan\delta}{\lambda_0 \epsilon_r'^{1/2}(\epsilon_r - 1)} \quad \text{dB/cm} \tag{3.7a}$$

Table 3.1 Microstrip substrates

Material	ϵ_r	$\tan\delta$	Temperature Range (°C)	Flexibility	Thermal Expansion Coefficient ($\times 10^{-6}$/°C)
Duroid®	2.22	0.0006	−60 to 200	Good	32
Fused silica	3.80	0.0001	−60 to 500	Very poor	0.55
Boron nitride	4.40	0.0003	−60 to 500	Poor	2
Alumina	9.6–10.1	0.00008	−60 to 500	Very poor	7.5
Sapphire	9.6–10.1	0.00008	−60 to 500	Very poor	8.3
Epsilam-10®	10.0–10.5	0.002	−60 to 150	Good	11–23

For $\sigma \neq 0$

$$\alpha_d = 4.34 \frac{\sigma \mu_0^{1/2}(\epsilon_r' - 1)}{\epsilon_0 \epsilon_r'^{1/2}(\epsilon_r - 1)} \tag{3.7b}$$

where ϵ_0 is the free space permittivity.

The dielectric loss α_d is usually very small compared with the conductor loss α_c. A variety of dielectric materials are given in Table 3.1 and the design parameters W/H, ϵ_r' and Z_0 are given in Appendix A3. Since the fields between the strip and the ground plane are not contained entirely in the substrate, the wave propagating mode along the strip is not transverse electromagnetic (TEM) but quasi-TEM. The phase velocity in the microstrip is given by

$$v_p = \frac{c}{\sqrt{\epsilon_r'}} \tag{3.8a}$$

where c is the velocity of light. The wavelength λ in the microstrip lines is given as

$$\lambda = \frac{v_p}{f} = \frac{c}{f\sqrt{\epsilon_r'}} = \frac{11.803}{\sqrt{\epsilon_r'}\, f\,(\mathrm{GHz})} \quad \text{in.} \tag{3.8b}$$

where f is the frequency.

3.2 LUMPED ELEMENTS

A lumped element is very much smaller than the wavelength at its operating frequency and exhibits negligible phase shift. Because of thin film technology the size of lumped elements can be reduced and their operation at frequencies up to 12 GHz is now possible. The MIC lumped elements are resistors, inductors, and capacitors.

(a) *Resistors.* Resistors are used in MICs for bias networks, terminations, and attenuators; sometimes, but not often, they are used in amplifier matching networks. The properties of good microwave resistors are the same as those of low frequency resistors: good stability, low temperature coefficient of resistance, and good dissipation capability. A typical thin film resistor is shown in Fig. 3.2. It consists of a thin resistive film deposited on an insulating substrate. Thin film resistor materials are aluminum, nichrome, titanium, tantalium, copper, gold and so forth with resistivity ranging from 30 to 1000 Ω/□.

(b) *Inductors.* Inductance can be realized by a metal ribbon of width W, thickness t and length l given as [2, pp. 148–158]:

$$L = 5.08\times 10^{-3} l \left(\ln\frac{l}{W+t} + 1.19 + 0.022\frac{W+t}{l} \right) \quad \text{nH/mil} \tag{3.9a}$$

If round wire is used (as bonding wire in the MIC) the inductance is

$$L = 5.08\times 10^{-3} l [\ln(l/d) + 0.386] \quad \text{nH/mil} \tag{3.9b}$$

where d is the diameter of the wire (all dimensions are in mils). The ribbon and round wire inductors are practical for inductances of 2–3 nH or less. For larger inductances, spiral inductors are necessary. The thin film spiral inductor shown in Fig. 3.3 can give several nanohenries of inductance and is given as

$$L = 0.03125 n^2 d_o, \qquad d_o = 5d_i = 2.5n(W+s) \tag{3.10}$$

where all dimensions are in mils, and n is the number of turns. The inductance of the spiral coil changes slightly with the thickness t of the conductor. For a single turn flat circular loop of circumference l, strip

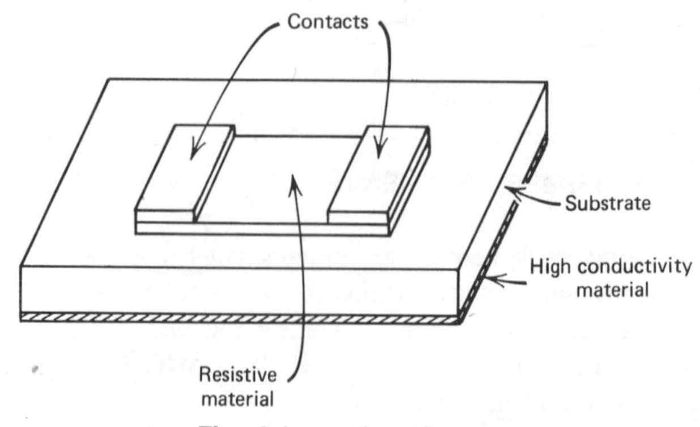

Fig. 3.2. A thin film resistor.

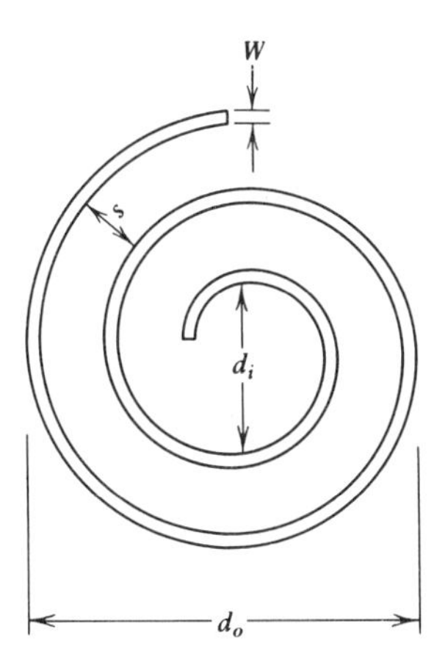

Fig. 3.3. A spiral inductor.

width W and thickness t, the inductance is given by

$$L = 5.08 \times 10^{-3} l \left(\ln \frac{t}{W+t} - 1.76 \right) \quad \text{nH/mil} \tag{3.11}$$

For the purpose of constructing rf chokes, a greater inductance per usable area can be achieved with square and rectangular spirals than with circular spirals. The expression for high frequency square spirals is given by

$$L = 8.5 A^{1/2} n^{5/3} \quad \text{nH} \tag{3.12}$$

where A is the surface area in square centimeters and n is the number of turns.

(c) *Capacitors.* The two types of capacitors that are used widely in MICs are the metal-oxide-metal capacitor shown in Fig. 3.4 and the interdigital capacitor shown in Fig. 3.5. The metal-oxide-metal capacitor has three layers; the bottom and top layers are electrodes that sandwich the middle dielectric layer. The capacitance can be approximated as $C = \epsilon_0 \epsilon_r Wl/H$ where $\epsilon_0 = 8.86 \times 10^{-12}$ F/cm is the permittivity of free space, ϵ_r is the

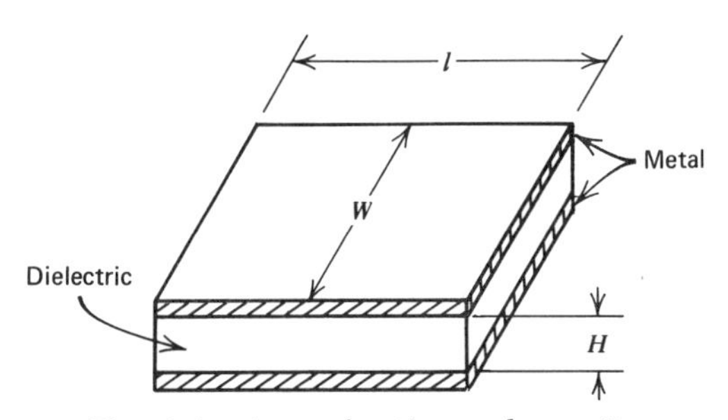

Fig. 3.4. A metal-oxide-metal capacitor.

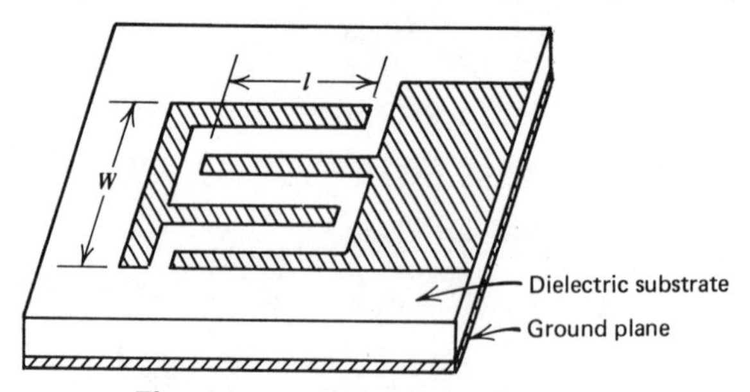

Fig. 3.5. An interdigitated capacitor.

relative dielectric constant of the dielectric material. The interdigital capacitor is fabricated on a single-layer structure as opposed to the three-layer structure of the metal-oxide-metal capacitor. It can be fabricated easily on substrates as microstrip lines with values between 0.1 and 15 pF. The large capacitors resonate at low frequencies; therefore only small interdigital capacitors (less than 2 pF) are practical at microwave frequencies (above 2 GHz). The capacitance can be approximated as [2, pp. 161–167]

$$C=\frac{\epsilon_r+1}{W}l[(N-3)A_1+A_2] \qquad \text{F/unit length} \tag{3.13}$$

where N is the number of fingers. The constants A_1 and A_2 represent the contribution of the interior and two exterior fingers and are a function of H/W. If the substrate is thick enough, A_1 and A_2 can be approximated by $A_1=0.225$ and $A_2=0.252$ pF/in. (actually, these values are for infinitely thick substrates).

3.3 MICROWAVE APPROXIMATION OF LUMPED-ELEMENTS (SEMILUMPED ELEMENTS)

Since truly lumped elements such as inductors with low loss are difficult to obtain at frequencies above 4 GHz, it is necessary to approximate lumped elements whenever appropriate with transmission lines that can be fabricated on microstrip substrates with low loss and much lower cost. These semilumped elements, such as series inductors, shunt inductors, and shunt capacitors, are short lengths of transmission lines that can be made to approximate the characteristics of the lumped elements by proper choice of their length and characteristic impedance. Consider a uniform lossless transmission line (UL) and a series inductor L_s shown in Fig. 3.6a and b with their scattering matrices $\mathbf{S}_1=[(S_{ij})_1]$ and $\mathbf{S}_2=[(S_{ij})_2]$ normalized to real characteristic impedances Z_{01}

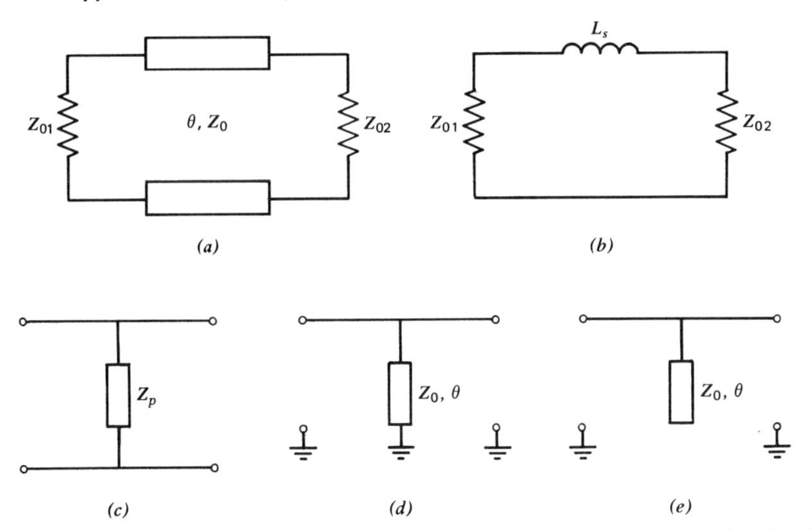

Fig. 3.6. (*a*) A UL normalized to Z_{01} and Z_{02}. (*b*) A series inductor normalized to Z_{01} and Z_{02}. (*c*) A shunt impedance Z_p. (*d*) A short-circuited UL. (*e*) An open-circuited UL.

and Z_{02} given as

$$(S_{11})_1 = \Big[\Big(\sqrt{Z_{02}/Z_{01}} - \sqrt{Z_{01}/Z_{02}}\Big)\cos\theta + j\Big(Z_0/\sqrt{Z_{01}Z_{02}} - \sqrt{Z_{01}Z_{02}}/Z_0\Big)\sin\theta\Big]/D_1$$

$$(S_{22})_1 = \Big[\Big(\sqrt{Z_{01}/Z_{02}} - \sqrt{Z_{02}/Z_{01}}\Big)\cos\theta + j\Big(Z_0/\sqrt{Z_{01}Z_{02}} - \sqrt{Z_{01}Z_{02}}/Z_0\Big)\sin\theta\Big]/D_1$$

$$(S_{12})_1 = (S_{21})_1 = 2/D_1$$

where

$$D_1 = \Big(\sqrt{Z_{02}/Z_{01}} + \sqrt{Z_{01}/Z_{02}}\Big)\cos\theta + j\Big(Z_0/\sqrt{Z_{01}Z_{02}} + \sqrt{Z_{01}Z_{02}}/Z_0\Big)\sin\theta \tag{3.14}$$

$$(S_{11})_2 = \Big(\sqrt{Z_{02}/Z_{01}} - \sqrt{Z_{01}/Z_{02}} + j\omega L_s/\sqrt{Z_{01}Z_{02}}\Big)/D_2$$

$$(S_{22})_2 = \Big(\sqrt{Z_{01}/Z_{02}} - \sqrt{Z_{02}/Z_{01}} + j\omega L_s/\sqrt{Z_{01}Z_{02}}\Big)/D_2$$

$$(S_{12})_2 = (S_{21})_2 = 2/D_2$$

where

$$D_2 = \left(\sqrt{Z_{02}/Z_{01}} + \sqrt{Z_{01}/Z_{02}}\right) + j\omega L_s/\sqrt{Z_{01}Z_{02}} \tag{3.15}$$

To approximate the series inductor L_s by the UL it is necessary that (3.14) and (3.15) are approximately equivalent. We observe that if θ is small ($\cos\theta \approx 1$ and $\sin\theta \approx \theta$) and Z_0 is large compared to $\sqrt{Z_{01}Z_{02}}$ then S_1 and S_2 are approximately equivalent if we let

$$\omega L_s = Z_0\theta \tag{3.16a}$$

Hence

$$\theta = \omega L_s/Z_0 \tag{3.16b}$$

In practice $Z_0 \geqslant 2\sqrt{Z_{01}Z_{02}}$ and $\theta \leqslant \pi/8$ yield good approximations, where θ is evaluated at the midband frequency of the operating band.

To find the UL approximation for the shunt inductor and capacitor, consider the shunt impedance Z_p in Fig. 3.6*c*. Its chain matrix is given as

$$\mathcal{C} = \begin{bmatrix} 1 & 0 \\ 1/Z_p & 1 \end{bmatrix} \tag{3.17}$$

where $Z_p = j\omega L_p$ for a shunt inductor L_p, $Z_p = -j/\omega C_p$ for a shunt capacitor C_p, $Z_p = jZ_0 \tan\theta$ for a short-circuited UL (Fig. 3.6*d*) and $Z_p = -jZ_0/\tan\theta$ for an open-circuited UL (Fig. 3.6*e*). If θ is small, then $\tan\theta \approx \theta$ and the shunt inductor L_p can be approximated by a short-circuited UL if we let

$$\theta = \omega L_p/Z_0 \tag{3.18a}$$

Furthermore, the shunt capacitor C_p can be approximated by an open-circuited UL if

$$\theta = \omega C_p Z_0 \tag{3.18b}$$

where θ is evaluated at the midband frequency of the band. In practice, $\theta \leqslant \pi/8$ yields good results.

3.4 DIRECTIONAL COUPLERS

A directional coupler is a four-port passive network that is used to transfer energy from one circuit to another. Its symbol is given in Fig. 3.7 and its

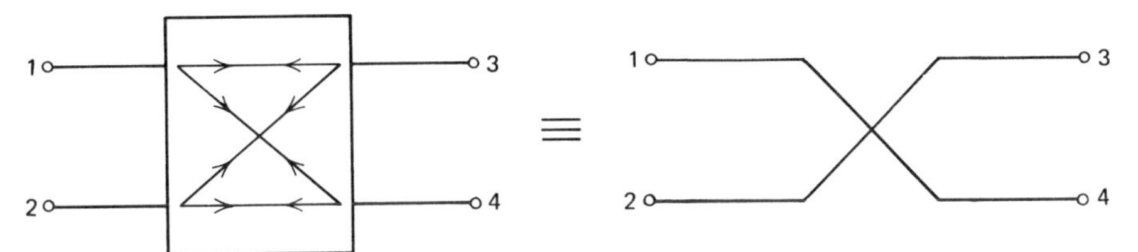

Fig. 3.7. The symbol of the directional coupler.

scattering matrix is given as

$$S = \begin{bmatrix} 0 & 0 & \alpha & j\beta \\ 0 & 0 & j\beta & \alpha \\ \alpha & j\beta & 0 & 0 \\ j\beta & \alpha & 0 & 0 \end{bmatrix} \tag{3.19}$$

where α and β are real numbers and satisfy the relationship

$$\alpha^2 + \beta^2 = 1 \tag{3.20}$$

Thus the wave incident at port 1 couples power to ports 3 and 4 but not port 2. Similarly, the wave incident at port 2 couples to ports 3 and 4 but not port 1. Therefore, ports 1 and 2 are isolated. The directional coupler can be made on a microstrip substrate using coupled lines as shown in Fig. 3.8, or by branch lines as in Fig. 3.9, where λ is evaluated at midband frequency.

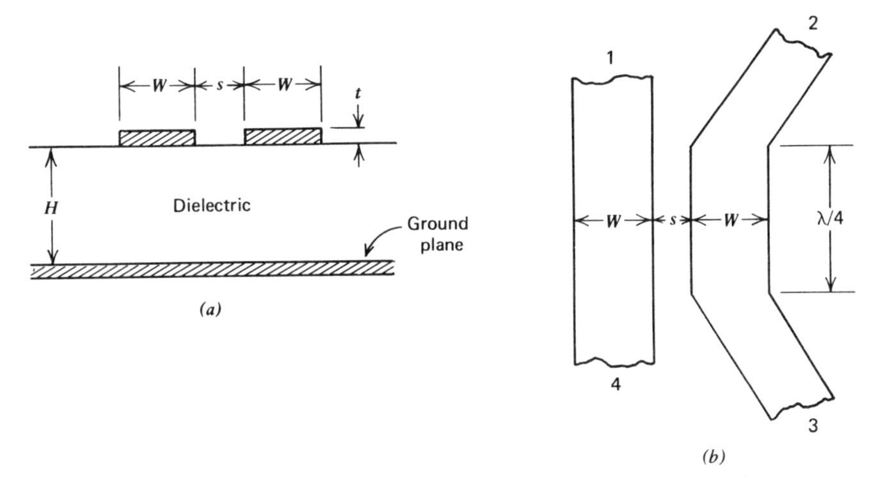

Fig. 3.8. (a) and (b) A parallel-coupled microstrip line coupler.

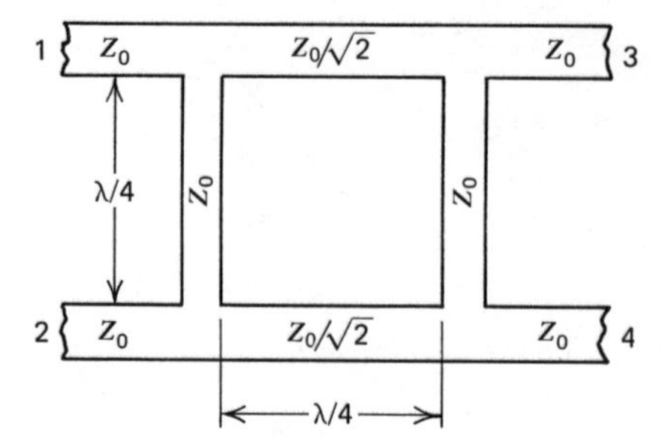

Fig. 3.9. A branched line 3 dB coupler.

3.5 CIRCULATORS AND ISOLATORS

An ideal three-port circulator is defined as a lossless network in which the energy incident on port 1 is routed to port 2 and none is routed to port 3. This also happens for port 2 and port 3 in that circular sequence. The symbol for the ideal circulator is shown in Fig. 3.10 and its scattering matrix is given as

$$\mathbf{S}=\begin{bmatrix} 0 & 0 & 1 \\ 1 & 0 & 0 \\ 0 & 1 & 0 \end{bmatrix} \tag{3.21}$$

A common application of the circulator is the isolator. Its symbol is shown in Fig. 3.11. Port 3 of the circulator is terminated in a matched load, and the isolator is a unilateral two-port network with transmission from port 1 to port 2 only. All the signal reflected from port 2 is absorbed by the load in port 3. The isolator scattering matrix is given as

$$\mathbf{S}=\begin{bmatrix} 0 & 0 \\ 1 & 0 \end{bmatrix} \tag{3.22}$$

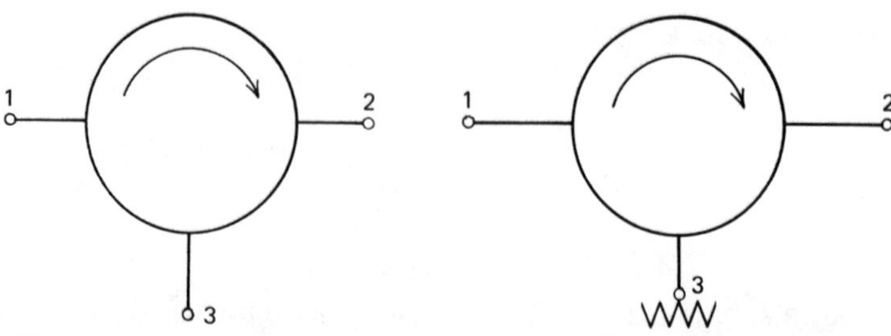

Fig. 3.10. The circulator.

Fig. 3.11. The isolator.

3.6 GALLIUM ARSENIDE FIELD-EFFECT TRANSISTORS (GaAs FET)

The field-effect transistor (FET) is a semiconductor device whose current is controlled by an electric field. It is a unipolar device because the current is carried by one type of carrier, the majority carrier of the semiconductor, which flows along the element under the applied electric field at its ends. This field is applied to the end of the semiconductor element by ohmic contacts, which are called the source and drain electrodes. The controlling of the electric field is done at a junction called the gate electrode. These gates can be metallic Schottky barriers (MESFET), *p-n* junctions (JFET), metals on oxide layers (MOSFET), or metals on insulating layers other than oxides (IGFET).

Most microwave FETs are fabricated on gallium arsenide (GaAs) instead of silicon, since GaAs has a higher electron bulk mobility and greater maximum

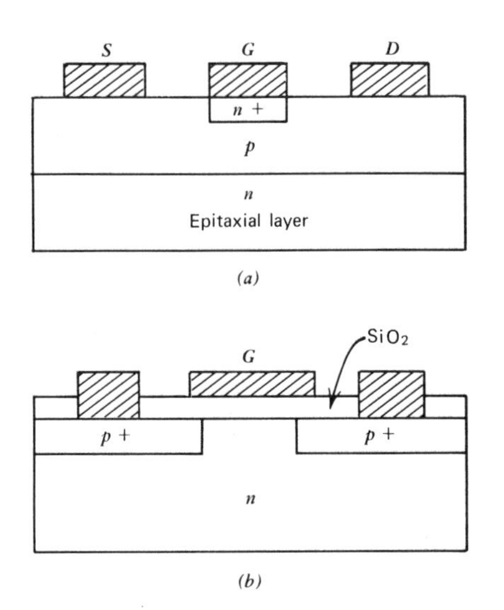

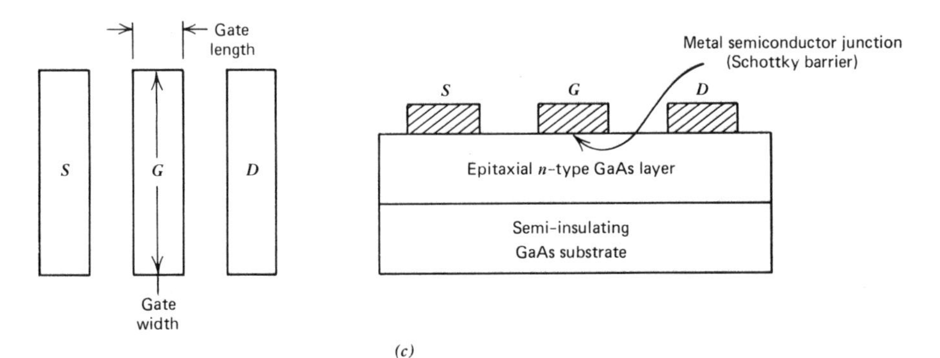

Fig. 3.12. (*a*) A JFET. (*b*) An IGFET. (*c*) A MESFET.

electron drift velocity than silicon. Therefore the parasitic resistances are smaller, the transconductance is larger, and the transit time of electrons in the high field region is shorter. These properties permit lower noise figure, higher gain and higher cutoff frequencies, all important characteristics for microwave transistors.

Today MESFETs, JFETs, and IGFETs are used at microwave frequencies. The cross sections of their structures are illustrated in Fig. 3.12*a–c*. At the present time GaAs MESFETs are the most popular GaAs FETs for microwave applications such as low noise, high gain and high power amplifications above 3 GHz. The reasons for this are easy realization on GaAs and the fact that the two critical dimensions—the gate length and channel thickness—can be accurately controlled by photolithography, ion implantation and molecular epitaxy.

3.6.1 GaAs MESFET Structure and Operation

In the fabrication of a GaAs MESFET, a very thin (less than 1 μ thick for low operating voltages and good noise and gain performance) epitaxial *n*-type GaAs layer (doped with either sulfur or tin) is grown on a semi-insulating GaAs substrate using vapor or liquid epitaxial techniques. The semi-insulating GaAs substrate material is produced by doping pure GaAs with chromium. Diffusion of impurities from this substrate during the epitaxial growth process of the active thin layer can degrade the electrical properties of the layer. For superior noise and gain performance, an intermediate semi-insulating buffer layer may be grown on the substrate prior to the growth of the active thin layer. The buffer layer is relatively pure compared to the substrate and acts as a barrier against the diffusion of impurities from the substrate into the active thin layer. Once the material has been grown, Au-Te-Ge source and drain areas are formed on the surface of the active thin layer and sintered to produced low resistance ohmic contacts. The gate control electrode is an area running between the source and the drain; the metal may be Cr-Ni-Au, Cr-Au, or Al-Ge, and forms a metal-to-semiconductor contact (Schottky barrier diode) to the GaAs. Standard photolithography is used to define the gate and this enables a short gate length (0.5 μ) to be achieved. For good microwave performance the source-gate spacing and the gate length should be small in order to minimize the source resistance and the electron transit time through the device.

In the FET, electrons flow from the source to the drain through the active thin layer when a positive voltage V_{DS} is applied to the drain with no gate-to-source voltage. For a small V_{DS} the active layer behaves like a linear resistor. For a larger V_{DS} the electron drift velocity does not increase at the same rate as the electric field and consequently the current-voltage characteristic falls below the initial resistor line. When V_{DS} reaches a large enough value, the electron drift velocity reaches its maximum and the electron flow starts to saturate. When the gate is shorted to the source and a small V_{DS} is applied, a depletion layer is formed in the active layer and restricts the electron flow from the source to the drain (higher source-drain resistance). When a negative voltage is applied between the gate and the source, the gate-to-channel junction is reverse biased, and the depletion layer is deeper. For small values of V_{DS} the active layer will act as a linear resistor, but this resistance will be larger due to a narrower cross section available to the electron flow than when the gate is shorted to the source (i.e., the gate-to-source voltage $V_{GS}=0$). As V_{DS} increases, the critical field is reached at a lower level than when $V_{GS}=0$ due to larger active layer resistance, and the electron flow will saturate for further increases in V_{DS}. When V_{GS} is negative enough, the depletion layer reaches the semi-insulating substrate and the electron flow from the source to the drain is cut off. In conclusion, the operation of the MESFET is controlled by the active thin layer, whose thickness can be varied by the depletion layer under the metal-to-semiconductor junction (gate). The depletion layer depth is the effect of the applied negative voltage

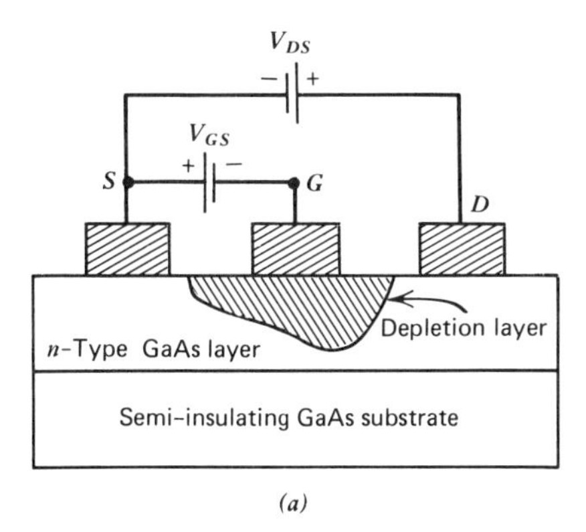

(a)

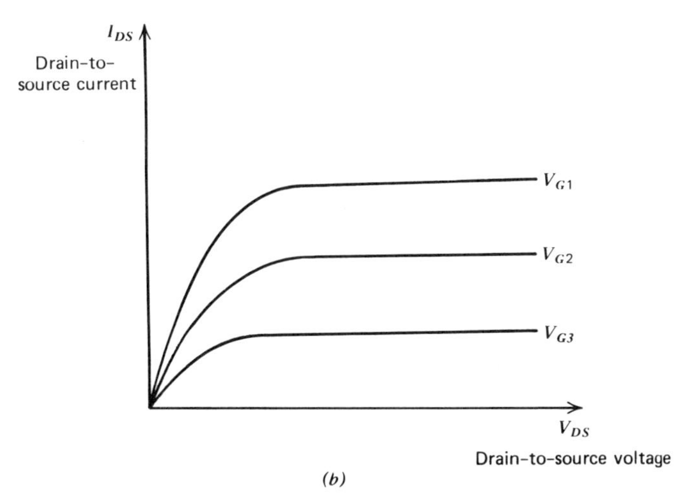

(b)

Fig. 3.13. (*a*) MESFET operation. (*b*) *I-V* characteristics with $V_{GN} < V_{G(N+1)} < 0$.

between the gate and the source. The current-voltage (I-V) characteristics resulting from this type of operation are shown in Fig. 3.13*a*–*b*.

3.6.2 Small-Signal Equivalent Circuit

The small-signal equivalent circuit for the GaAs MESFET can be represented accurately by lumped elements up to 12 GHz for the extrinsic and 14 GHz for the intrinsic FET as is shown in Fig. 3.14 for common source operation. In the intrinsic model, the depletion layer capacitance under the gate electrode is denoted by the source-gate capacitance C_i and its charging resistance in the channel by R_i; C_f represents the fringing capacitance between the drain and the gate, and R_o shows the effect of drain-source channel resistance. R_S, R_g, R_d, L_s, L_i, C_o, and L_o represent the extrinsic parasitic elements. The transconductance $g_m = g_{m0}e^{-j\omega\tau_0}$ where g_{m0} is independent of frequency and τ_0 is a phase delay. This delay corresponds to the time required for electrons to traverse the gate length at the scattering-limited velocity. For a 0.5 μ gate GaAs MESFET, $\tau_0 \approx L_g/v_s = 0.5\times10^{-4}$ cm$/2\times10^7$ cm sec$^{-1} = 2.5$ psec.

The activity of the FET is determined by its maximum frequency of oscillation $\omega_{\max}$ where the FET becomes passive. Consider the intrinsic model of the FET in Fig. 3.14. Its admittance matrix $\mathbf{Y} = [Y_{ij}]$ is given as

$$\mathbf{Y} = \begin{bmatrix} \dfrac{\omega^2\tau_i C_i}{1+\omega^2\tau_i^2} + j\omega(C_i + C_f) & \dfrac{g_m}{1+\omega^2\tau_i^2} - j\omega(g_m\tau_i + C_f) \\ -j\omega C_f & G_o + j\omega C_f \end{bmatrix} \tag{3.23}$$

where $\tau_i = C_i R_i$ and $G_o = 1/R_o$. When the FET is passive we get $\mathbf{Y} + \mathbf{Y}^{T*} \geqslant \mathbf{0}$,

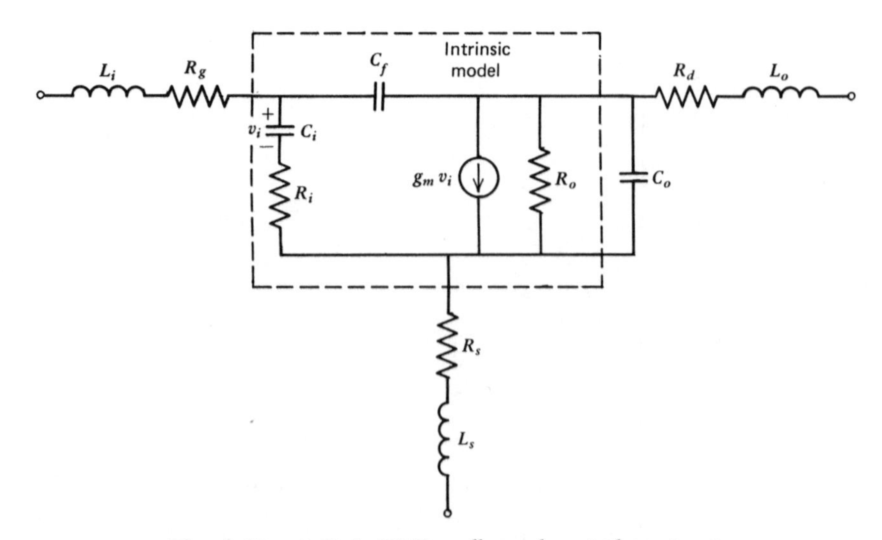

Fig. 3.14. A GaAs FET small-signal equivalent circuit.

hence

$$\mathbf{Y}+\mathbf{Y}^{T*}=\begin{bmatrix} \dfrac{2\omega^2\tau_i C_i}{1+\omega^2\tau_i^2} & \dfrac{g_m}{1+\omega^2\tau_i^2}-j\omega g_m\tau_i \\ \dfrac{g_m}{1+\omega^2\tau_i^2}+j\omega g_m\tau_i & 2G_o \end{bmatrix} \geqslant \mathbf{0} \tag{3.24}$$

implies that the determinant of $\mathbf{Y}+\mathbf{Y}^{T*}$ is nonnegative, that is,

$$\frac{4\omega^2\tau_i C_i G_o}{1+\omega^2\tau_i^2}-\frac{g_m^2}{\left(1+\omega^2\tau_i^2\right)^2}-\omega^2 g_m^2\tau_i^2 \geqslant 0 \tag{3.25}$$

In practice $\omega_{\max}^{-1} \gg \tau_i$, and (3.25) can be approximated as

$$4\omega_{\max}^2\tau_i C_i G_o - g_m^2 - \omega_{\max}^2 g_m^2 \tau_i^2 \geqslant 0 \tag{3.26}$$

which yields

$$\omega_{\max} \approx g_m\left(4\tau_i C_i G_o - g_m^2\tau_i^2\right)^{-1/2} \tag{3.27}$$

For a GaAs MESFET with a 1 μ gate length, the maximum frequency of oscillation is about 30–40 GHz.

For the extrinsic FET, the parasitic elements reduce $\omega_{\max}$ considerably. The maximum available power gain $G_{a,\max}$ of the extrinsic model can be approximated by [3]

$$G_{a,\max} \approx \alpha\left(\frac{\omega_T}{\omega}\right)^2 \tag{3.28a}$$

where

$$\omega_T = g_m/(C_i + C_f) \tag{3.28b}$$

and

$$\alpha=\left[4G_o\left(R_i+R_s+\tfrac{1}{2}\omega_T L_s\right)+2\omega_T C_f\left(R_i+R_s+\omega_T L_s\right)\right]^{-1} \tag{3.28c}$$

From (3.28a) we see that the FET has a gain roll-off of about 6 dB/octave. This is true for most microwave GaAs MESFETs. Furthermore, in the operating range, the feedback capacitance C_f is very small compared to C_i, and the reverse transmission scattering parameter S_{12} of the GaAs FET is very small in magnitude compared to S_{21}; hence $S_{12}=0$ can be assumed in the design if the FET is unconditionally stable in the operating frequency range. Such a unilateral model can be seen in Fig. 3.15 for a common source configuration. This kind of model

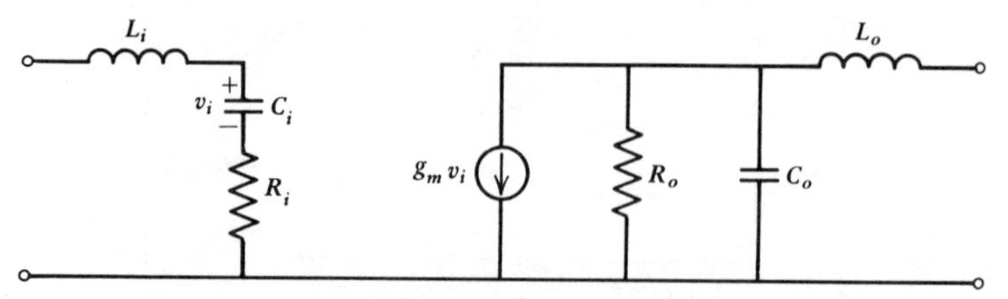

Fig. 3.15. A GaAs FET simplified unilateral model.

adequately represents practical FETs over a broad frequency band (more than an octave bandwidth).

3.6.3 Noise

The noise behavior of the FET can be represented by two noise sources connected across the input and the output ports as shown in Fig. 3.16 for the intrinsic FET. The current generator i_{nd} represents the channel noise, and its mean square can be expressed by [4]

$$\overline{i_{nd}^2} = 4kT_0 \Delta f g_{m0} P \tag{3.29}$$

where k is Boltzmann's constant, T_0 is the device temperature, Δf is the bandwidth, g_{m0} is the magnitude of the low frequency transconductance and P is a factor depending upon the biasing conditions and the device geometry. For zero drain voltage, i_{nd} characterizes the thermal noise generated by R_o, that is, $P = 1/g_m R_o$. For positive drain voltages, the noise generated in the channel is larger than the thermal noise generated by R_o.

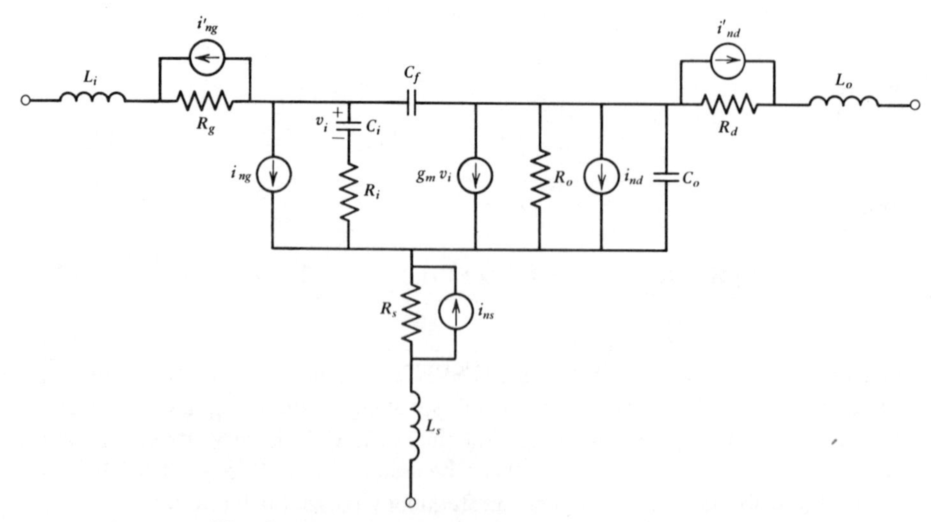

Fig. 3.16. A GaAs FET noise equivalent circuit.

The noise voltage fluctuations along the channel will induce noise fluctuations on the gate, since the gate is capacitively coupled to the channel. This can be represented by the noise current i_{ng} from gate to source electrodes as

$$\overline{i_{ng}^2} = 4kT_0 \Delta f \omega^2 C_i^2 Q / g_{m0} \tag{3.30}$$

where Q is a factor depending upon the FET geometry and bias conditions. Since the two noise currents i_{nd} and i_{ng} have the same origin, they are correlated by a factor C that depends on the biasing conditions and can be expressed as

$$jC = \frac{\overline{i_{nd} i_{ng}^*}}{\sqrt{\overline{i_{nd}^2}\ \overline{i_{ng}^2}}} \tag{3.31}$$

where $\overline{i_{nd} i_{ng}^*} = j4kT_0 \Delta f \omega C_i V$ (V is a bias dependent factor). For the extrinsic FET, the noise currents of the parasitic elements R_g, R_s and R_d can be expressed as

$$\overline{i_x^2} = 4kT_0 \Delta f / R_x, \qquad x = g, s, d \tag{3.32}$$

which correspond to the normal thermal noise.

The factors P, Q, and V can be calculated in terms of the drain-gate, source-gate, and pinch-off voltages, and the saturation field [4]; hence $C = V/\sqrt{PQ}$ as seen from (3.29)–(3.31). The minimum noise figure F_m of the intrinsic FET can be expressed by

$$F_m = 1 + 2\sqrt{PQ(1 - C^2)}\ \frac{f}{f_T} + 2g_{m0} R_i P \left(1 - C\sqrt{\frac{P}{Q}}\right) \left(\frac{f}{f_T}\right)^2 \tag{3.33}$$

where $f_T = g_{m0}/(2\pi C_i)$. For a half-micron gate length GaAs MESFET, F_m can be as low as 0.7 dB at 4 GHz. In practical GaAs MESFETs, the gate, source, and drain parasitic resistances generate thermal noise themselves and thus increase the minimum noise figure.

3.6.4 Power GaAs MESFET

The GaAs MESFET is not limited to small-signal amplifications or low noise applications above 4 GHz. High power high-efficiency amplification has been demonstrated at frequencies up to 14 GHz; power of 3 W at 8 GHz and 1 W at 14 GHz have been achieved for package GaAs MESFETs, and these figures will be higher in the future. A high power FET can be considered to be basically composed of many small-signal FETs connected in parallel. It is necessary, then, to obtain the desired output power by parallel operation of the FETs and maintaining the power gain as much as possible. It is also necessary to widen the gate width and suppress the increase of gate parasitic resistance and inductance as much as possible. For example, a 500 μ gate width FET delivered 100 mW at 10 GHz; then for a 1 W FET the total required gate width is about 5000 μ. The gate length of power GaAs FETs is usually 1.5–2.0 μ, the source-gate and drain-gate separations are about 2 μ and the width of the source and drain stripes are about 5 μ. A cross sectional diagram of the power GaAs MESFET is shown in Fig. 3.17.

To fabricate, first the n-type layer outside the active area is etched down to the semi-insulating substrate. An inlaid n^+ is grown beneath the source and drain regions. Interdigitated source and drain electrodes are formed on the inlaid n^+. Next a thin layer of SiO_2 is deposited by chemical vapor deposition and is etched to open the gate windows. The SiO_2 film serves as the insulating material between the overlaid gate and source stripes at crossovers, determining the gate length. It also prevents the source and drain electrodes from touching the gate because of misalignment and acts as a passivating material, protecting the GaAs surface from various chemical etchants. An Al film is then deposited on the SiO_2 film and etched to make the overlaid gate and the Schottky barrier gates themselves. Finally the source and drain windows are opened by etching the SiO_2 film.

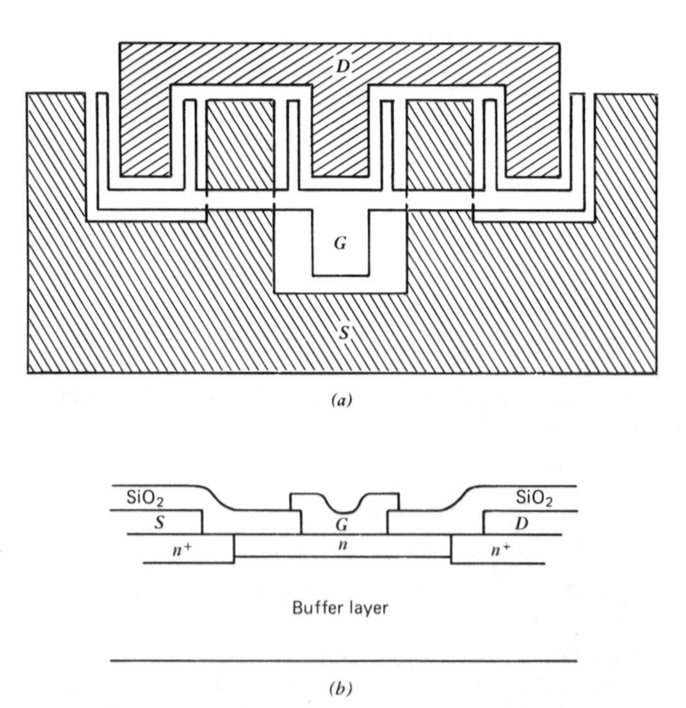

Fig. 3.17. (a) Layout of the planar power GaAs MESFET. The gate branches are interconnected with a metal line that crosses over the source. (b) Cross-sectional view of a single gate cell.

3.6.5 Biasing Circuits

The design of bias circuits for microwave GaAs FETs is as important as the design of matching networks, since the cost per decibel of microwave gain or noise figure is so high that rf performance cannot be sacrified by poor dc bias design. Depending upon the applications—low noise low power; low noise higher gain; Class A high power; and Class B or AB high power high efficiency—the optimum dc operating points can be selected. Low noise amplifiers operate at relatively low drain-source voltage V_{DS} and current I_{DS}, usually $I_{DS} = 0.15 I_{DSS}$ (I_{DSS} is the drain-source current when $V_{GS} = 0$) as shown by point A on Fig. 3.18. For higher gain the bias point is moved upward to a higher I_{DS} level of about $0.9 I_{DSS}$ with the same V_{DS} as at point B. As the FET output power is increased, the bias point must be shifted to a higher voltage and lower current to maintain linear operation in Class A as shown at point C for $V_{DS} \approx 8-10$ V and $I_{DS} \approx 0.5 I_{DSS}$. For higher efficiency, I_{DS} must be decreased and the bias is shifted down to point D. Various bias configurations are shown in Fig. 3.19*a–e*. These types of bias circuits can be used for low noise, high gain, high power, and high efficiency. The bias circuit in Fig. 3.19*a* uses two supply voltages; the negative gate voltage V_G must be turned on first to avoid operating beyond the safe region. This configuration gives the lowest source inductance and hence lowest noise figure and higher gain. The bias circuits in Fig. 3.19*b–c* employ only one type of supply, either positive or negative. Since the gate must be negatively biased before any drain voltage is applied, V_S must be applied first, before V_D is turned on in the circuit shown in Fig. 3.19*b*, and V_G must be turned on before V_S in the circuit shown in Fig. 3.19*c*. These types of bias circuits need very good microwave bypass capacitors at the source; this can cause a problem at higher frequencies since any small series source impedance would cause a noise figure degradation. Furthermore, this can cause low frequency instability

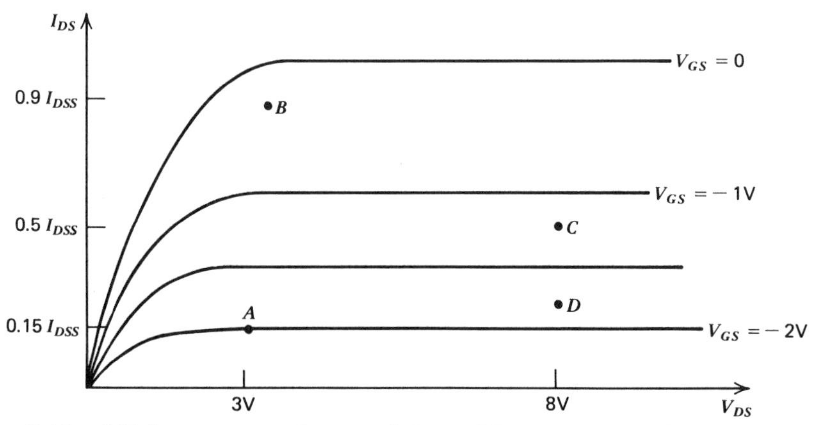

Fig. 3.18. *I-V* characteristics of a typical GaAs FET. Point *A* is for low noise, *B* is for high gain, *C* is for Class A, and *D* is for Class B or AB.

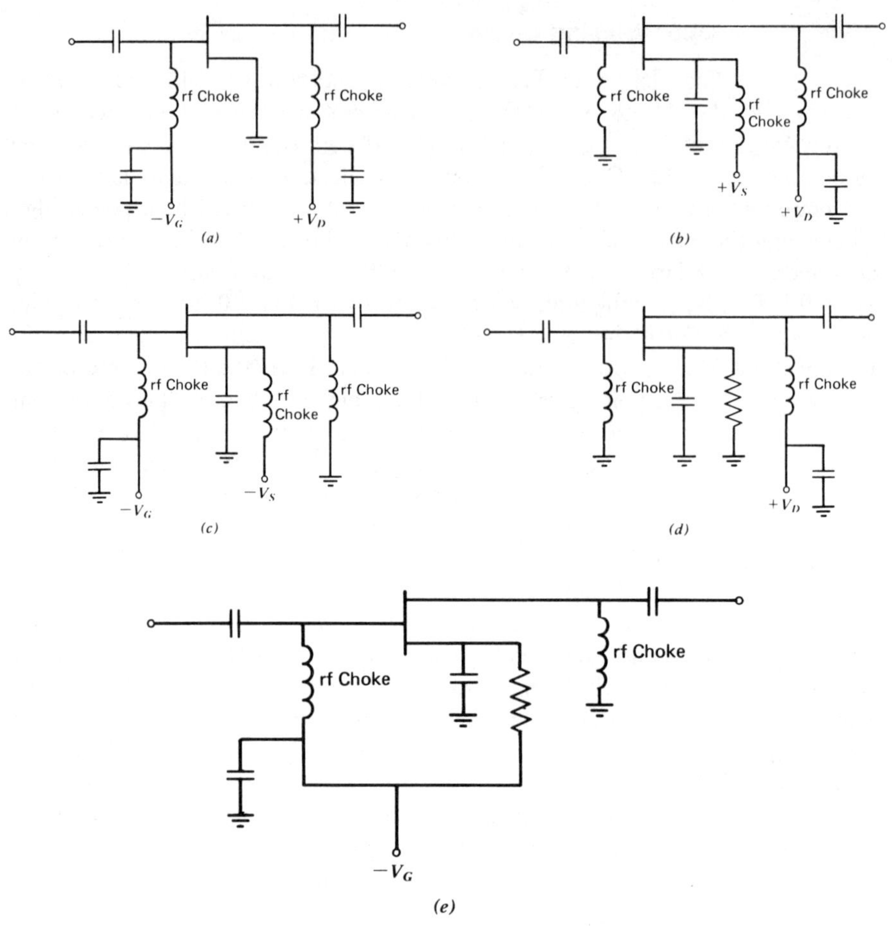

Fig. 3.19. (a)–(e) Basic bias circuits for GaAs FET amplifiers.

and give rise to bias oscillations. The bias circuits in Fig. 3.19d–e require only one power source; this type of bias energizes the gate, drain, and source simultaneously and can avoid the turn-on or turn-off problems of the two-supply bias circuits. Since a source resistor must now be used in parallel with a bypass capacitor, the efficiency of the amplifier is reduced because it draws part of the dc power supply. The advantage of this type of bias is the transient protection from the source resistance. The bias point can be adjusted by varying the value of the source resistor. The transient protection when using the two-supply bias circuits can be done by using a long RC time constant in the positive supply and a short RC time constant in the negative supply.

It is also necessary to decouple the bias circuits near the FET with a low inductance capacitor shunted with a Zener diode. In addition to limiting transients, the Zener diode gives protection against overvoltage and reverse biasing. Such a bias protection circuit is shown in Fig. 3.20.

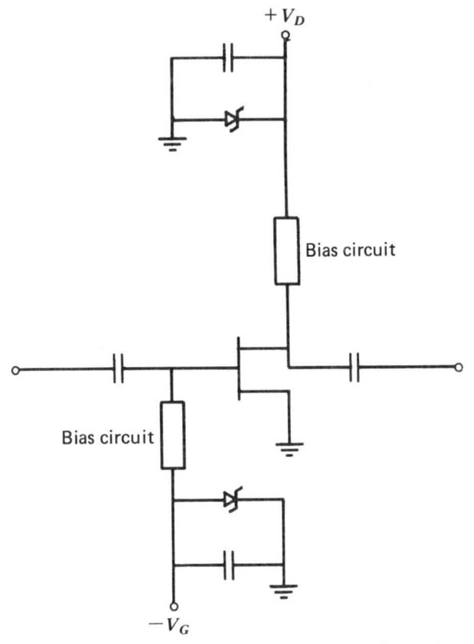

Fig. 3.20. Transient protection in a bias circuit.

Besides the above passive bias circuit, the active bias circuit shown in Fig. 3.21 can be used. In Fig. 3.21 the quiescent point is controlled by R_2 and R_3. R_2 is adjusted for proper V_{DS} and R_3 is adjusted for proper I_{DS}. The best bias circuit for GaAs FET amplifiers, therefore, must satisfy the following conditions. The FET source must have a good ground, transient protection must be provided, and when an FET in a chain fails it must not affect the other FETs.

3.7 SILICON BIPOLAR TRANSISTORS

Although the GaAs FET has higher gain, lower noise figure and higher power at frequencies above 4 GHz, the bipolar transistor still dominates at the lower frequency range. Silicon is the preferred semiconductor for microwave bipolar transistors. The fabrication technology is basically the same as that of lower frequency transistors except that attention must be paid to the processes that limit the frequency performance, such as the emitter width, the emitter-to-base contact spacing, and the collector area. The maximum oscillation frequency is inversely proportional to the emitter width and emitter-to-base spacing. Most microwave silicon bipolar transistors use the planar process and all are *n-p-n*. The critical device dimensions are etched into layers of SiO_2. These etched patterns are called the geometry of the transistor. The process begins on an *n*-type epitaxially grown silicon layer that has resistivities from 0.5 to 2 Ω/cm. The epitaxial layer is from 2 to 5 μ thick and is supported by a heavily doped n^+

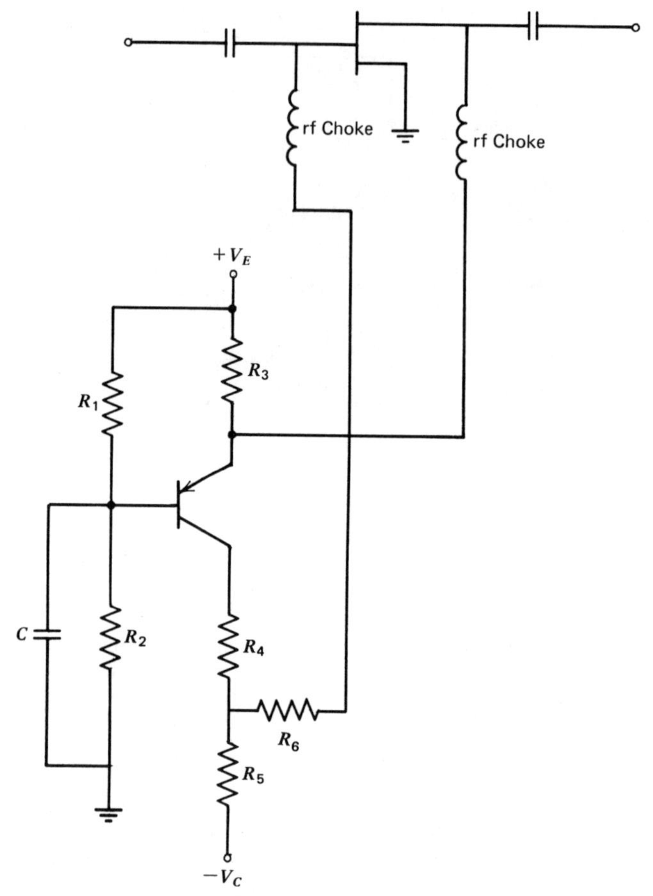

Fig. 3.21. An active bias circuit for a GaAs FET. R_2 and R_3 can be adjusted to provide proper V_{DS} and I_D.

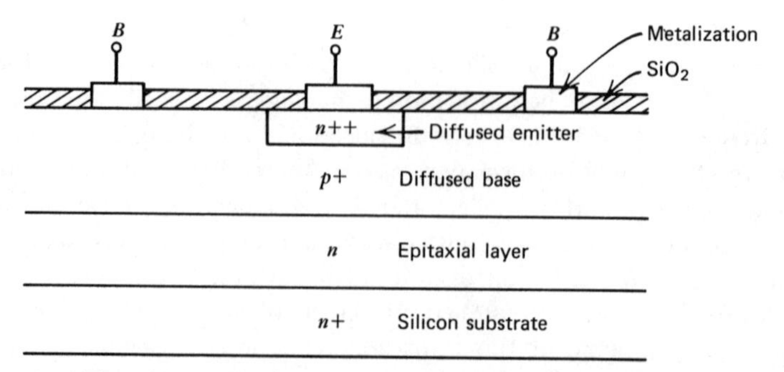

Fig. 3.22. Cross section of a silicon bipolar transistor.

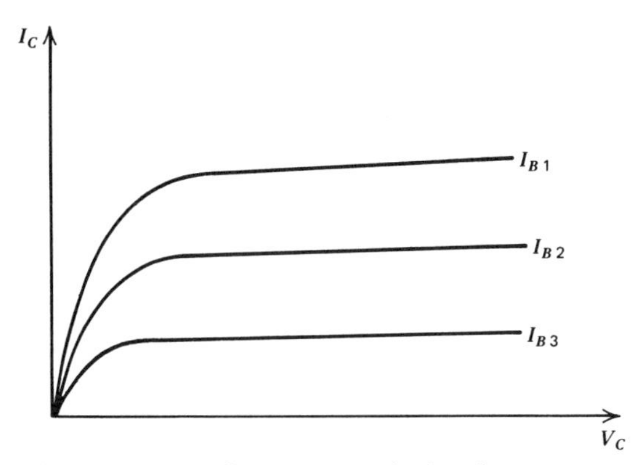

Fig. 3.23. *I-V* characteristic of a bipolar transistor.

substrate, which forms the collector contact. A thermally grown oxide layer of approximately 0.5 μ is then formed on the surface of the n-layer. Windows are opened in it by photoresistant exposure and etching to allow a p^+ diffusion for the base area. Through the open window, a heavily doped p-type diffusion is made to provide low resistance contacts to the base region. The base area is then cut into the oxide; and a lightly doped p-type diffusion is performed through the base area opening and is connected to the p^+ region. Additional SiO_2 deposition on the base area is also provided for emitter masking. Windows are then opened in the base oxide to form emitter contacts. The process is completed by diffusing a shallow heavily doped n-type layer into the emitter opening. The contact metallization, which may be either aluminum based or gold based, is evaporated and the contact pattern is defined and etched. Ohmic contacts are obtained by sintering at about 400°C for a few minutes. This process is called "diffused planar" technology. A cross section of a single emitter strip in an interdigitated transistor structure is shown in Fig. 3.22 (the term "interdigitated" means that the base and emitter areas appear as interleaved fingers, separated closely from each other). The I–V characteristics of bipolar transistors are shown in Fig. 3.23.

3.7.1 Small-Signal Equivalent Circuit

The small-signal lumped equivalent circuit model for a common emitter microwave package transistor is shown in Fig. 3.24. As in the case of the GaAs FET, the maximum available gain of the transistor can be approximated as

$$G_{a,\max} \approx \left(\frac{f_{\max}}{f} \right)^2 \tag{3.34}$$

where $f_{\max}$ is the maximum frequency of oscillation. Like GaAs FETs, microwave bipolar transistors have an intrinsic gain roll-off of 6 dB/octave.

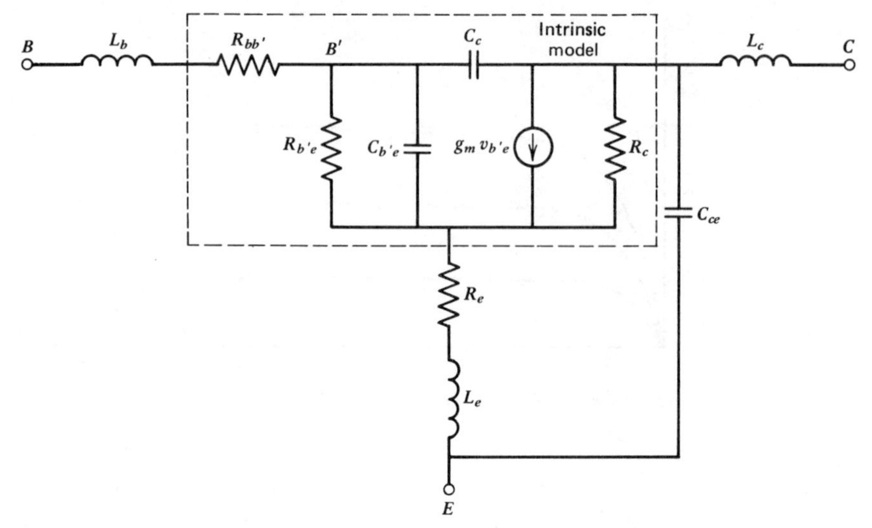

Fig. 3.24. Small-signal equivalent circuit for a common emitter microwave transistor.

3.7.2 Noise

Bipolar transistors have been used in low noise amplifications at frequencies up to 4 GHz although their noise figures are higher than those of the GaAs MESFETs. One of the two main sources of noise in microwave bipolar transistors is the thermal noise caused by thermal agitation of the current carrier in the ohmic resistance of the bulk material of the transistor. This consists of the base resistance, the emitter series resistance, and the collector series resistance. In general the base resistance $R_{bb'}$ dominates the others; hence the thermal noise source can be represented by a noise voltage e_{nb} in series with $R_{bb'}$ as [5]

$$\overline{e_{nb}^2} = 4kT_0 R_{bb'} \Delta f \tag{3.35}$$

The other noise source is the shot noise caused by the fluctuations in the dc base current and the dc collector current, which can be expressed as noise current sources at the input and output

$$\overline{i_{nb}^2} = 2qI_b \Delta f \tag{3.36}$$

$$\overline{i_{nc}^2} = 2qI_c \Delta f \tag{3.37}$$

where q is the electron charge.

These three noise sources are uncorrelated, that is, $\overline{e_{nb}^* i_{nb}} = 0$, $\overline{e_{nb}^* i_{nc}} = 0$, and $\overline{i_{nb}^* i_{nc}} = 0$. The optimum noise figure F_m of the microwave transistor is approximated by

$$F_m = 1 + h\left(1 + \sqrt{1 + 2/h}\right) \tag{3.38}$$

where

$$h = 0.04 I_C R_{bb'} (f/f_T)^2 \tag{3.39}$$

In (3.39), f_T is the unity gain frequency of the transistor.

3.7.3 Power Bipolar Transistors

The important properties of power bipolar transistors are the power output, power gain, and efficiency. The power output is determined by the current and voltage handling capability, which are limited by the emitter

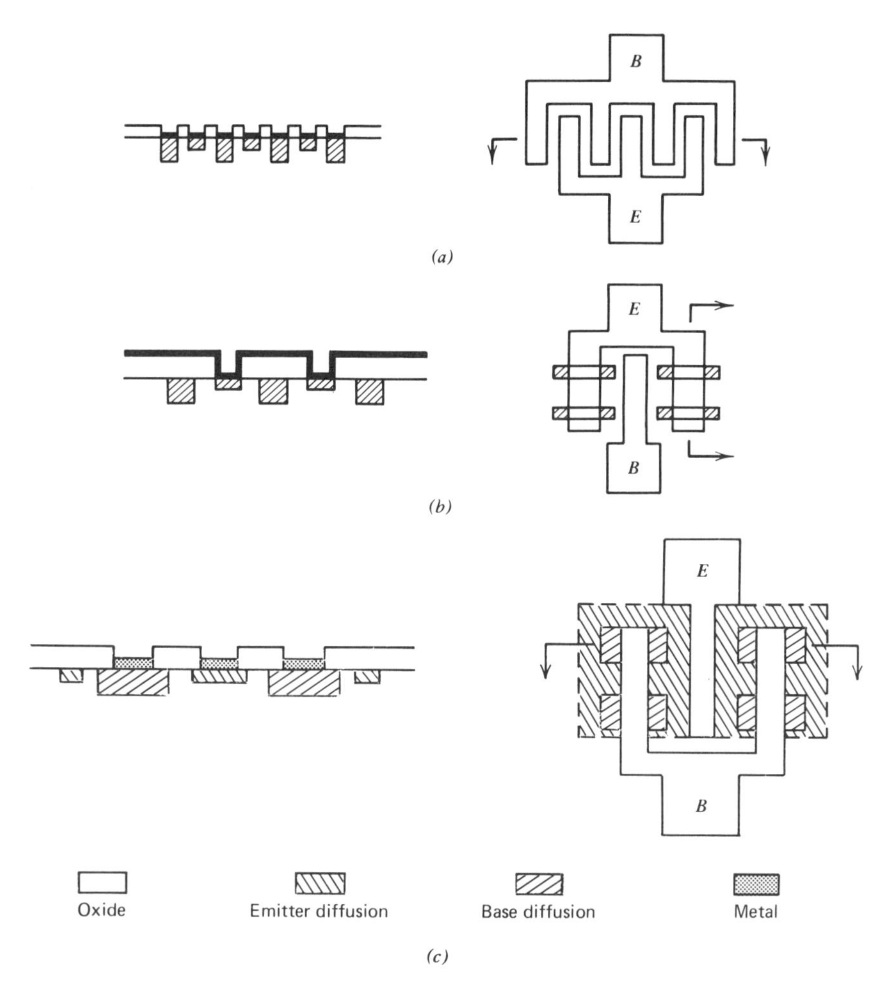

Fig. 3.25. Three geometries of the power transistor. (*a*) Interdigitated. (*b*) Overlay. (*c*) Mesh.

periphery and the breakdown voltage. The gain is determined by the maximum frequency of oscillation f_{max}, and the efficiency is determined by heat dissipated within the device. There are three basic designs for microwave bipolar transistors. The first is interdigitated, as previously explained. The second is overlay, that is, the emitter sites consist of square, rectangular, or round patterns and are surrounded by a diffused base grid. One finger of the emitter metallization contact can make contact with others so that the emitter metallization overlays the base grid with oxide as insulating material. The base grid is connected by narrow base metal fingers. The third design is mesh, that is, the inverse of the overlay; the individual sites are the p^+ base contacts and the emitter is a continuous n^+ diffused grid. The three structures are shown in Fig. 3.25*a–c*.

3.7.4 Biasing Circuits

In the design of microwave transistor amplifiers, the biasing networks must be able to hold the operating point constant over a given temperature range. The transistor temperature sensitive parameters are found to

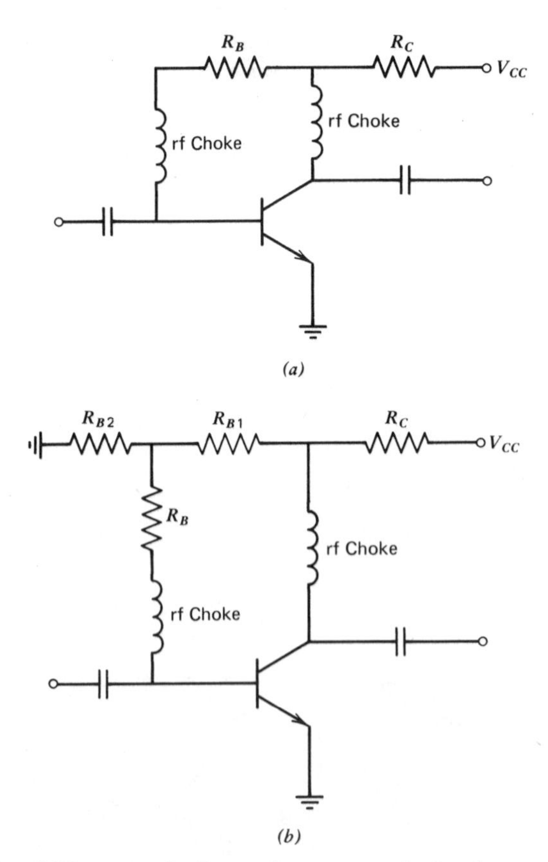

Fig. 3.26. (*a*)–(*b*) Passive bias circuits for bipolar transistors.

be $V_{B'E}$, I_{CB_0} and β. $V_{B'E}$ is the voltage across the junction capacitance and has a negative coefficient of about -2 mV/°C. I_{CB_0} is the leakage current flowing through the reverse bias *p-n* junction. It is linearly proportional to the rise in temperature. And β is dc current gain and is linearly proportional to the rise in temperature at the rate of about 5%/°C. The two most commonly used passive bias circuits that provide good stability are given in Fig. 3.26*a*–*b*. The circuits in Fig. 3.26*b* with the addition of R_{B1} and R_{B2} in the voltage feedback circuit allow the use of smaller resistance values and make hybrid integrated circuits suitable.

Example 3.1

Consider the bias circuit in Fig. 3.26*a*. Assume that the operating point is $V_{CE}=5$ V, $I_C=5$ mA and the available supply is $V_{CC}=15$ V. Also assume that $I_{CB_0}=0$, $V_{BE}=0.7$ V and $\beta=50$. Then the base current $I_B=I_C/\beta=5/50=0.1$ mA. The base resistor R_B can be calculated as $R_B=(V_{CE}-V_{BE})/I_B=(5-0.7)/0.1=43$ kΩ and the collector resistor R_C is found to be $R_C=(V_{CC}-V_{CE})/(I_C+I_B)=(15-5)/(5+0.1)=2$ kΩ.

Example 3.2

Consider the bias circuit in Fig. 3.26*b* and using the same operating conditions as in Example 3.1, with the base current $I_B=0.1$ mA, we can let $V_B=1.7$ V. This yields $R_B=(V_B-V_{BE})/I_B=(1.7-0.7)/0.1=10$ kΩ.

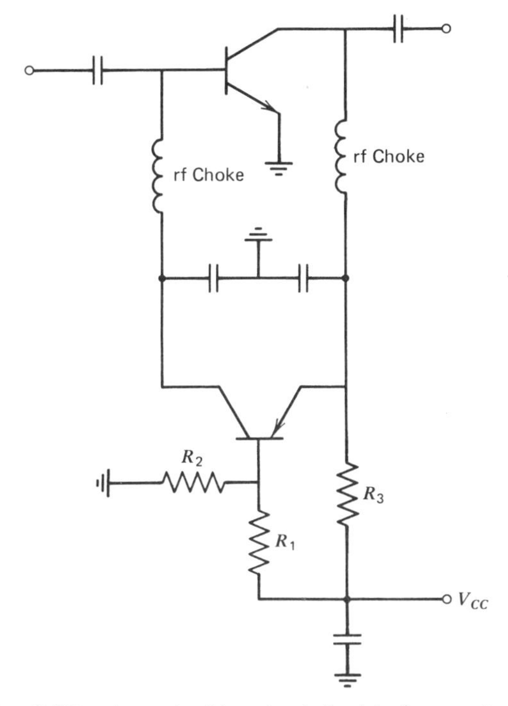

Fig. 3.27. An active bias circuit for bipolar transistors.

Assume the current flowing through R_{B2} is 1.7 mA then $R_{B2} = V_B/1.7 = 1.7/1.7 = 1\ \text{k}\Omega$, $R_{B1} = (V_{CE} - V_B)/(I_B + 1.7) = 1.83\ \text{k}\Omega$, and $R_C = 2.2\ \text{k}\Omega$.

From the standpoint of temperature compensation, passive bias circuits can provide satisfactory results and are inexpensive to build. But when the dc current gain variation is large, automatic compensation can be made effectively only by an active bias circuit. Furthermore, the active bias circuit can provide better operating point stability, especially for low noise or high power amplifications. A typical active bias circuit is shown in Fig. 3.27. The bias operating point can be selected by adjusting the values of the resistors R_1, R_2, and R_3. The active bias circuit is actually a feedback loop that senses the collector current of the microwave transistor and adjusts the base current to hold the collector current fixed. The control transistor of the bias circuit is used as a *p-n-p* type switch. The voltage divider R_1 plus R_2 is chosen to fix the V_{CE} of the microwave transistor, and the current flowing through its collector is determined by R_3. Equilibrium will be reached when the base plus collector current of the microwave transistor is equal to the current in R_3. The low frequency parasitic oscillations of the active bias circuit can be suppressed using adequate low frequency capacitors.

3.8 AVALANCHE TRANSIT-TIME MICROWAVE DIODES

The use of negative resistance in a reverse-biased *p-n* junction to generate microwave power was predicted by Read [6]. The negative resistance is the result of a delay between an applied rf voltage and the avalanche breakdown current in a properly designed diode. These negative resistance diodes can be employed in both small-signal and high power amplifiers and in oscillators. They are usually called IMPATT diodes (*IMP*act ionization Avalanche *T*ransit *T*ime). The IMPATT diode is a p^+-n-n^+ three-layer junction shown in Fig. 3.28 and its

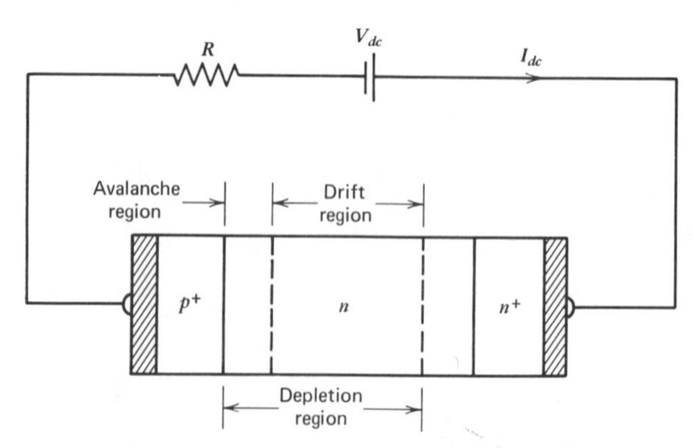

Fig. 3.28. Representation of an IMPATT diode.

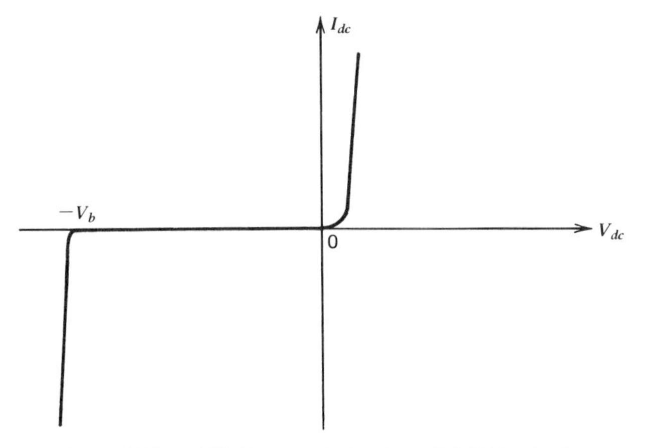

Fig. 3.29. *I-V* characteristic of an IMPATT diode.

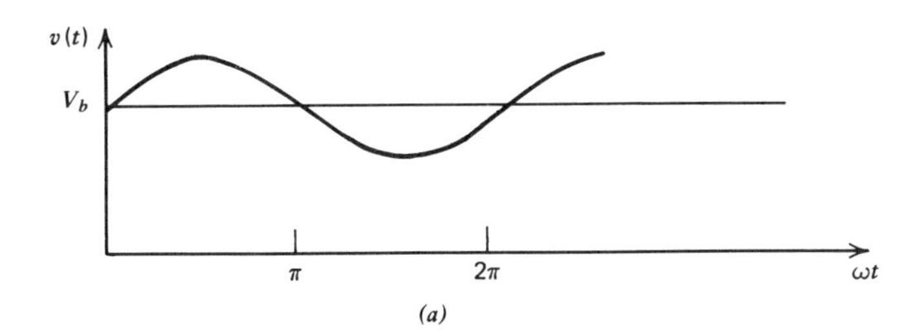

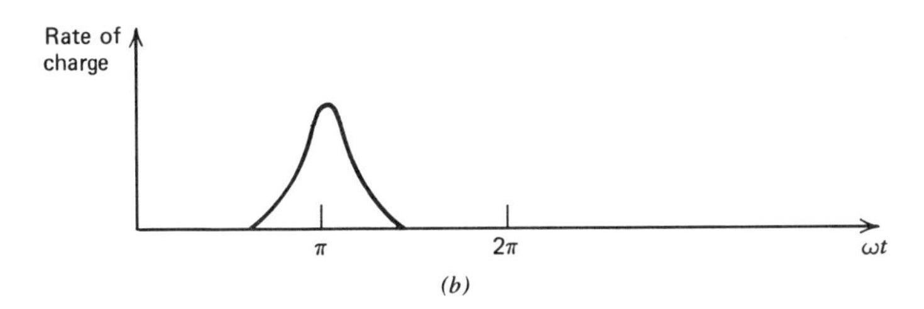

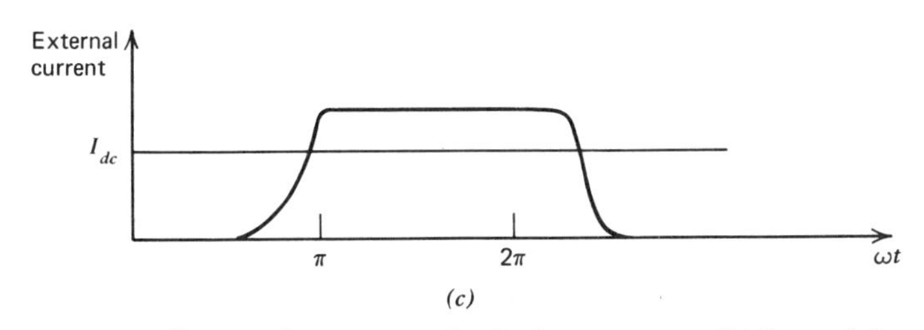

Fig. 3.30. (*a*) Oscillation voltage across the depletion region. (*b*) Rate of charge in the avalanche region. (*c*) External current due to the flow of electrons across the drift region.

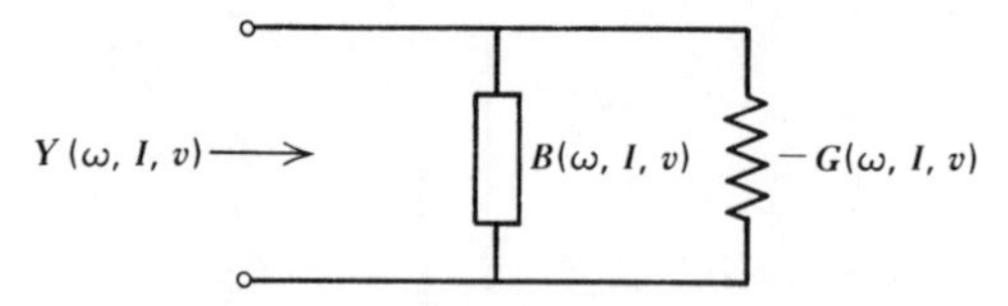

Fig. 3.31. A general model of IMPATT diodes with frequency, bias current, and input signal level dependent elements.

I-V characteristic is shown in Fig. 3.29. In the forward-bias mode the current increases rapidly as the applied voltage reaches 0.5 V. In the reverse-bias mode, a very small current flows through the junction (saturation current) until a breakdown voltage V_b is reached. IMPATT diodes operate in the breakdown region with a dc reverse voltage greater than V_b and a large reverse current flowing through the junction.

When a reverse-bias voltage is applied to the p^+-n-n^+ junction, a depletion region is formed in the n-type region and acts as a nonlinear capacitor if the reverse-bias voltage is smaller than the breakdown voltage. Furthermore, in this case the saturation current is formed by a small number of electrons flowing from the p^+ region into the avalanche region. When the reverse-bias voltage exceeds the breakdown voltage, the avalanche process occurs and additional electrons are created in the avalanche region and flow into the drift region, causing a large current to flow with little increase in the applied reverse-bias voltage, typically 70–100 V. The bias current for IMPATT diodes ranges from 20 mA for small-signal to 200 mA for large-signal amplification. The microwave signal properties of IMPATT diodes can be explained as follows:

(a) Let $v(t) = V_{dc} + V_0 \sin \omega t$ be a voltage appearing across the diode in the breakdown region (V_{dc} is the dc breakdown voltage) as in Fig. 3.30*a*. Excess charge builds up in the avalanche region when $\omega t = \pi/2$, reaches its maximum rate when $\omega t = \pi$ and returns to normal when $\omega t = 3\pi/2$, all during the positive half cycle of $v(t)$ as in Fig. 3.30*b*.

(b) During the avalanche process, the flow of electrons from the avalanche region into the drift region induces a current in the external circuit as in Fig. 3.30*c*, while the voltage $v(t)$ undergoes the negative half-cycle, that is, the diode exhibits negative resistance.

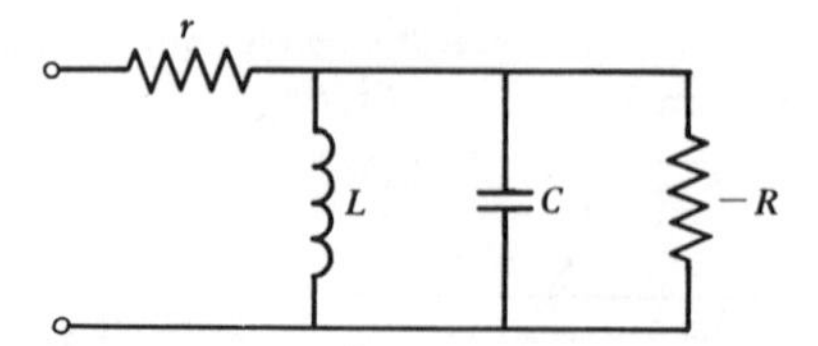

Fig. 3.32. A general equivalent circuit of IMPATT diodes with series loss resistance.

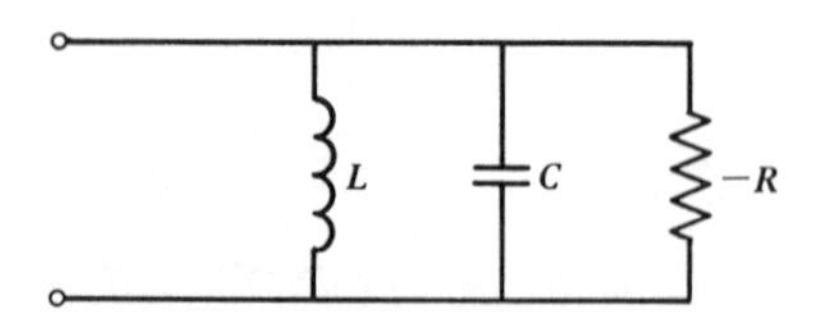

Fig. 3.33. A dissipationless shunt-resonant equivalent circuit.

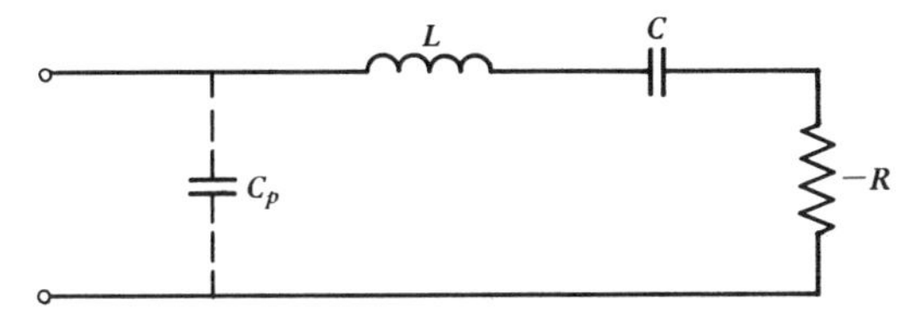

Fig. 3.34. A dissipationless series-resonant equivalent circuit.

The IMPATT diode can be modeled by various kinds of equivalent circuits depending upon the level of signal and bandwidth. A general model is shown in Fig. 3.31, where the diode admittance is a combination of negative conductance and capacitive susceptance. Both are functions of frequency, input signal level, and bias current. For most classes of IMPATT diodes, an equivalent circuit that includes the series loss resistance is given in Fig. 3.32. In many cases the dissipationless model of Fig. 3.33 can be employed if r is small enough to be ignored, or for ease of design with the correction made in the final computer optimization process. The series-resonant model shown in Fig. 3.34 can also be used in many cases; the package capacitor C_p can also be included as a shunt element. These models can be derived from the measured data fairly accurately over octave bandwidths. In many cases of larger bandwidths computer approximation is necessary to obtain an accurate design model.

3.9 TRANSFERRED ELECTRON DIODES

The transferred electron, or Gunn, diode is a one-port device that exhibits negative resistance at microwave frequencies because of the transfer of electrons, upon the application of an electric field, from a low mass, high mobility energy band to another energy band with higher effective mass and lower mobility. Gunn diodes are constructed by putting ohmic contacts between a slab of n-type gallium-arsenide as shown in Fig. 3.35. Usually n^+-type buffer layers are employed to provide simpler ohmic metal contacts for the diode. When the bias voltage applied across the ohmic contacts exceeds a threshold value, the diode can exhibit a differential negative resistance over a broad range

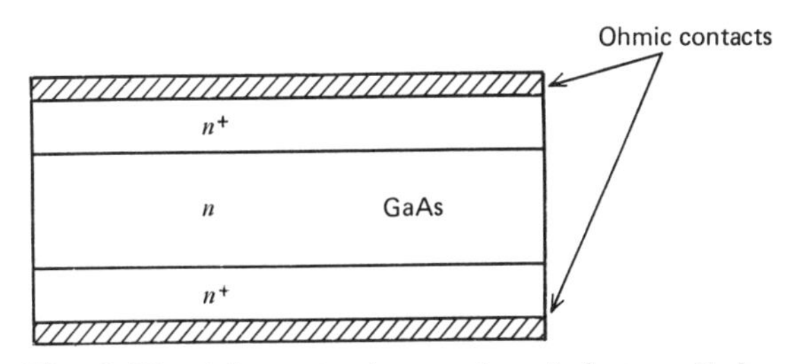

Fig. 3.35. Schematic of a transferred electron diode.

of frequency in the order of one octave. In the stable negative resistance mode, the diode exhibits an rf impedance with a negative real part for frequencies close to the transit time frequency. For a slab of GaAs of length 10 μ, the transit time frequency is approximately 10 GHz. The stable Gunn diode rf impedance can be modeled either by parallel or series $-RLC$ circuits as in Fig. 3.33 or Fig. 3.34. Gunn diodes employed in amplification mode require stable bias with dc voltage two or three times the threshold voltage of the diodes.

3.10 CONCLUDING REMARKS

An introduction to microwave integrated circuits has been presented in this chapter. Practical equations for the design of microstrip lines were given in detail, including calculation of Z_0 for a given ratio W/H and calculation of W/H for a given Z_0. Useful tables of microstrip line designs for frequently used MIC substrates are given in Appendix A3. Design equations for microwave lumped elements such as inductors and capacitors were also discussed, together with their approximations by uniform lossless transmission lines. These approximations are particularly useful in the design of practical amplifiers from 600 MHz to 15 GHz. The selection of suitable substrates will enable the designer to control the size of the distributed networks fairly well. To familiarize the reader with microwave amplifier building blocks, the coupler, the circulator, and the isolator were introduced. All play indispensable roles in the design of single-ended or balanced amplifiers in subsequent chapters. For active devices such as GaAs FETs, bipolar transistors, and IMPATT diodes, which are currently being used extensively in telecommunications and electronic countermeasures, comprehensive treatment was given on device structure and operating principles, especially the bias networks for transistors. Basic practical bias networks of GaAs FETs and bipolar transistors were discussed. Modifications to fit individual requirements based upon this discussion can be made with more sophisticated bias networks employing voltage, current regulators, and protection time-delay networks.

PROBLEMS

3.1. Generate a computer program to compute the ratio W/H of microstrip lines in terms of Z_0 from (3.2).

3.2. Show that if l/W is large in (3.9a) then the inductance per unit length is approximately 10 nH/cm.

3.3. Consider the general transmission line shown in Fig. P3.3 with length l, characteristic impedance $Z_0=[(r+j\omega L)/(g+j\omega C)]^{1/2}$ and propagation constant $\gamma=[(r+j\omega L)(g+j\omega C)]^{1/2}$ where the series elements L and r and the shunt elements g

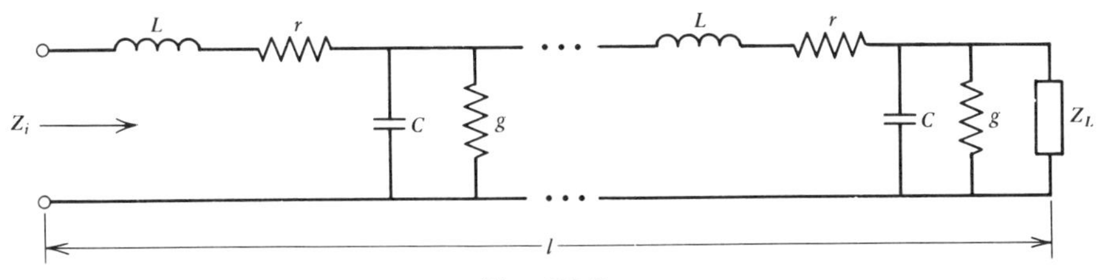

Fig. P3.3.

and C are all per unit length of line. The input impedance Z_i is given as

$$Z_i = Z_0 \frac{Z_L \cosh \gamma l + Z_0 \sinh \gamma l}{Z_0 \cosh \gamma l + Z_L \sinh \gamma l}$$

For a very short length ($\gamma l \ll 1$) of short-circuited ($Z_L = 0$) line, show that $Z_i \approx rl + j\omega Ll$. Also show that for a very short l and $Z_L = \infty$, $Z_i \approx rl/3 + g/\omega^2 Cl - 1/j\omega Cl$.

3.4. Find the scattering matrix of two cascaded 3 dB couplers as shown in Fig. P3.4.

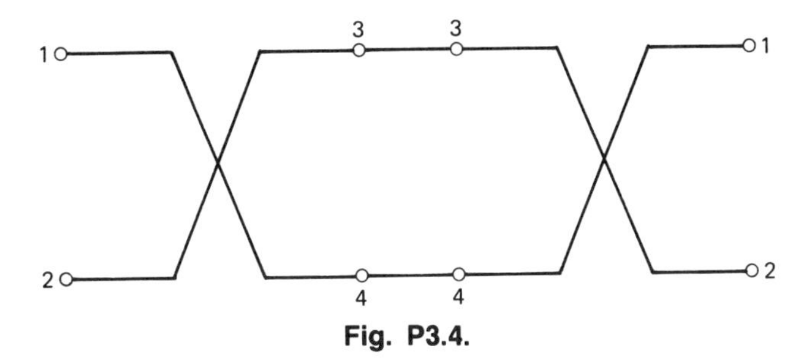

Fig. P3.4.

3.5. Let the 3 dB couplers in Fig. P3.5 be normalized to Z_0 and let port 4 of the first coupler be terminated in Z_0 Ω, and let a_1 be the incident wave into port 1. Show that the reflected waves b_2 and b_3 at ports 2 and 3 of the second coupler are given

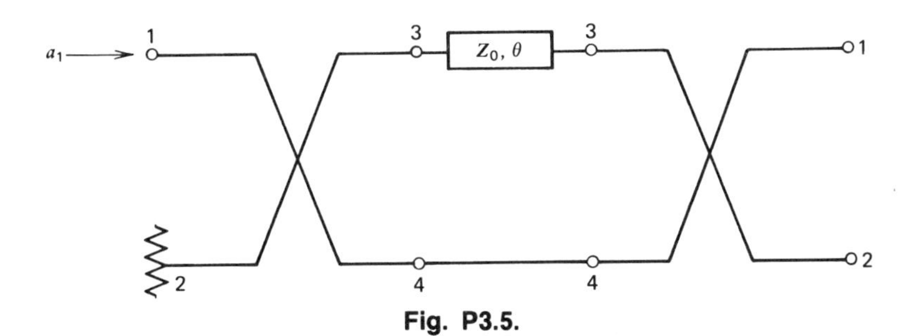

Fig. P3.5.

as

$$b_1 = \tfrac{1}{2}(1 - e^{i\theta})\, a_1$$

$$b_2 = -\frac{i}{2}(1 + e^{i\theta})\, a_1$$

Compute $|b_2|^2$ and $|b_3|^2$ for $\theta = 0°$, 70.5°, 90°, 109.5° and 180°.

3.6. Compute the overall scattering matrix of the network shown in Fig. P3.6 assuming the scattering matrices $\mathbf{S}_A$ and $\mathbf{S}_B$ of N_A and N_B and the circulators are all normalized to Z_0.

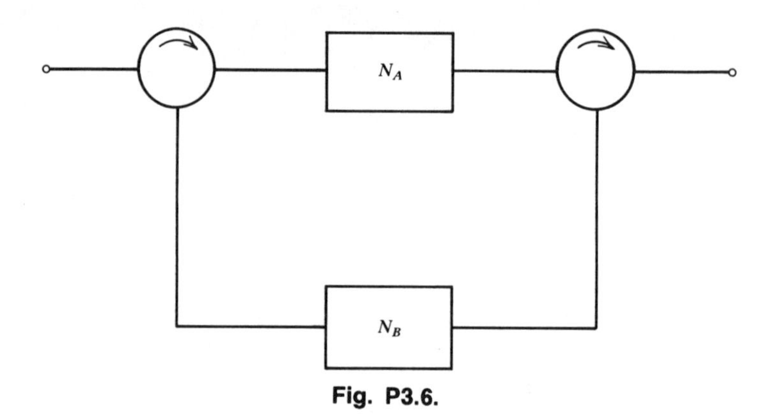

Fig. P3.6.

3.7. Calculate the maximum oscillation frequency $\omega_{\max}$ of the GaAs FET in Fig. 3.14 assuming $R_g = 0$, $R_s = 0$, and $R_d = 0$.

3.8. Calculate the transducer power gain of the GaAs FET in Fig. 3.15.

3.9. Calculate the maximum oscillation frequency $\omega_{\max}$ of the intrinsic bipolar transistor in Fig. 3.24.

3.10. Approximate a shunt inductor $L = 2$ nH and a shunt capacitor $C = 0.7$ pF by uniform lossless transmission line at 4 GHz.

REFERENCES

1. I. J. Bahl and D. K. Trivedi, "A designer's guide to microstrip lines," *Microwaves*, **16**, no. 12, 174–182, 1977.
2. L. Young, *Advances in Microwaves*. New York: Academic Press, 1974.
3. M. Fukuta, K. Sumuya, K. Suzuki, and H. Ishikawa,"GaAs microwave power FET," *IEEE Trans. Electron Devices*, **ED-23**, no. 4, 388–394, 1976.
4. W. Baechtold, "Noise behavior of GaAs FETs with short gate length," *IEEE Trans. Electron Devices*, **ED-19**, no. 5, 674–680, 1972.

5. H. Fukui, "The noise performance of microwave transistors," *IEEE Trans. Electron Devices*, **ED-13**, no. 3, 329–341, 1966.
6. M. J. Howes and D. V. Morgan, *Microwave Devices*. New York: Wiley, 1976.
7. E. O. Hammerstad, "Equations for microstrip circuit design," *Proc. Europ. Microwave Conference, Hamburg, Germany*, 268–272, 1975.
8. C. A. Liechti, "Microwave field-effect transistors—1976," *IEEE Trans. Microwave Theory Tech.*, **MTT-24**, no. 6, 279–300, 1976.
9. M. V. Schneider, "Microstrip lines for microwave integrated circuits," *Bell Syst. Tech. J.*, **48**, 1421–1444, 1969.
10. H. A. Wheeler, "Transmission lines properties of parallel strips separated by a dielectric sheet," *IEEE Trans. Microwave Theory Tech.*, **MTT-13**, no. 3, 172–185, 1965.
11. G. D. Vendelin, "Five basic bias designs for GaAs FET amplifiers," *Microwaves*, **17**, no. 2, 40–42, 1978.
12. M. H. White and M. O. Thurston, "Characterization of microwave transistors," *Solid-State Electronics*, **13**, 522–542, 1970.
13. P. Wolf, "Microwave properties of Schottky-barrier FETs," *IBM J. Research Development*, **14**, 125–141, 1970.
14. Microwave transistor bias considerations. Hewlett-Packard application note 944-1.
15. Microwave power generation and amplification using IMPATT diodes. Hewlett-Packard application note 935.
16. H. Statz, H. Hauss and R. A. Pucel, "Noise characteristics of GaAs FETs," *IEEE Trans. Electron Devices*, **ED-21**, no. 9, 549–562, 1974.
17. R. A. Pucel, D. J. Masse, and C. F. Krumm, "Noise performance of GaAs FETs," *IEEE J. Solid-State Circuits*, **SC-11**, no. 2, 243–255, 1976.
18. R. D. Kasser, "Noise factor contours for field-effect transistors at moderately high frequencies," *IEEE Trans. Electron Devices*, **ED-19**, no. 2, 164–171, 1972.
19. H. F. Cook, "Microwave transistors: theory and design," *Proc. IEEE*, **59**, no. 8, 1163–1181, 1971.
20. J. A. Archer, "Design and performance of small-signal microwave transistors," *Solid-State Electronics*, **15**, 249–258, 1972.
21. C. S. Aitchison et al., "Lumped-circuit elements at microwave frequencies," *IEEE Trans. Microwave Theory Tech.*, **MTT-19**, no. 12, 928–937, 1971.
22. G. D. Alley, "Interdigital capacitors and their application to lumped-element microwave integrated circuits," *IEEE Trans. Microwave Theory Tech.*, **MTT-18**, 1028–1032, 1970.

4

Microwave Amplifiers: Narrowband Design

Microwave amplifiers employing semiconductor devices can be classified into two types: transistor amplifiers (using GaAs FETs and bipolar transistors) and reflection amplifiers (using negative resistance diodes such as IMPATT, Gunn, and tunnel diodes). The GaAs FETs are used for low noise, high gain, and linear medium power (Class A) amplifications from 2 to 20 GHz with research toward 40 GHz applications under way. Silicon bipolar transistors are dominant in applications requiring low noise, high gain, and high power (Classes A, B, AB, and especially Class C) from hf to 3 GHz. The IMPATT diodes are used for high power amplifications at frequencies as low as 6 GHz and as high as 100 GHz both in CW and pulsed modes. The Gunn amplifier is finding a place as a wideband medium power amplifier at high frequencies, above 10 GHz, and as driver for traveling wave tube amplifiers or IMPATT amplifiers. The tunnel diode amplifier is also a candidate for a low noise amplifier above 20 GHz. The general configurations for the transistor amplifier and the reflection amplifier are shown in Fig. 4.1 and Fig. 4.2 with the appropriate matching networks, which match the device impedances to the appropriate source or load resistances. In the following sections, the basic concepts in the design of microwave amplifiers for narrowband applications (less than 10% bandwidth) are analysed for both transistor amplifiers and negative resistance reflection amplifiers.

4.1 TRANSISTOR AMPLIFIERS: BASIC CONCEPTS

There are three basic designs of transistor amplifiers: maximum gain, low noise, and high power amplifications. In each mode of operation, the FET or bipolar transistor presents different reflection coefficients to its source and load resistances (50 Ω in general), hence each requires different design approaches.

4.1.1 Maximum Gain Design

In the design of narrowband amplifiers for maximum gain, the input and output matching networks must transform the source and the load

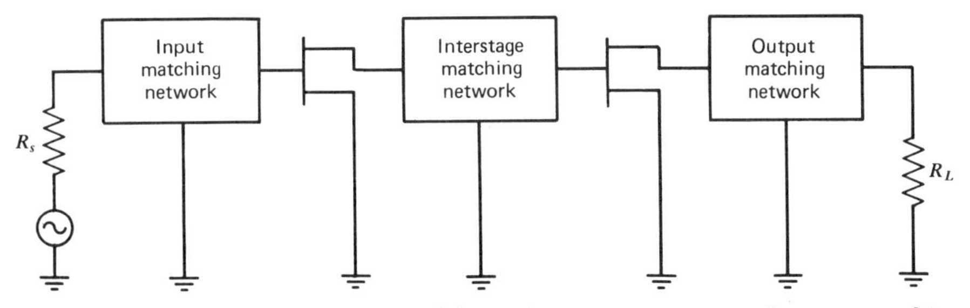

Fig. 4.1. A two-stage GaAs FET amplifier with input, interstage, and output matching networks.

reflection coefficients Γ_s and Γ_L to the matched counterparts Γ_{sM} and Γ_{LM} given in (2.47a) and (2.47b), respectively, when the transistor is unconditionally stable, that is, $K>1$ and $|\Delta|<1$ where K and Δ are given by (2.51c) and (2.51d) respectively. Let $\mathrm{S}=[S_{ij}]$ be the scattering matrix of the transistor, normalized to the real characteristic impedance Z_0 (50 Ω in practice); the unconditional stability criteria yields

$$1-|S_{22}|^2>|S_{12}S_{21}| \tag{4.1a}$$

$$1-|S_{11}|^2>|S_{12}S_{21}| \tag{4.1b}$$

as seen from (A.17) and (A.19) in Appendix A1. From (4.1) it can be shown that $B_1=1+|S_{11}|^2-|S_{22}|^2-|\Delta|^2>0$ and $B_2=1+|S_{22}|^2-|S_{11}|^2-|\Delta|^2>0$ (Appendix A4), and therefore the matched source and load reflection coefficients Γ_{sM} and Γ_{LM} together with the associated maximum transducer power gain are given from (2.47) and (2.50) as

$$\Gamma_{sM}=C_1^*\left[B_1-\left(B_1^2-4|C_1|^2\right)^{1/2}\right]\Big/\left(2|C_1|^2\right) \tag{4.2a}$$

$$\Gamma_{LM}=C_2^*\left[B_2-\left(B_2^2-4|C_2|^2\right)^{1/2}\right]\Big/\left(2|C_2|^2\right) \tag{4.2b}$$

$$G_{\max}=\left|\frac{S_{21}}{S_{12}}\right|\left(K-\sqrt{K^2-1}\right) \tag{4.2c}$$

where C_1 and C_2 are given in (2.48c) and (2.48d).

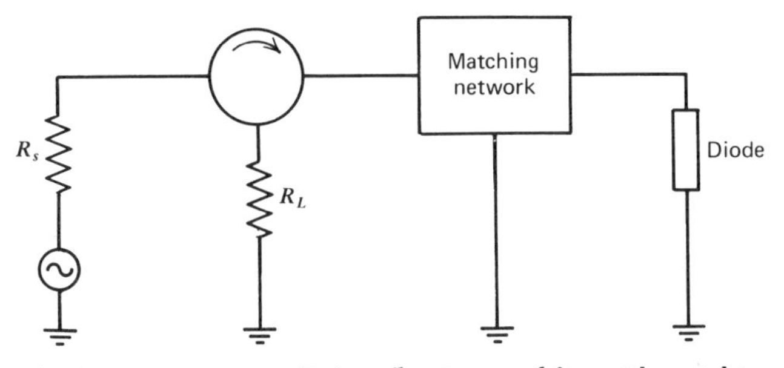

Fig. 4.2. A negative resistance diode reflection amplifier with matching network.

From (4.2c) we see that for $K>1$ and $K \to 1$, the maximum transducer power gain $G_{max} \to |S_{21}/S_{12}|$. This quantity is called the maximum stable gain of the transistor, that is,

$$\mathrm{MSG} = \left| \frac{S_{21}}{S_{12}} \right| \tag{4.3}$$

When the transistor is potentially unstable and $K<1$ then G_{max} in (4.2c) does not exist. In this case the power gain G_p must be used for the design. The power gain G_p is defined as the ratio of the power delivered to the load to the input power, that is, it is independent of the source impedance. From Section 2.2, using the generalized scattering matrix we see that

$$G_p = \frac{|b_2|^2}{|a_1|^2 - |b_1|^2} = \frac{|b_2|^2/|a_1|^2}{1-\left(|b_1|^2/|a_1|^2\right)} = \frac{|S'_{21}|^2}{1-|S'_{11}|^2} \tag{4.4}$$

where $\mathbf{S}' = [S'_{ij}]$ is the scattering matrix of the transistor normalized to the source and load impedance Z_s and Z_L, respectively. We note that G_p is independent of Z_s, by definition. Hence to simplify computations we let $\Gamma_s = 0$ when using (4.4). From (2.35) we get, for $\Gamma_s = 0$

$$S'_{21} = \frac{1}{A_2} \frac{S_{21}\left(1-|\Gamma_L|^2\right)}{1-\Gamma_L S_{22}} \tag{4.5a}$$

$$S'_{11} = \frac{(1-\Gamma_L S_{22}) S_{11} + \Gamma_L S_{12} S_{21}}{1-\Gamma_L S_{22}} \tag{4.5b}$$

where

$$A_2 = (1-\Gamma_L^*)|1-\Gamma_L|^{-1}\left(1-|\Gamma_L|^2\right)^{1/2} \tag{4.5c}$$

Substituting (4.5) into (4.4) yields

$$G_p = \frac{|S_{21}|^2\left(1-|\Gamma_L|^2\right)}{|1-\Gamma_L S_{22}|^2 - |(1-\Gamma_L S_{22}) S_{11} + \Gamma_L S_{12} S_{21}|^2} \tag{4.6}$$

Rearranging the terms in the denominator of the right side of (4.6) we get

$$G_p = \frac{|S_{21}|^2\left(1-|\Gamma_L|^2\right)}{1-|S_{11}|^2 + |\Gamma_L|^2\left(|S_{22}|^2 - |\Delta|^2\right) - 2\,\mathrm{Re}(C_2 \Gamma_L)} \tag{4.7}$$

where $C_2 = S_{22} - \Delta S_{11}^*$.

From (4.7) we see that G_p is a function of the two-port S-parameters and the load only. If one selects a source that matches the two-port at the input, then the input power and the maximum available input power are equal and the transducer power gain G and the power gain G_p are equal. If the source does not match the input impedance of the two-port, then the input power will be less than the maximum available input power and G will be less than G_p.

Now let $g_p = G_p/|S_{21}|^2$ denote the normalized power gain, that is,

$$g_p = \frac{(1-|\Gamma_L|^2)}{1-|S_{11}|^2+|\Gamma_L|^2(|S_{22}|^2-|\Delta|^2)-2\,\mathrm{Re}(C_2\Gamma_L)} \tag{4.8}$$

Then it will be shown that for a given g_p, (4.8) defines a circle that is the locus of all Γ_L that results in the same g_p. For various values of g_p, this defines a family of constant power gain circles of the transistor. To show this, let $D_2 = |S_{22}|^2 - |\Delta|^2$ then we get

$$\frac{1}{g_p} = -D_2 + \frac{B_2 - 2\,\mathrm{Re}(C_2\Gamma_L)}{1-|\Gamma_L|^2} \tag{4.9}$$

where $B_2 = 1-|S_{11}|^2 + D_2$. Rearranging the terms in (4.9) yields

$$\frac{B_2 - 2\,\mathrm{Re}(C_2\Gamma_L)}{1-|\Gamma_L|^2} = \frac{1+g_p D_2}{g_p} = X$$

or

$$X^2|\Gamma_L|^2 - 2X\,\mathrm{Re}(C_2\Gamma_L) + |C_2|^2 = X^2 - XB_2 + |C_2|^2$$

$$|X\Gamma_L - C_2^*|^2 = X^2 - XB_2 + |C_2|^2$$

or

$$\left|\Gamma_L - \frac{C_2^*}{X}\right|^2 = 1 - \frac{B_2}{X} + \frac{|C_2|^2}{X^2} \tag{4.10a}$$

Further rearranging terms on the right side of (4.10a) yields

$$\left|\Gamma_L - \frac{C_2^*}{X}\right|^2 = \frac{1-2K|S_{12}S_{21}|g_p + |S_{12}S_{21}|^2 g_p^2}{g_p^2 X^2} \tag{4.10b}$$

where K is the stability factor given previously.

Equation (4.10) defines a circle with center C_p and radius r_p given by

$$C_p = \frac{C_2^*}{X} = \frac{g_p(S_{22}^* - \Delta^* S_{11})}{1 + g_p(|S_{22}|^2 - |\Delta|^2)} \tag{4.11a}$$

$$r_p = \frac{\left(1 - 2K|S_{12}S_{21}|g_p + |S_{12}S_{21}|^2 g_p^2\right)^{1/2}}{\left|1 + g_p(|S_{22}|^2 - |\Delta|^2)\right|} \tag{4.11b}$$

If the load reflection coefficient Γ_L is selected to be in the stable region of the Smith chart, that is, it satisfies the load stability criteria defined by (2.52) and at the same time lies on the constant power gain circle, then the amplifier is stable and can be designed to have the transducer power gain $G = G_p = |S_{21}|^2 g_p$ if the source reflection coefficient is selected by the cascade-load formula (2.46a), that is,

$$\Gamma_s = \left(\frac{S_{11} - \Gamma_L \Delta}{1 - \Gamma_L S_{22}} \right)^* \tag{4.12}$$

The following examples illustrate the maximum gain design concept.

Example 4.1

This example demonstrates the design of a maximum gain GaAs FET amplifier using the HFET-1101 at 6 GHz whose S-parameters normalized to 50 Ω at this frequency are

$$S_{11} = 0.614\angle -167.4^\circ \qquad S_{21} = 2.187\angle 32.4^\circ$$

$$S_{12} = 0.046\angle 65^\circ \qquad S_{22} = 0.716\angle -83^\circ$$

First, the FET must be checked for its stability at this frequency. Using (2.51) we get $|\Delta| = |0.34195\angle 113.16^\circ| = 0.34195 < 1$ and $K = 1.1296 > 1$. Thus the FET is unconditionally stable and (4.2) can be used to compute the conjugate match source and load for maximum gain. We have $B_1 = 0.747$, $B_2 = 1.019$, $C_1 = 0.37\angle -169.75^\circ$ and $C_2 = 0.507\angle -84.47^\circ$. Subsequently we obtain $G_{max} = 14.58$ dB, $\Gamma_{sM} = 0.868\angle 169.75^\circ$ and $\Gamma_{LM} = 0.9\angle 84.47^\circ$. With the computed values of Γ_{sM} and Γ_{LM}, it is now possible to synthesize the input and output matching networks. We will use the conventional single-stub impedance matching on the Smith chart.

Input Matching Network

From $\Gamma_{sM} = 0.868\angle 169.75^\circ$, the conjugate matched input immittances Z_{sM} and Y_{sM} are located on the Smith chart in Fig. 4.3. A circular arc whose center is the center of the Smith chart is drawn through Y_{sM} and the upper half of the unit resistance circle at point A. The distance from $Z = 0$ to the

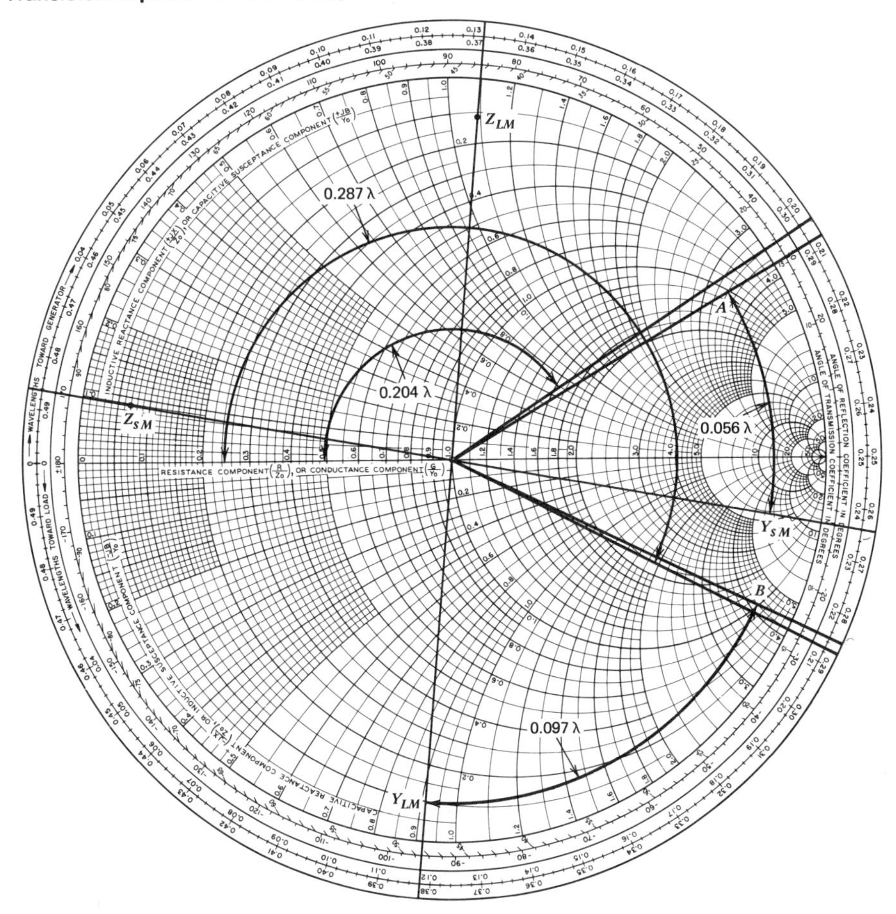

Fig. 4.3. An input and output matching network design for Example 4.1.

line drawn from the center of the Smith chart and in tangent with the reactive circle $+j3.4$ passing through point A is the length of the input open stub in parallel with the 50 Ω source (0.204λ in length). The distance from point A to Y_{sM} is the length of the input line that connects the open stub to the input of the FET (0.056λ in length). The two lines have the same characteristic impedance of 50 Ω.

Output Matching Network

From $\Gamma_{LM} = 0.9\angle 84.47°$, the conjugate matched output immittances Z_{LM} and Y_{LM} are located in Fig. 4.3. A circular arc is drawn through Y_{LM} and point B of the lower half of the unit resistance circle. The distance from $Z=0$ to the line drawn from the center of the Smith chart and in tangent with the reactive circle $-j4.2$ passing through point B, which is 0.287λ, is the length of the output shunt open stub. The distance from point B to Y_{LM}, which is

0.097λ, determines the length of the output line that connects the output of the FET to the output shunt open stub. The two lines have the same characteristic impedance of 50 Ω.

The complete amplifier with input and output matching networks is shown in Fig. 4.4. The capacitors C_B are dc bias blocking capacitors and must be selected to have very low loss at 6 GHz, and low impedance compared to 50 Ω. Suppose now we want to build the matching networks on the microstrip substrate Duroid® with relative dielectric constant $\epsilon_r = 2.22$ and thickness $H = 0.031$ in. Then from Appendix A3, the 50 Ω microstrip line will have width $W = 3.1H = 0.096$ in. and the effective dielectric constant $\epsilon_r' = 1.91$. Hence according to (3.8b) we obtain $\lambda = 1.423$ in. and hence $l_1 = 0.204\lambda = 0.29$ in., $l_2 = 0.056\lambda = 0.08$ in., $l_3 = 0.138$ in., and $l_4 = 0.409$ in. Although the amplifier is unconditionally stable at 6 GHz, the designer must check stability at other frequencies within the working bandwidth of the amplifier.

In this example, the GaAs FET is unconditionally stable at 6 GHz, and conjugate matching is possible for maximum gain. When the GaAs FET is

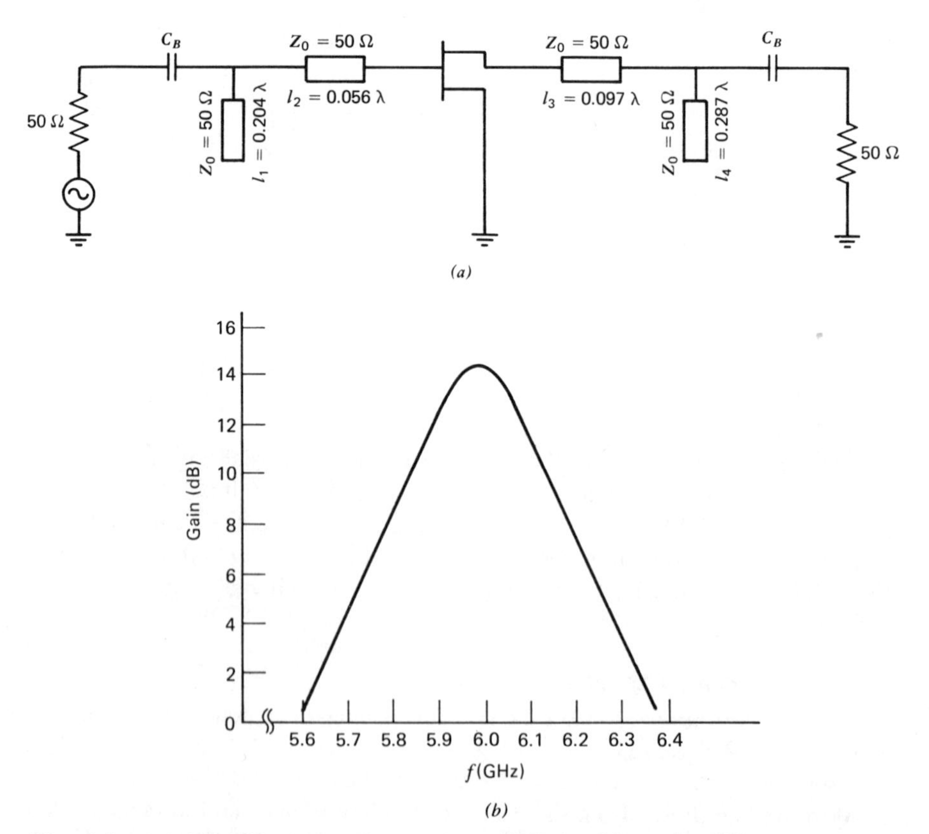

Fig. 4.4. (*a*) A 6 GHz maximum gain GaAs FET amplifier. (*b*) The gain response of the 6 GHz GaAs FET amplifier.

potentially unstable, the design must incorporate the stability condition as illustrated in the next example.

Example 4.2

At 2 GHz, the GaAs FET HFET-1101 biased for maximum gain has the following scattering parameters:

$$S_{11} = 0.894\angle -60.6^\circ \qquad S_{21} = 3.122\angle 123.6^\circ$$

$$S_{12} = 0.021\angle 62.4^\circ \qquad S_{22} = 0.781\angle -27.6^\circ$$

Using (2.51) we get $|\Delta| = |0.697\angle -83.1^\circ| = 0.697 < 1$, but $K = 0.618 < 1$. Thus according to the stability criteria in (2.51), the FET is potentially unstable. We

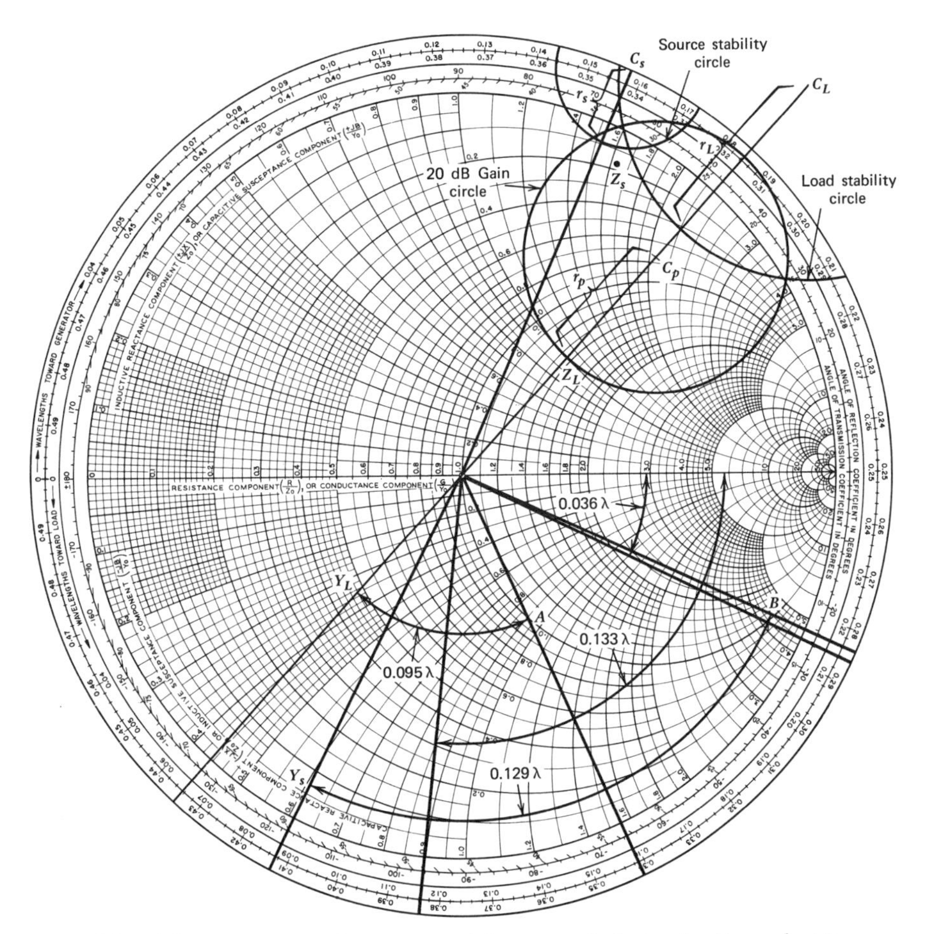

Fig. 4.5. An input and output matching network design for Example 4.2.

note that at this frequency, the maximum stable gain is given by (4.3) as MSG=21.93 dB. In order to select a stable load, the load stability circle is plotted on the Smith chart in Fig. 4.5. Its center C_L and its radius r_L are given by (2.52) as $C_L = 1.367\angle 46.9°$ and $r_L = 0.5$. Suppose that we want to design an amplifier with 20 dB gain (1.93 dB less than the maximum stable gain); thus let $G_p = 20$ dB. Hence the normalized power gain g_p is given by $g_p = G_p/|S_{21}|^2 = 100/(3.122)^2 = 10.26$. From (4.11), the 20 dB constant power gain circle with center $C_p = 0.766\angle 46.9°$ and radius $r_p = 0.346$ is plotted on the Smith chart in Fig. 4.5. The region of the Smith chart outside the load stability circle is the stable load region. Any load in this region and on the 20 dB constant power gain circle will make the amplifier stable and yield a gain of 20 dB. Select the load reflection coefficient $\Gamma_L = 0.42\angle 46.9°$ that satisfies the above conditions, then according to (4.12), the source reflection coefficient that together with the selected load yield 20 dB of transducer power gain is given as $\Gamma_s = 0.915\angle 62.6°$. This source must be a stable source, indeed it is outside the source stability circle

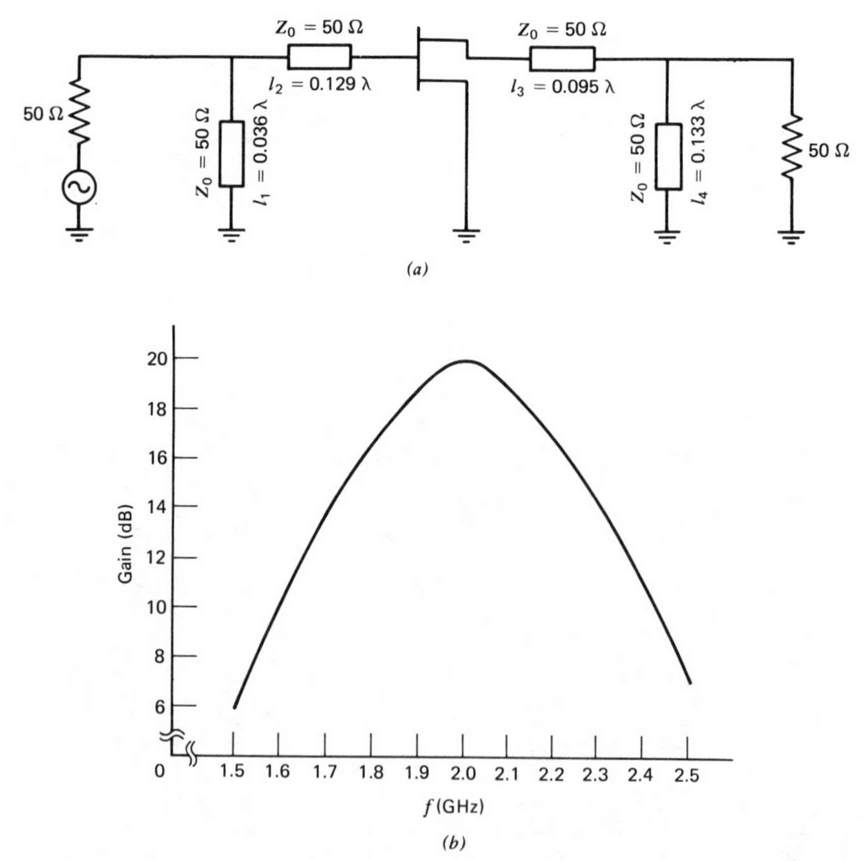

Fig. 4.6. (*a*) The 2 GHz GaAs FET amplifier. (*b*) Gain response of the 2 GHz GaAs FET amplifier.

with center $C_s = 1.136\angle 68.4°$ and radius $r_s = 0.198$ [from Equation (2.53)]. Proceeding as in Example 4.1, the input and output matching circuits are given in Fig. 4.6*a*. Since at 2 GHz the wavelength is three times longer than at 6 GHz, short stubs are employed instead of open stubs. If Duroid® substrate is used as the microstrip substrate, $\lambda = 4.27$ in., hence $l_1 = 0.153$ in., $l_2 = 0.55$ in., $l_3 = 0.406$ in., and $l_4 = 0.568$ in. The amplifier gain response is shown in Fig. 4.6*b*.

4.1.2 Low Noise Design

In a low noise amplifier, the front stage of the amplifier has to have an optimum noise source reflection coefficient Γ_m in order to achieve its minimum noise figure F_m. For a given transistor, a minimum noise figure F_m is associated with it when a source of reflection coefficient Γ_m is connected to its input, together with a weighing coefficient R_n called the equivalent noise resistance. For an arbitrary source reflection coefficient connected to the transistor, its noise figure F is given from (2.59b) as

$$F = F_m + 4r_n \frac{|\Gamma_s - \Gamma_m|^2}{|1+\Gamma_m|^2(1-|\Gamma_s|^2)} \tag{4.13}$$

where $r_n = R_n/Z_0$ (Z_0 is the normalized characteristic impedance of the transistor, which is 50 Ω in practice). It is clear from (4.13) that if $\Gamma_s = \Gamma_m$ then $F = F_m$. The purpose of the design of the low noise amplifier is to provide a matching network that is capable of transforming Γ_s to Γ_m. The noise equation (4.13) defines a family of circles called constant noise figure circles. Each circle when plotted on the Smith chart represents the locus of source impedances that give the same noise figures. Let $F = F_i$, a given noise figure, and define N_i as

$$N_i = \frac{|\Gamma_s - \Gamma_m|^2}{1-|\Gamma_s|^2} = \frac{F_i - F_m}{4r_n}|1+\Gamma_m|^2 \tag{4.14}$$

From (4.14) we get

$$|\Gamma_s - \Gamma_m|^2 = (\Gamma_s - \Gamma_m)(\Gamma_s^* - \Gamma_m^*) = N_i - N_i|\Gamma_s|^2 \tag{4.15}$$

Rearranging the terms in (4.15) yields

$$|\Gamma_s|^2 + |\Gamma_m|^2 - 2\,\mathrm{Re}(\Gamma_s\Gamma_m^*) = N_i - N_i|\Gamma_s|^2$$

$$|\Gamma_s|^2(1+N_i) + |\Gamma_m|^2 - 2\,\mathrm{Re}(\Gamma_s\Gamma_m^*) = N_i \tag{4.16}$$

Multiplying both sides of (4.16) by $1+N_i$ yields

$$|\Gamma_s|^2(1+N_i)^2 + |\Gamma_m|^2 - 2(1+N_i)\,\mathrm{Re}(\Gamma_s\Gamma_m^*) = N_i^2 + N_i(1-|\Gamma_m|^2)$$

and

$$|\Gamma_s(1+N_i)-\Gamma_m|^2=N_i^2+N_i(1-|\Gamma_m|^2)$$

or equivalently,

$$\left|\Gamma_s-\frac{\Gamma_m}{1+N_i}\right|^2=\frac{N_i^2+N_i(1-|\Gamma_m|^2)}{(1+N_i)^2} \tag{4.17}$$

Equation (4.17) defines a circle with center C_F and radius r_F given as

$$C_F=\frac{\Gamma_m}{1+N_i} \tag{4.18a}$$

$$r_F=\frac{1}{1+N_i}\sqrt{N_i^2+N_i(1-|\Gamma_m|^2)} \tag{4.18b}$$

From (4.14) and (4.18) we note that $N_i=0$ when $F_i=F_m$, which gives $C_F=\Gamma_m$ and $r_F=0$, that is, the center of the F_m circle with zero radius is located at Γ_m on the Smith chart. From (4.18a) we see that the centers of other constant noise figure circles lie along the Γ_m vector on the Smith chart.

In the design of low noise amplifiers, it is important to note that the noise figure contribution of the subsequent stages can be significant if the gain of the first stage is low, as seen from (2.62b). Therefore it is not always necessary to minimize the first stage noise figure if the sacrifice of gain is too great because the minimum noise figure and maximum gain cannot be obtained simultaneously. Therefore it is better sometimes to compromise between a first stage noise figure and gain to obtain a lower overall noise figure for a cascade amplifier or a receiving system. This is illustrated in the following examples.

Example 4.3

Consider the two amplifiers A_1 and A_2 in cascade as shown in Fig. 4.7, with noise figure F_i and gain G_i, $i=1,2$, respectively. Let $F_1=1.5$ dB, $G_1=10$ dB, and $F_2=2.5$ dB. Then the overall noise figure F is given by (2.62b) as

$$\begin{aligned} F=F_1+\frac{F_2-1}{G_1}&=\log^{-1}\left(\frac{1.5}{10}\right)+\frac{\log^{-1}(2.5/10)-1}{\log^{-1}(10/10)} \\ &=1.49 \end{aligned}$$

or in terms of decibels, $F=1.733$ dB. Now if $F_1=1.6$ dB and $G_1=13$ dB with F_2 unchanged, then $F=1.716$ dB. Thus the overall amplifier has lower noise figure in the second case even though the first stage noise figure of the first case is lower than that of the second case.

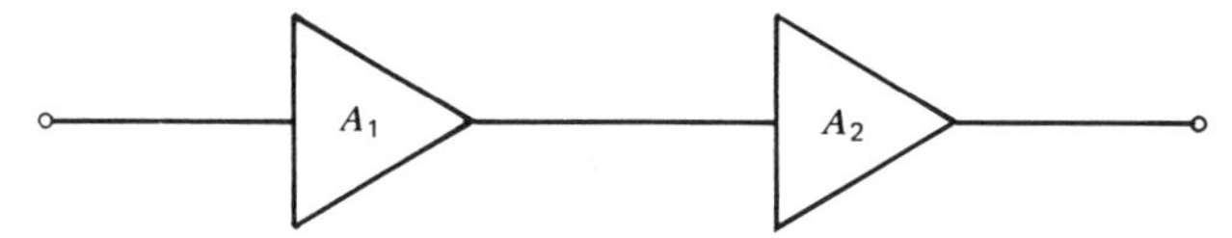

Fig. 4.7. A cascade amplifier.

Example 4.4

This example demonstrates the design of a low noise GaAs FET amplifier using the HFET-1101 at 6 GHz whose noise parameters at this frequency are given as $F=2.2$ dB, $\Gamma_m=0.575\angle 138°$, and $R_n=6.64\ \Omega$. The HFET-1101 scattering parameters at the minimum noise figure bias are

$$S_{11}=0.674\angle -152° \qquad S_{21}=1.74\angle 36.4°$$

$$S_{12}=0.075\angle 6.2° \qquad S_{22}=0.60\angle -92.6°$$

Using (2.51) we have $|\Delta|=|0.385\angle 134.3°|=0.385<1$ and $K=1.275>1$. Thus the GaAs FET is unconditionally stable at this bias condition. With the minimum noise figure reflection coefficient Γ_m as the source, the load that produces the maximum gain at the minimum noise bias is given by the cascade-load formula in (2.46b) as

$$\Gamma_L=\left(\frac{S_{22}-\Gamma_s\Delta}{1-\Gamma_s S_{11}}\right)^*=\left(\frac{S_{22}-\Gamma_m\Delta}{1-\Gamma_m S_{11}}\right)^*=0.602\angle 104°$$

Substituting Γ_m and Γ_L into (2.43) yields the amplifier gain $G=9.73$ dB. The graphical method of previous examples can be used to synthesize the input and output matching networks. Here we offer another approach.

Input Matching Network

The impedance Z_m, corresponding to $\Gamma_m=0.575\angle 138°$ is $Z_m=15.32+j17.6$. Hence the admittance $Y_m=1/Z_m=0.028-j0.032$. It is seen that Y_m can be realized by paralleling a conductance of 0.028 mhos and a succeptance of 0.032 mhos. To realize the succeptance, we note that an open-circuited stub is seen to have an input admittance $Y=jY_0\tan\theta$ as in (2.9) and if $\theta=2\pi l/\lambda=3\pi/4$ (i.e., $l=3\lambda/8$), then $Y=-jY_0$. Thus an open-circuited stub that is three-eighths of a wavelength long looks like a shunt element with admittance $-jY_0$. For our case, $Y=-jY_0=-j0.032$ and thus $Y_0=0.032$ or $Z_0=1/Y_0=31.25\ \Omega$ is the characteristic impedance of the open-circuited stub. Since the source generator is 50 Ω and in order to realize the parallel conductance element of 0.028 mhos, a quarter-wave transformer ($l=\lambda/4$) of characteristic impedance $Z_0=\sqrt{(50)(1/0.028)}=42.26\ \Omega$ is employed. The input matching network is shown in Fig. 4.8.

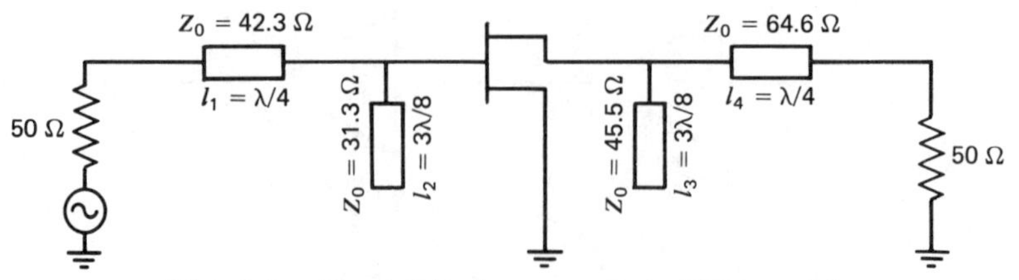

Fig. 4.8. The 6 GHz low noise GaAs FET amplifier.

Output Matching Network

The output impedance Z_L, corresponding to Γ_L, is $Z_L = 19.28 + j35.32$. Hence the admittance $Y_L = 1/Z_L = 0.012 - j0.022$. As above the output matching network can be realized with a three-eighths of a wavelength long open-circuited stub of characteristic impedance of 45.45 Ω and a quarter-wave transformer of characteristic impedance 64.55 Ω as shown in Fig. 4.8.

Example 4.5

This example is used to demonstrate the design of a low noise amplifier at 2 GHz using a GaAs FET potentially unstable at this frequency. We use an HFET-1101 whose noise parameters and whose scattering parameters at the minimum noise figure bias are given as $F_m = 1.25$ dB, $\Gamma_m = 0.730\angle 60°$, and $R_n = 19.4\ \Omega$. And

$$S_{11} = 0.935\angle -51.9° \qquad S_{21} = 2.166\angle 128.3°$$

$$S_{12} = 0.045\angle 54.6° \qquad S_{22} = 0.733\angle -30.5°$$

From the S-parameters data, $|\Delta| = |0.7\angle -74.4°| = 0.7 < 1$ and $K = 0.405 < 1$. Thus the FET is potentially unstable. To check whether the optimum noise source reflection coefficient Γ_m is a stable source at 2 GHz, the source stability circle is plotted on the Smith chart as in Fig. 4.9, with center $C_s = 1.126\angle 61.4°$ and radius $r_s = 0.252$. From Fig. 4.9 it is seen that Γ_m is a stable source reflection coefficient. To achieve the maximum gain with Γ_m as a source, the load reflection coefficient Γ_L must be selected by the cascade-load formula (2.46b), that is,

$$\Gamma_L = \left(\frac{S_{22} - \Gamma_m \Delta}{1 - \Gamma_m S_{11}} \right)^* = 0.831\angle 44.3°$$

This load reflection coefficient Γ_L falls inside the load stability circle with center $C_L = 2.622\angle 77.4°$ and radius $r_L = 2.06$ and hence is not appropriate. The selection of stable load reflection coefficients must be such that they can produce gain as high as possible and can be practically realizable on microstrip substrates. Based upon these criteria, we select a new stable load reflection coefficient

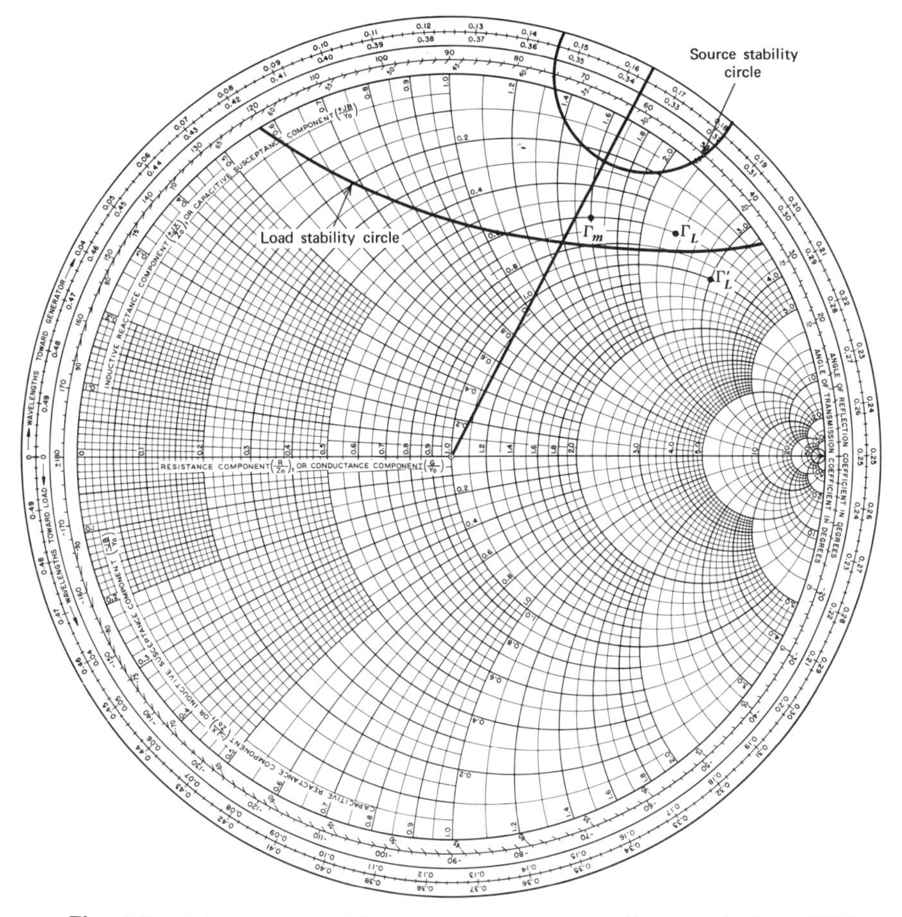

Fig. 4.9. A low noise amplifier design using a potentially unstable GaAs FET.

$\Gamma_L' = 0.831\angle 34°$ that lies outside the load stability circle as in Fig. 4.9. Substituting Γ_m and Γ_L' into (2.43) yields the amplifier transducer power gain $G = 17$ dB (the maximum stable gain is MSG = 16.8 dB and the maximum power gain given by Problem 2.17 is $G_{p,\max} = 17.74$ dB). Using the procedure in Example 4.4, the input and output matching networks that realize Γ_m and Γ_L' are presented in Fig. 4.10.

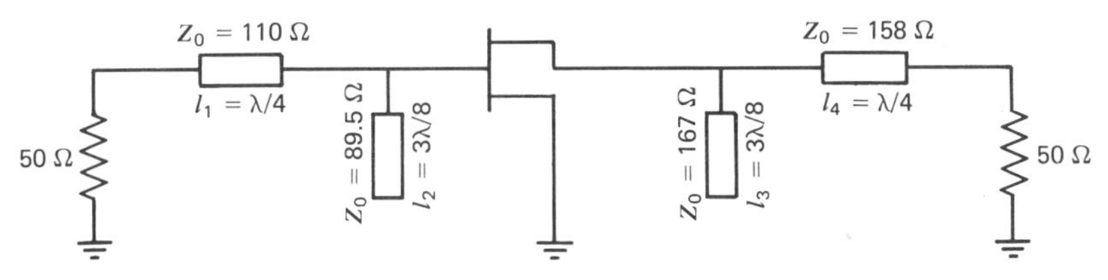

Fig. 4.10. A 2 GHz low noise GaAs FET amplifier.

Example 4.6

This example shows how to generate the noise circles. Consider the bipolar transistor HXTR-6105 at 2 GHz whose noise parameters are $F_m = 2.25$ dB, $\Gamma_m = 0.429\angle 173°$ and $R_n = 5.04\ \Omega$. Suppose we want to plot the 3 dB noise figure circle. Then $F_i = 3$ dB and using (4.14) we get

$$N_i = \frac{2-1.68}{4(5.04/50)}|1+0.429\angle 173°|^2 = 0.261$$

Substituting $N_i = 0.261$ into (4.18) yields

$$C_F = 0.34\angle 173°$$

$$r_F = 0.42$$

In many cases, the designer of the low noise amplifier has to compromise noise figure and gain in order to achieve a low overall system noise figure, as shown before in Example 4.3. Such design can be accomplished systematically using constant noise figure circles and constant available power gain circles as described in the subsequent discussion.

It is known for a two-port network that the available power gain G_a is defined as the ratio of the maximum power available to the load to the maximum power available from the source, that is, it is independent to the load impedance. From Section 2.2, using the generalized scattering matrix we obtain

$$G_a = \frac{\text{maximum power available to load}}{\text{maximum power available at source}} \cdot \frac{\text{power delivered to load}}{\text{power delivered to load}}$$

$$= \frac{\text{power delivered to load}}{\text{maximum power available at source}}$$

$$\cdot \frac{1}{\text{power delivered to load/maximum power available to load}}$$

Since the power delivered to the load is equal to the maximum power available to the load minus the power reflected from the load, and since the ratio of the load reflected power to the load maximum available power is $|S'_{22}|^2$ as shown in Section 2.2, we conclude that

$$G_a = \frac{|S'_{21}|^2}{1-|S'_{22}|^2} \tag{4.19}$$

Proceeding in a similar way to the derivation of the power gain G_p we arrive to the following expression

$$G_a = \frac{|S_{21}|^2(1-|\Gamma_s|^2)}{1-|S_{22}|^2+|\Gamma_s|^2(|S_{11}|^2-|\Delta|^2)-2\,\mathrm{Re}(C_1\Gamma_s)} \tag{4.20}$$

where $C_1 = S_{11} - \Delta S_{22}^*$.

Note that G_a and G_p can be interchanged by interchanging S_{22} for S_{11}, Γ_s for Γ_L and C_1 for C_2. Hence if we let $g_a = G_a/|S_{21}|^2$ denote the normalized available power gain, then (4.20) represents a circle that is the locus of all Γ_s that results in the same g_a. The center C_a and radius r_a of this circle are obtained from (4.11) by the appropriate interchange mentioned above.

$$C_a = \frac{g_a(S_{11}^* - \Delta^* S_{22})}{1+g_a(|S_{11}|^2-|\Delta|^2)} \tag{4.21a}$$

$$r_a = \frac{\left(1-2K|S_{12}S_{21}|g_a+|S_{12}S_{21}|^2g_a^2\right)^{1/2}}{\left|1+g_a(|S_{11}|^2-|\Delta|^2)\right|} \tag{4.21b}$$

By definition it is seen that $G_{p,\max} = G_{a,\max} = G_{\max} = |S_{21}/S_{12}|(K-\sqrt{K^2-1})$ when the transistor is unconditionally stable. When potential unstability exists, $G_{a,\max}$ and $G_{\max}$ become infinite. After some Γ_s is chosen, the correct value of Γ_L to maximize G_a and to realize the two-port transducer power gain $G = G_a$ is given by the cascade-load formula (2.46b)

$$\Gamma_L = \left(\frac{S_{22}-\Gamma_s\Delta}{1-\Gamma_s S_{11}}\right)^* \tag{4.22}$$

The constant available power gain circle and the constant noise figure circle can be plotted together on the Smith chart and their intersections, if they exist, yield the input reflection coefficient Γ_s that provides the desired noise figure and gain.

Example 4.7

This example demonstrates the design of a GaAs FET amplifier that compromises between noise figure and gain at 6 GHz. Typical scattering parameters and noise parameters are given as

$$S_{11} = 0.641\angle -171.3^\circ \qquad S_{21} = 2.058\angle 28.5^\circ$$

$$S_{12} = 0.057\angle 16.3^\circ \qquad S_{22} = 0.572\angle -95.7^\circ$$

and

$$F_m = 2.9\ \mathrm{dB},\ R_n = 9.42\ \Omega \qquad \text{and} \qquad \Gamma_m = 0.542\angle 141^\circ$$

From the above data $|\Delta|=|0.302\angle 109.8°|=0.302$ and $K=1.509$. Hence the GaAs FET is unconditionally stable. The maximum gain is given as $G_{\max}=G_{a,\max}=G_{p,\max}=11.36$ dB; the gain with the associated minimum noise figure F_m is 9.33 dB. We decide that a compromise between noise figure and gain is necessary if an input reflection coefficient Γ_s exists.

Let $F_i=3.15$ dB and $G_a=10.55$ dB. The F_i-circle and the G_a-circle have centers C_F and C_a and radii r_F and r_a as shown in Fig. 4.11.

$$C_F=0.507\angle 141° \qquad r_F=0.217$$

$$C_a=0.691\angle 177.3° \qquad r_a=0.243$$

These two circles intersect at two points. The source reflection coefficients Γ_s and Γ_s' at these two points provide an amplifier with 3.15 dB noise figure and

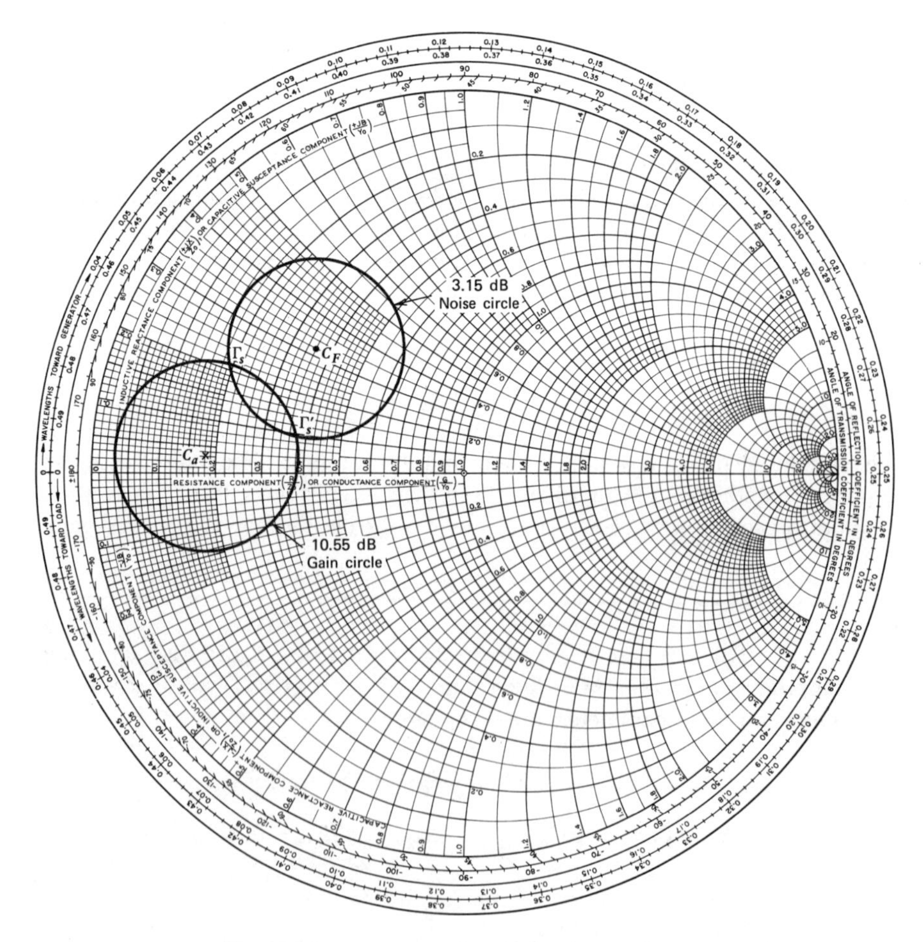

Fig. 4.11. A noise and gain compromise design.

10.55 dB gain. The selection of the source reflection coefficients Γ_s and Γ_s' must be such that they and their corresponding load reflection coefficients can be practically realizable on microstrip substrates. Select $\Gamma_s = 0.68\angle 156°$. Using (4.22) we obtain $\Gamma_L = 0.67\angle 105°$.

4.1.3 High Power Design

The maximum gain and low noise amplifiers described in the previous two sections are small-signal, low power amplifiers and are often used in receiving systems where the signal level is low enough that distortion is negligible and efficiency is not a major concern. For these types of amplifiers, the equivalent circuits of the small-signal S-parameters completely predict their behavior. In contrast, transmitting systems require the amplifications of high-level signals to furnish considerable signal power to the load such as an antenna. The amplifiers used for these purposes are called power amplifiers or large-signal amplifiers, and since they are biased at a high current level, their efficiency is of major importance. Furthermore, since the input signal level is large, distortion becomes a problem because the transistor parameters now vary appreciably over the signal cycle. For these amplifiers, small-signal S-parameters are of limited value for the design of matching networks, but are useful for stability analysis, and since large-signal S-parameters are ill-defined and are not often available, large-signal input and output impedances that are readily measured using the conventional substitution method to be discussed later, are indispensable in the design of matching networks for microwave transistor power amplifiers.

For power amplifiers, the input signal level is often high, and consequently the output current is either in the cutoff or saturation region during a portion of the input signal cycle. This leads to the classification of power amplifiers into three basic modes of operation: Class A, Class B, and Class C as illustrated in Fig. 4.12*a–c* where the input signal is assumed to be sinusoidal. If the output current flows for 360° of the input voltage cycle, the amplifier is operated in the Class A mode. If the output current flows for 180° of the input voltage cycle, the amplifier is designated a Class B amplifier and is biased at cutoff. If the output current flows for less than 180° of the input voltage cycle, the amplifier is called a Class C amplifier and is biased beyond cutoff. The typical power amplifier with output matching network shown in Fig. 4.13 is used for the following analysis.

(a) *Class A amplifiers.* As shown in Fig. 4.12*a*, the transistor in the Class A mode is biased as a current source that conducts all the time. The dc component of the current source is forced through the rf choke while the ac component is forced to the load through the dc blocking capacitor. The tuned circuit is employed to tune out the output capacitor of the transistor. When the ac component of the current source passes through the load R_L, it generates a collector output voltage the dc component of which is V_{cc}. If this signal is a sine wave of amplitude V_0 then the output power P_o is

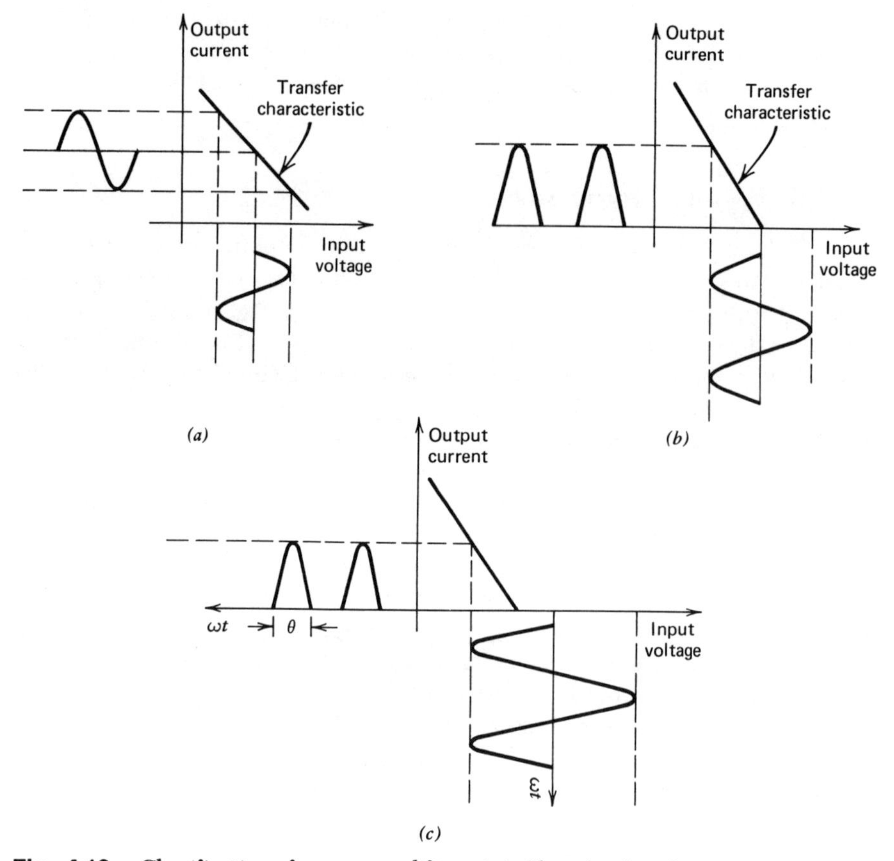

Fig. 4.12. Classification of power amplifiers. (*a*) Class A. (*b*) Class B. (*c*) Class C.

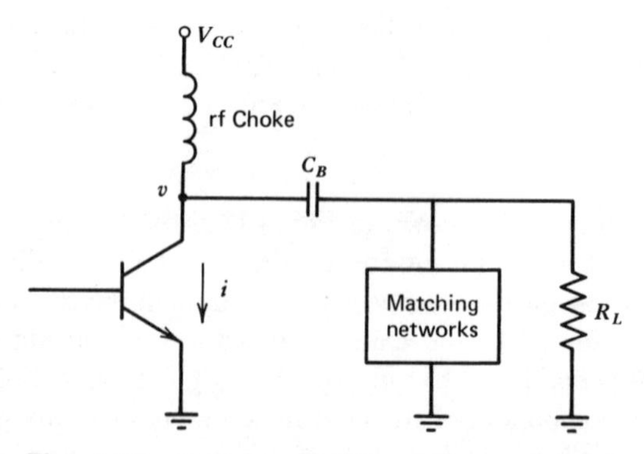

Fig. 4.13. A power amplifier with tuned circuit.

expressed as

$$P_o = V_0^2/2R_L \leqslant V_{cc}^2/2R_L \tag{4.23}$$

The maximum P_o occurs when $V_0 = V_{cc}$, the bias supply voltage. Since the quiescent collector current is $I_c = V_{cc}/R_L$ and the quiescent collector voltage is V_{cc}. The input power into the collector of the transistor is

$$P_i = V_{cc}^2/R_L \tag{4.24}$$

Thus for sine wave amplification, the efficiency of the Class A amplifier is given as

$$\eta = \frac{P_o}{P_i} = \frac{V_0^2/2R_L}{V_{cc}^2/R_L} = \frac{V_0^2}{2V_{cc}^2} \leqslant 50\% \tag{4.25}$$

If other waveforms are amplified, the collector efficiency can be higher than 50% as in the case of a square wave, where the efficiency can approach 100%.

(b) *Class B amplifiers.* In Class B operation, as shown in Fig. 4.12*b*, the transistor acts as a current source during half of the signal cycle and is cut off during the other half period. For a sine wave input signal, the output current is half the sine wave that contains inherent harmonics. Therefore a parallel-tuned circuit that presents a low impedance to the harmonics to pass them to ground is necessary. We note that the output power P_o of class B amplifiers is the same as that of Class A amplifiers, but the input power is different because of a different load current. For sine wave amplifications, the average current for half the sine wave flowing through the load is $(2/\pi)V_0/R_L$. Hence the input power is

$$P_i = V_{cc} \cdot 2V_0/\pi R_L \tag{4.26}$$

The collector efficiency for sine wave amplification is

$$\eta = \frac{P_o}{P_i} = \frac{V_0^2/2R_L}{2V_0V_{cc}/\pi R_L} = \frac{\pi V_0}{4V_{cc}} \leqslant \frac{\pi}{4} = 78.5\% \tag{4.27}$$

Again, the amplification of different waveforms yields different efficiency and can reach 100% for square waves.

(c) *Class C amplifiers.* The Class C amplifier is biased below cutoff. Hence if a sine wave input signal is applied, the collector current flows for less than 180° of a cycle. The angle over which the current flows is called the conduction angle. This results in a higher efficiency than Class A or Class B amplifiers, but the input-output characteristic is no longer linear. If the input is sinusoidal, the input power is

$$P_i = \frac{1}{2\pi}\int_{-\pi}^{\pi} V_{cc} i(t)\, d(\omega t) = \frac{2}{2\pi}\int_{0}^{\pi} V_{cc} i(t)\, d(\omega t) \tag{4.28}$$

where $i(t)$ is the collector current. If $i(t)$ is taken to be a portion of a sine wave within the conduction angle θ, then $i(t)$ can be written as

$$i(t) = I_0(\cos \omega t - \cos \theta/2) \tag{4.29}$$

Hence

$$P_i = \frac{V_{CC} I_0}{\pi}\left[\sin\frac{\theta}{2} - \left(\frac{\theta}{2}\right)\cos\left(\frac{\theta}{2}\right)\right] \tag{4.30}$$

Part of the input power P_i is the collector dissipation P_d and the remaining power is the output power P_o. We note that the dissipation power P_d at the

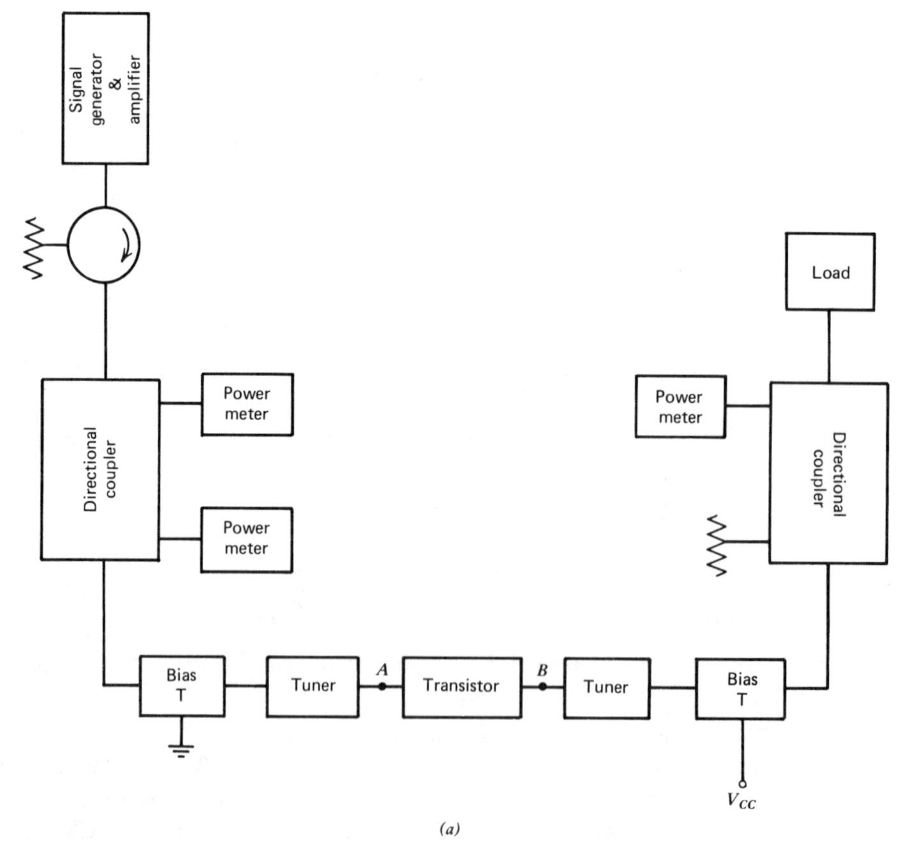

Fig. 4.14. (a) A bipolar transistor amplifier test circuit. (b) A GaAs FET amplifier test circuit.

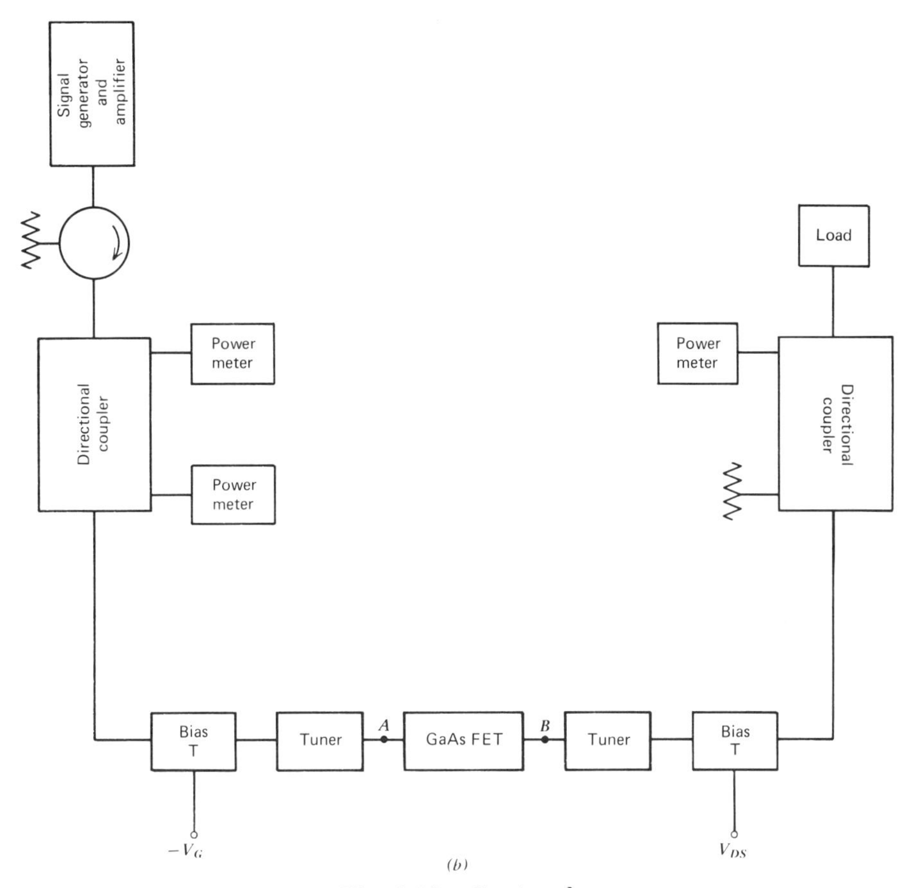

Fig. 4.14. *Continued*

collector is

$$P_d = \frac{1}{2\pi}\int_{-\pi}^{\pi}(V_{cc} - V_{cc}\cos\omega t)\, i(t)\, d(\omega t) \tag{4.31}$$

and hence

$$P_o = P_i - P_d = \frac{V_{cc} I_0}{4\pi}(\theta - \sin\theta) \tag{4.32}$$

Thus for sine wave amplification, the maximum collector efficiency of a Class C amplifier with conduction angle θ is

$$\eta = \frac{P_o}{P_i} = \frac{\theta - \sin\theta}{4\sin\theta/2 - 2\theta\cos\theta/2} \tag{4.33}$$

when $\theta=\pi$, $\eta=\pi/4=78.5\%$, which is the sine wave amplification efficiency of Class B amplifiers. Using L'Hospital's rule in (4.33) and letting $\theta \to 0$ then $\eta \to 100\%$. However as $\theta \to 0$, the output power $P_o \to 0$ too for a finite $i(t)$ and thus in Class C amplifiers, θ is made as small as possible to increase efficiency while still producing the required power output.

As mentioned above, small-signal S-parameters are of little use in the design of power amplifiers except for stability analysis. Only large-signal input and output impedances with power output and gain data provide the designer with the necessary information for the design of matching networks and predict amplifier performance. The large-signal input impedance and output impedance are referred to the actual transistor terminal input and output impedances when operating in a matched amplifier at the desired power output level and dc supply voltage. The matched condition occurs when the input and output matching networks of the test transistor provide a conjugate match to its input and output such that the input and output of the overall test amplifier are 50 Ω impedances. Most microwave bipolar transistors are used in Class A, B, and C amplifiers, whereas most GaAs FETs are used in Class A amplifiers. A test circuit for large-signal impedances and power measurement is shown in Fig. 4.14 for both bipolar transistor amplifiers and GaAs FET amplifiers. The input tuner is used to tune for zero reflected power, thus indicating that the input network matches the transistor input impedance to 50 Ω. The output tuner is then tuned for maximum output power with a given input power. After the test amplifier has been properly tuned for maximum output power, the test transistor is disconnected from the test circuit at points A and B. The impedance looking toward the generator at point A and toward the load at point B are the matched large-signal source and load impedances. The conjugate of these two impedances represent the transistor's large-signal input and output impedances, respectively. The connections shown in Fig. 4.14 are for common emitter bipolar transistors and for common source GaAs FETs. It should also be noted that other connections such as common base, and so forth are possible and sometimes desirable.

Example 4.8

This example is used to illustrate the design of a Class A linear power amplifier using the bipolar transistor HP-35853E at 1.5 GHz biased at 15 V and 100 mA. The corresponding scattering parameters and the large-signal input and output matching reflection coefficients Γ_s and Γ_L for output power at 1 dB gain compression $P_o=26.4$ dBm (see Appendix A5) are given as follows:

$$S_{11}=0.67\angle 144^\circ \qquad S_{21}=2.7\angle 66^\circ$$

$$S_{12}=0.11\angle 60^\circ \qquad S_{22}=0.31\angle -177^\circ$$

$$\Gamma_s=0.86\angle -146^\circ \qquad \Gamma_L=0.52\angle 110^\circ$$

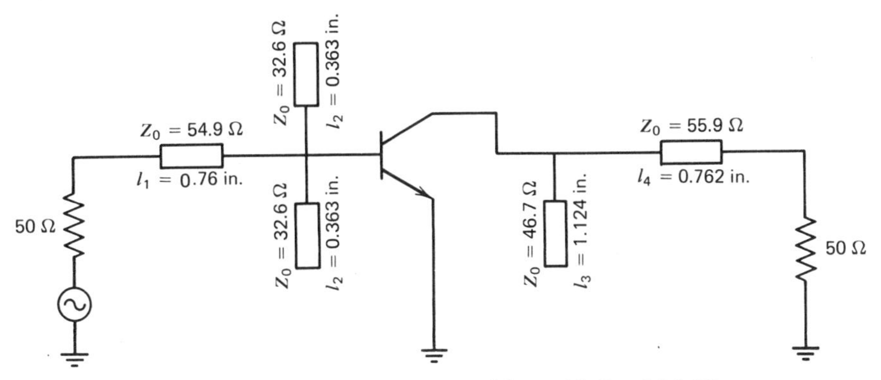

Fig. 4.15. A 1.5 GHz power amplifier with $P_o = 24.6$ dBm.

At this bias, the stability parameters Δ and K are computed as $|\Delta| = |0.388\angle -22°| = 0.388 < 1$ and $K = 1.02$. Thus the transistor is unconditionally stable. The inherent maximum transducer power gain is calculated from (4.2c) as $G_{max} = 13$ dB. Substituting Γ_s and Γ_L into (2.43) yields the actual gain under this optimum linear output power situation as $G = 10.3$ dB. Therefore in order to obtain an output power $P_o = +26.4$ dBm at 1 dB gain compression, the input power must be $P_i = +17.1$ dBm.

From $\Gamma_s = 0.86\angle -146°$, the input impedance Z_s is given as $Z_s = 4.113 - j15.192$ and consequently $Y_s = 1/Z_s = 0.0166 + j0.0613$. We note that an open-circuited stub has an input admittance $Y = jY_0 \tan\theta$ as seen from (2.9) and if $\theta = 2\pi l/\lambda = \pi/4$ ($l = \lambda/8$), then $Y = jY_0$. Thus an open-circuited stub that is 1/8 wavelength long looks like a shunt element with admittance $+jY_0$. For this case, $Y = jY_0 = j0.0613$, thus $Y_0 = 0.0613$ and $Z_0 = 1/0.0613 = 16.3\ \Omega$ is the characteristic impedance of the stub. To match the parallel conductance of 0.0166 mhos to the source admittance of $\frac{1}{50}$ mhos, a quarter-wave transformer of characteristic impedance $Z_0 = \sqrt{50(1/0.0166)} = 54.88\ \Omega$ is employed.

For the output matching network, $\Gamma_L = 0.52\angle 110°$ and hence $Z_L = 22.43 + j30.05$ and $Y_L = 0.016 - j0.0214$. As shown in the previous examples, the output matching network comprises a three-eighths wavelength long open-circuited stub and a quarter wavelength long line with 46.7 Ω and 55.9 Ω characteristic impedances, respectively.

Assume an alumina substrate with $\epsilon_r = 10$ and $H = 0.025$ in. is used as microstrip line substrate for the input and output matching networks. Then from Appendix A3, $W = 0.051$ in., $\epsilon_r' = 7.36$ for 32.6 Ω; $W = 0.019$ in., $\epsilon_r' = 6.69$ for 54.88 Ω; $W = 0.026$ in., $\epsilon_r' = 6.89$ for 46.7 Ω and $W = 0.018$, $\epsilon_r' = 6.67$ for 55.9 Ω. Here we will parallel two 32.6 Ω open-circuited stubs of the input matching network to obtain the 16.3 Ω stub. The complete circuit is shown in Fig. 4.15.

4.2 NEGATIVE RESISTANCE REFLECTION AMPLIFIERS

In this section we will study the design of circulator reflection amplifiers using negative resistance devices such as IMPATT, Gunn, and tunnel diodes. The negative resistance device is assumed to have impedance $Z_d(j\omega)$ such that $\text{Re } Z_d(j\omega) < 0$ over the operating frequency band. The general configuration of this type of amplifier is shown in Fig. 4.16. Let the circulator C be ideal with all the ports having real characteristic impedance Z_0. Then its scattering matrix $\mathbf{S}_c$ normalized to Z_0 is given as

$$\begin{bmatrix} b_1 \\ b_2 \\ b_3 \end{bmatrix} = \mathbf{S}_c \begin{bmatrix} a_1 \\ a_2 \\ a_3 \end{bmatrix} = \begin{bmatrix} 0 & 0 & 1 \\ 1 & 0 & 0 \\ 0 & 1 & 0 \end{bmatrix} \begin{bmatrix} a_1 \\ a_2 \\ a_3 \end{bmatrix} \tag{4.34}$$

where a_i and b_i $(i=1,2,3)$ are the incident and reflected waves. Note that if the generator impedance $Z_s = Z_0$ and the load impedance $Z_L = Z_0$ then $b_1 = 0$ and $a_3 = b_L = 0$. Let Γ_1 denote the reflection coefficient at the input of network N normalized to Z_0, that is, $\Gamma_1 = (Z_1 - Z_0)/(Z_1 + Z_0)$ where Z_1 is the corresponding impedance. Then we have

$$a_2 = \Gamma_1 b_2 \tag{4.35a}$$

Since

$$a_L = b_3 = a_2 \qquad \text{and} \qquad b_2 = a_1$$

$$b_3 = \Gamma_1 a_1 \tag{4.35b}$$

Thus the transducer power gain from port 1 to port 3 is given by

$$G = \left| \frac{b_3}{a_1} \right|^2 = |\Gamma_1|^2 \tag{4.36}$$

Now let $\mathbf{S} = [S_{ij}]$ be the scattering matrix of the lossless reciprocal matching network N, normalized with respect to Z_0 at port 2 of the circulator and to Z_d at

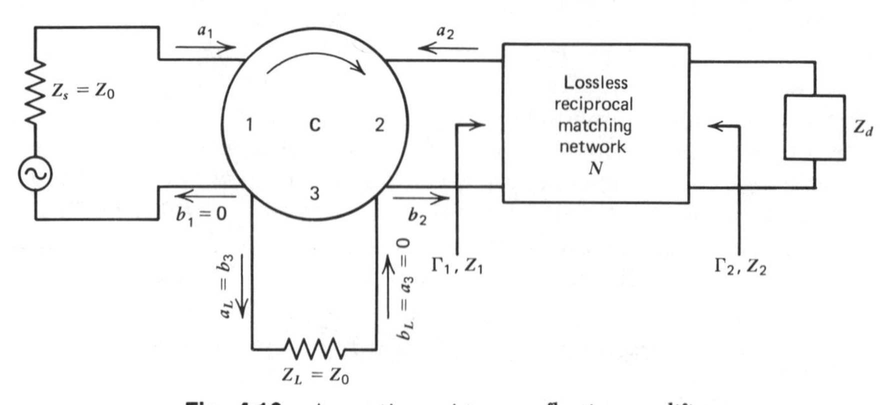

Fig. 4.16. A negative resistance reflection amplifier.

the device's port. Also let Γ_2 be the output reflection coefficient of N, at the device's port normalized to Z_d, that is, $\Gamma_2 = (Z_2 - Z_d^*)/(Z_2 + Z_d)$ where Z_2 is the corresponding output impedance. Then it is seen that $\Gamma_1 = S_{11}$ and $\Gamma_2 = S_{22}$ as explained in Chapter 2. The lossless and reciprocal properties of N yield

$$\mathbf{S}^{T*}\mathbf{S} = \mathbf{1}_2 \tag{4.37a}$$

and

$$\mathbf{S}^T = \mathbf{S} \tag{4.37b}$$

From (4.37) we get

$$|S_{11}|^2 + |S_{21}|^2 = |S_{22}|^2 + |S_{21}|^2 = 1 \tag{4.38a}$$

$$S_{11}^* S_{21} + S_{21}^* S_{22} = 0 \tag{4.38b}$$

This implies that $|S_{11}| = |S_{22}|$ and hence $|\Gamma_1| = |\Gamma_2|$. Therefore the transducer power gain from port 1 to port 3 of the amplifier is also expressed by

$$G = |\Gamma_2|^2 = \left| \frac{Z_2 - Z_d^*}{Z_2 + Z_d} \right|^2 \tag{4.39}$$

Since Z_2 is the impedance of N at the device's port when the other port is loaded in Z_0, Z_2 is passive, that is, $\operatorname{Re} Z_2 > 0$. Furthermore because $\operatorname{Re} Z_d = \operatorname{Re} Z_d^* < 0$, we have $G > 1$. Thus using negative resistance device one can amplify an input rf signal.

For the narrowband design in this section, let $Z_d = -R_d + jX_d$ where $R_d > 0$, $X_d < 0$ for a series resonant device and $X_d > 0$ for a parallel resonant device. Also let $Z_2 = R_2 + jX_2$ ($R_2 > 0$ and X_2 can be either negative or positive). Then substituting into (4.39) we get

$$G = \frac{(R_2 + R_d)^2 + (X_2 + X_d)^2}{(R_2 - R_d)^2 + (X_2 + X_d)^2} \tag{4.40}$$

If X_2 is selected such that $X_2 = -X_d$, then G is given by

$$G = \frac{(R_2 + R_d)^2}{(R_2 - R_d)^2} \tag{4.41}$$

For amplification to be possible, it is necessary that $R_2 > R_d$; if $R_2 = R_d$, then G becomes infinite and oscillation occurs.

In high power amplifications, the large-signal output power depends not only on the gain but also on the diode dc bias current I_{dc} and the peak rf current I_d through the diode, which determines the negative resistance $-R_d$ of the diode.

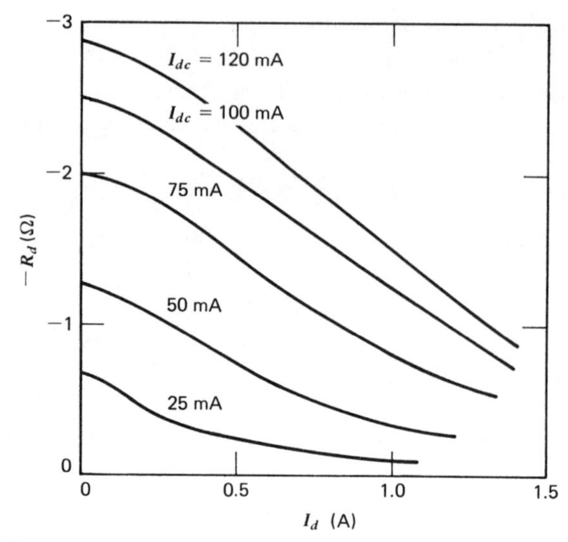

Fig. 4.17. Diode negative resistance $-R_d$ as a function of its rf current amplitude I_d for the IMPATT diode HP-5082-0401 at 11 GHz.

A typical graph of $-R_d$ versus I_d of an IMPATT diode is shown in Fig. 4.17. Referring to the graph, we see that if R_2 is selected to be 3 Ω and the diode is biased at $I_{dc} \leqslant 120$ mA, then R_2 can be used for amplifier design. If $R_2 = 2$ Ω and $I_{dc} = 100$ mA, then oscillation will occur with $I_d = 0.5$ A. In this case the diode is capable of delivering an rf power $P_d = \frac{1}{2}|I_d|^2 R_d = \frac{1}{2}(0.5)^2(-2) = -0.25$ W. The minus sign indicates that the power is generated by the diode; this is also called the oscillation power. Let $P_A = P_o - P_i$ be the added power of the amplifier where P_i and P_o are the input and output power, respectively. We note that

$$P_A = P_o - P_i = |b_3|^2 - |a_1|^2 = |a_1|^2(G-1) \tag{4.42}$$

We see that for P_A to be finite as $G \to \infty$, it is necessary that $|a_1| \to 0$ at the same time. In this case $P_A \to -P_d$, the oscillation power, that is, $-P_d = P_{A,\max}$. Consider Fig. 4.17. Let $I_{dc} = 100$ mA and $R_d = -2$ Ω. Then $I_d = 0.5$ A and $P_{A,\max} = -P_d = 0.25$ W, and the corresponding $G = 25$ or 14 dB with $R_2 = 3$ Ω. More power could be obtained at a lower gain. For example, at $I_{dc} = 100$ mA and $R_d = -1.2$ Ω, we get $I_d = 1$ A and $P_{A,\max} = 0.6$ W. The gain is now $G = 7.4$ dB.

Example 4.9

This example is used to illustrate the design of a large-signal negative resistance amplifier at 10 GHz referring to the following data. The bias current $I_{dc} = 100$ mA, the equivalent impedance $Z_d = -2 - j31$ (Ω), the output

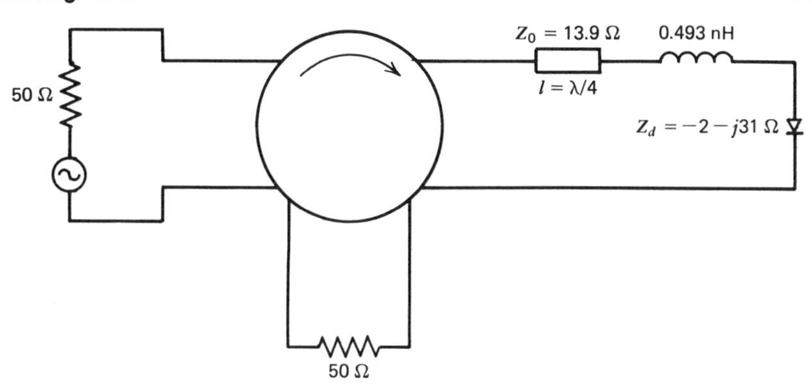

Fig. 4.18. A 10 GHz IMPATT diode amplifier.

power obtained is 0.7 W at a gain of 10 dB. Since $G=10$, using (4.41) we get

$$R_2 = \frac{G^{1/2}+1}{G^{1/2}-1} R_d = 3.85\ \Omega$$

In order to resonate the reactance $-j31$ of Z_d, an inductance of $L=31/2\pi f= 31/(2\pi)(10^{10})=0.493$ nH is put in series with the diode. Assume that $Z_0=50$ Ω, then a quarter wavelength line is necessary to transform Z_0 to R_2, which has the characteristic impedance $\sqrt{50(3.85)}=13.9$ Ω. The circuit is shown in Fig. 4.18.

Although negative resistance reflection amplifiers employing circulators are the most common, others employing hybrids to obtain twice the output power at the expense of using two diodes are also possible (Problem 6.2).

4.3 CONCLUDING REMARKS

We have presented various techniques for designing microwave transistor amplifiers and negative resistance reflection amplifiers depending upon the applications such as maximum gain, low noise, and high power amplifications. For the maximum gain amplifier design, two cases were considered. First, if the transistor is unconditionally stable, the maximum transducer power gain is obtainable if the source and load are conjugately matched and can be obtained from existing formulas. Second, if the transistor is potentially unstable, maximum transducer power gain does not exist and graphical method on the Smith chart must be employed using the power gain concept. In this case the load must be chosen to lie outside the stability circle for a selected constant gain circle.

For the low noise amplifier design, matching for minimum noise figure, in general, does not yield maximum gain. Therefore the designer, in some circumstances, has to make a trade-off between noise figure and gain. Such a compromise can be made by using constant noise figure and constant available power gain circles.

For the large-signal (power) amplifier design, the large-signal input and output matching impedances of the transistor strongly depend upon the signal level and the bias, and can be obtained by the substitution measurement method. Small-signal S-parameters at the same bias conditions can be used to determine the stability of the amplifier.

The design of negative resistance amplifiers depends on resonating out the reactance part of the diode impedance and then matching the negative resistance part to the load, which depends upon a particular output power and gain at the frequency of interest. The maximum added power of a negative resistance amplifier is also its oscillation power.

In summary, even the design of narrowband microwave amplifiers described here is at a single frequency. It can be used for less than 10% bandwidth with moderate accuracy if the device parameters within the bandwidth do not vary considerably (the percentage bandwidth is determined by $(\Delta\omega/\omega_0)$ 100% where ω_0 is the midband frequency and $\Delta\omega$ is the bandwidth).

PROBLEMS

4.1. A GaAs FET at 4 GHz has the following scattering parameters:

$$S_{11}=0.398\angle-81^\circ \qquad S_{12}=0.118\angle 78^\circ$$

$$S_{21}=1.38\angle 84^\circ \qquad S_{22}=0.75\angle-35^\circ$$

Show that the GaAs FET is unconditionally stable. Design a maximum gain amplifier using this FET.

4.2. A GaAs FET at 2 GHz has the following scattering parameters:

$$S_{11}=0.804\angle-44^\circ \qquad S_{12}=0.076\angle 70^\circ$$

$$S_{21}=1.549\angle 125^\circ \qquad S_{22}=0.832\angle-19^\circ$$

Show that it is potentially unstable. Plot the source and load stability circles. Design an amplifier with 12 dB gain with this FET.

4.3. A bipolar transistor at 4 GHz has the following scattering and noise parameters:

$$S_{11}=0.522\angle 169^\circ \qquad S_{12}=0.049\angle 23^\circ$$

$$S_{21}=1.681\angle 26^\circ \qquad S_{22}=0.839\angle-67^\circ$$

$$F_m=2.5\text{ dB} \qquad \Gamma_m=0.475\angle 166^\circ \qquad R_n=3.5\ \Omega$$

Design an amplifier using alumina substrate with $\epsilon_r = 10$ and $H = 0.025$ in. to achieve the above noise figure.

4.4. Using the same transistor in Problem 4.3 to design an amplifier with $F_i = 2.9$ dB and $G = 12.7$ dB using the same type of microstrip substrate.

4.5. A GaAs FET is biased for linear power at 4 GHz ($I_D \approx 50\%\ I_{DSS}$). The input power is $P_i = +5$ dBm, the output power at 1 dB compression is $P_o = +15.5$ dBm and the large-signal input and output source and load reflection coefficients are $\Gamma_s = 0.732\angle 120°$, $\Gamma_L = 0.390\angle 81°$ with the following S-parameters:

$$S_{11} = 0.75\angle -116.5° \qquad S_{12} = 0.05\angle 30°$$

$$S_{21} = 2.458\angle 74.5° \qquad S_{22} = 0.632\angle -59.3°$$

Note that Γ_s is almost identical to S_{11}^* whereas Γ_L is very much different from S_{22}^*. For linear power GaAs FET amplifiers (Class A), this happens in general. Design this power amplifier using a substrate with $\epsilon_r = 10$ and $H = 0.050$ in.

4.6. The input reflection detuning factor δ is defined as the fractional change in the two-port input reflection coefficient with respect to the fractional change in the load reflection coefficient, that is,

$$\delta = \frac{\partial \Gamma_i / \Gamma_i}{\partial \Gamma_L / \Gamma_L}$$

where Γ_i is given in (2.46a). Show that in terms of the two-port S-parameters δ can be expressed as

$$\delta = \frac{S_{12} S_{21} \Gamma_L}{(1 - S_{22}\Gamma_L)(S_{11} - \Delta\Gamma_L)}$$

For a given value of δ show that

$$|\Gamma_L| = \left| \beta \pm \left| \beta^2 - \frac{S_{11}}{S_{22}\Delta} \right|^{1/2} \right|$$

where

$$\beta = \frac{\Delta + S_{11}S_{22} + S_{12}S_{21}\delta^{-1}}{2S_{22}\Delta}$$

This is the limiting value of $|\Gamma_L|$ for a given δ.

4.7. Find the detuning factors for amplifiers in Problems 4.1 and 4.3.

4.8. Using the cascade-load formula in (1.21), derive (4.36).

4.9. Show that the transducer power gain $G \leqslant |\Gamma_2|^2$ for the amplifier in Fig. P4.9.

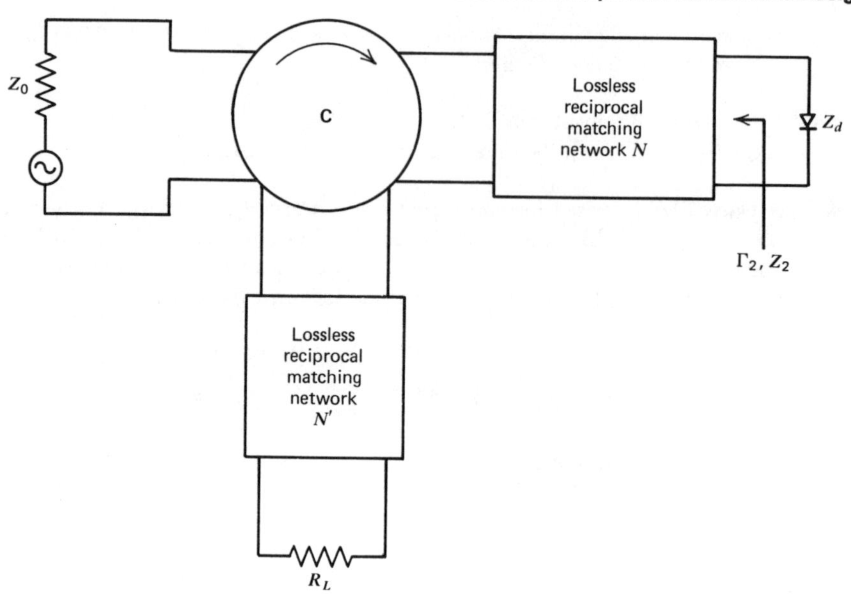

Fig. P4.9. A general configuration for a negative resistance amplifier.

4.10. Explain why $\Gamma_L = 0.42\angle 46.9°$ in Example 4.2 is selected. Show the gain curve if one or two other points are selected.

REFERENCES

1. S-Parameters... circuit analysis and design. Hewlett-Packard application note 95, Sept. 1968.
2. S-Parameter design. Hewlett-Packard application note 154, April 1972.
3. G. J. Wheeler, *Introduction to Microwaves*. Englewood Cliffs, NJ: Prentice-Hall, 1963.
4. R. E. Collins, *Foundations for Microwave Engineering*. New York: McGraw-Hill, 1966.
5. D. T. Paris and F. K. Hurd, *Basic Electromagnetic Theory*. New York: McGraw-Hill, 1969.
6. A. M. Cowley, Design and application of Silicon IMPATT diodes. *Hewlett-Packard J.*, **21**, no. 9, 2, 1970.
7. Microwave power generation and amplification using IMPATT diodes. Hewlett-Packard application note 935, March 1972.
8. R. S. Carson, *High Frequency Amplifiers*. New York: Wiley, 1975.
9. H. A. Atwater, *Introduction to Microwave Theory*. New York: McGraw-Hill, 1962.

5

Microwave Amplifiers: Broadband Design

In Chapter 4 we introduced design techniques for microwave amplifiers with narrow bandwidth (less than 10%) using the device scattering parameters at a single frequency, which is usually at the center of the band. This results in unpredictable bandwidth and ripple. Also, in many applications such as in communication or surveillance systems that require broader bandwidth amplifiers, the previous design techniques are not applicable. The purpose of this chapter is to study various broadband matching techniques for transistor and negative resistance amplifiers. At microwaves frequencies, the reverse transmission scattering parameter S_{12} of the transistors, especially the GaAs FETs, is very small. Hence the unilateral model ($S_{12}=0$) can be used with minimal corrections for the unilateral approximations, and these are usually incorporated in the final optimization procedure. With the unilateral model, the transistor input and output matching networks can be designed separately to match the input impedance of the transistor to its generator resistance (50 Ω in practice) and its output impedance to its load resistance (50 Ω in practice). Furthermore, since the available gain of microwave transistors typically decreases with increasing frequency by approximately 6 dB/octave, it is necessary to design equalizers that can compensate the gain roll-off to obtain an overall amplifier flat response with specified bandwidth ripple.

Unlike the situation with narrowband transistor amplifiers, the success of a broadband amplifier design depends heavily on the transistor's model for high gain, low noise, and large-signal amplifications.

5.1 PRACTICAL UNILATERAL LUMPED MODEL OF UNCONDITIONALLY STABLE MICROWAVE TRANSISTORS

The microwave transistor in package and chip form can be modeled by the circuits in Figs. 5.1*a* and *b*, respectively. The unilateral model is useful only when the transistor is unconditionally stable, although it also represents the potentially unstable transistor. In order to determine accurately whether

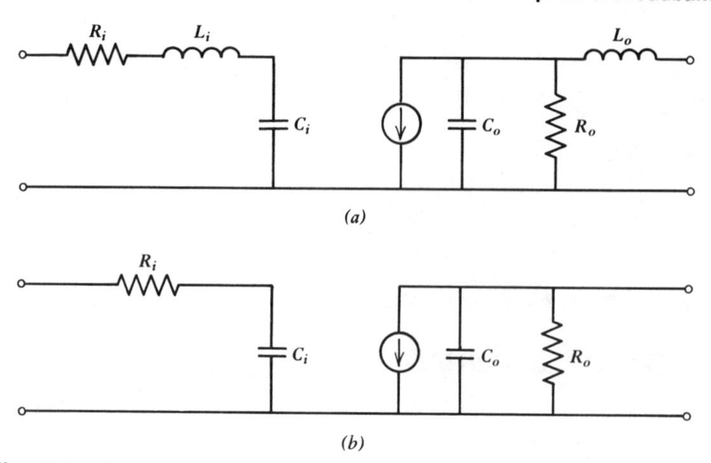

Fig. 5.1. A microwave transistor unilateral model (a) Package. (b) Chip.

a transistor can be considered unilateral from the standpoint of transducer power gain, the designer can use the unilateral figure of merit u defined as follows. From (2.44) we obtain

$$\frac{G}{G_u} = \frac{|1-\Gamma_s S_{11}|^2 |1-\Gamma_L S_{22}|^2}{|(1-\Gamma_s S_{11})(1-\Gamma_L S_{22}) - \Gamma_s \Gamma_L S_{12} S_{21}|^2} = \frac{1}{|1-U|^2} \tag{5.1}$$

where

$$U = \Gamma_s \Gamma_L S_{12} S_{21} / (1-\Gamma_s S_{11})(1-\Gamma_L S_{22})$$

Since

$$\frac{1}{(1+|U|)^2} < \frac{1}{|1-U|^2} < \frac{1}{(1-|U|)^2}$$

we obtain

$$\frac{1}{(1+|U|)^2} < \frac{G}{G_u} < \frac{1}{(1-|U|)^2} \tag{5.2}$$

For unconditionally stable transistors $S_{11} < 1$ and $S_{22} < 1$. Hence if $\Gamma_s = S_{11}^*$ and $\Gamma_L = S_{22}^*$, then (5.2) becomes

$$\frac{1}{(1+u)^2} < \frac{G}{G_{u,\max}} < \frac{1}{(1-u)^2} \tag{5.3a}$$

Table 5.1 The GaAs FET HFET-1102 S-parameters

f (GHz)	S_{11}	S_{21}	S_{12}	S_{22}	S_{21} (dB)	K
4	$0.72\angle -116°$	$2.6\angle 75.6°$	$0.028\angle 56.7°$	$0.73\angle -54°$	8.29	1.29
5	$0.66\angle -142°$	$2.4\angle 53.8°$	$0.034\angle 62°$	$0.72\angle -68°$	7.57	1.29
6	$0.62\angle -167°$	$2.2\angle 32.5°$	$0.045\angle 64.3°$	$0.72\angle -83°$	6.79	1.15
7	$0.59\angle 169°$	$2.0\angle 12.1°$	$0.060\angle 61.6°$	$0.71\angle -100°$	5.96	0.98

where

$$u=\frac{|S_{11}||S_{22}||S_{12}||S_{21}|}{\left(1-|S_{11}|^2\right)\left(1-|S_{22}|^2\right)} \tag{5.3b}$$

is defined as the unilateral figure of merit. Equations (5.3a) and (5.3b) give the maximum error range of the transducer power gain when employing the unilateral model. At microwave frequency, this range is typically ± 0.5 dB.

To show that the unilateral lumped model of microwave transistors in Figs. 5.1*a–b* is indeed valid, we verify it below by modeling the input and output impedances of a number of GaAs FETs and bipolar transistors currently used in industry over a broadband of frequency.

Consider the Hewlett-Packard GaAs FET HFET-1102, whose scattering parameters are shown in Table 5.1. Although the GaAs FET is potentially unstable above 7 GHz, it can be employed in the design of a cascade amplifier (to be discussed later) whose overall stability factor K can be greater than 1. The HFET-1102 can be modeled by the unilateral circuit in Fig. 5.1*a* with $R_i=12.3$ Ω, $L_i=0.748$ nH, $C_i=0.791$ pF, $R_o=240.3$ Ω, $C_o=0.38$ pF and $L_o=0.3$ nH. The input and output reflection coefficients Γ_i and Γ_o of this model are shown in Table 5.2. It is seen that Γ_i and Γ_o approximate S_{11} and S_{22} fairly well over a 3 GHz bandwidth.

The models of other microwave transistors shown here include the Nippon Electric Company GaAs FET NE-38806, NE-24406, the Hewlett-Packard bipolar transistor HXTR-6104 and GaAs FET HFET-1000 in chip form.

The series $R_iL_iC_i$ input impedance of the unilateral model can also be used to approximate the conjugate, Γ_m^*, of the optimum noise source reflection coefficient Γ_m in (4.13). This is very important in the design of broadband low noise

Table 5.2 The input and output reflection coefficients of the HFET-1102 unilateral model

f (GHz)	Γ_i	Γ_o
4	$0.703\angle -113.3°$	$0.680\angle -56°$
5	$0.639\angle -141°$	$0.694\angle -70°$
6	$0.609\angle -167°$	$0.710\angle -83°$
7	$0.607\angle 170°$	$0.728\angle -96°$

Table 5.3 The GaAs FET NE-38806 S-parameters

f (GHz)	S_{11}	S_{21}	S_{12}	S_{22}	S_{21} (dB)	K
4	$0.78\angle -93°$	$3.03\angle 79°$	$0.058\angle 28°$	$0.55\angle -63°$	9.63	0.92
5	$0.71\angle -118°$	$2.80\angle 56°$	$0.054\angle 13.9°$	$0.52\angle -82°$	8.95	1.26
6	$0.68\angle -141°$	$2.51\angle 38°$	$0.046\angle 14°$	$0.56\angle -100°$	8.00	1.51
7	$0.68\angle -164°$	$2.14\angle 16.6°$	$0.046\angle 31°$	$0.62\angle -119°$	6.62	1.39
8	$0.69\angle 176°$	$1.77\angle -9°$	$0.054\angle 46°$	$0.68\angle -137°$	4.96	1.05

Table 5.4 The input and output reflection coefficients of the NE-38806 unilateral model with $R_i = 10.5\ \Omega$, $L_i = 0.65$ nH, $C_i = 0.64$ pF, $R_o = 164\ \Omega$, $C_o = 0.428$ pF, and $L_o = 0.474$ nH

f (GHz)	Γ_i	Γ_o
4	$0.797\angle -94°$	$0.553\angle -67°$
5	$0.732\angle -118°$	$0.569\angle -84°$
6	$0.685\angle -141°$	$0.590\angle -102°$
7	$0.659\angle -164°$	$0.620\angle -119°$
8	$0.654\angle -176°$	$0.650\angle -136°$

Table 5.5 The GaAs FET NE-24406 S-parameters

f (GHz)	S_{11}	S_{21}	S_{12}	S_{22}	S_{21} (dB)	K
5	$0.85\angle -121°$	$2.14\angle 65°$	$0.023\angle 23°$	$0.74\angle -77°$	6.62	1.09
6	$0.79\angle -144°$	$1.95\angle 46°$	$0.018\angle 16°$	$0.71\angle -93°$	5.81	2.38
7	$0.76\angle -164°$	$1.80\angle 33.4°$	$0.024\angle 33°$	$0.70\angle -108°$	5.10	2.07
8	$0.74\angle 179°$	$1.67\angle 20°$	$0.033\angle 52°$	$0.73\angle -124°$	4.50	1.47

Table 5.6 The input and output reflection coefficients of the NE-24406 unilateral model with $R_i = 5.93\ \Omega$, $L_i = 0.56$ nH, $C_i = 0.72$ pF, $R_o = 299\ \Omega$, $C_o = 0.387$ pF, and $L_o = 0.49$ nH

f (GHz)	Γ_i	Γ_o
5	$0.832\angle -123°$	$0.722\angle -73.6°$
6	$0.806\angle -144.3°$	$0.730\angle -91°$
7	$0.792\angle -163.7°$	$0.743\angle -108°$
8	$0.788\angle 179°$	$0.760\angle -125°$

Table 5.7 The bipolar transistor HXTR-6104 S-parameters

f (GHz)	S_{11}	S_{21}	S_{12}	S_{22}	S_{21} (dB)	K
2.0	$0.50\angle -151°$	$2.93\angle 69°$	$0.05\angle 31°$	$0.72\angle -43°$	9.34	1.16
2.5	$0.50\angle -169°$	$2.45\angle 55°$	$0.05\angle 31°$	$0.69\angle -51°$	7.78	1.46
3.0	$0.49\angle 175°$	$2.12\angle 42°$	$0.06\angle 33°$	$0.68\angle -57°$	6.53	1.43
3.5	$0.54\angle 165°$	$1.87\angle 29°$	$0.06\angle 35°$	$0.65\angle -68°$	5.44	1.58
4.0	$0.52\angle 156°$	$1.67\angle 19°$	$0.06\angle 37°$	$0.68\angle -76°$	4.45	1.68

Table 5.8 The input and output reflection coefficients of the HXTR-6104 unilateral model with $R_i = 16.5\ \Omega$, $L_i = 0.89$ nH, $C_i = 3.52$ pF, $R_o = 257.4\ \Omega$, $C_o = 0.526$ pF, and $L_o = 0.611$ nH

f (GHz)	Γ_i	Γ_o
2.0	$0.524\angle -151°$	$0.680\angle -39°$
2.5	$0.506\angle -170°$	$0.680\angle -49°$
3.0	$0.504\angle 176°$	$0.684\angle -60°$
3.5	$0.511\angle 163°$	$0.688\angle -70°$
4.0	$0.523\angle 152°$	$0.695\angle -81°$

Table 5.9 The GaAs FET HFET-1000 S-parameters

f (GHz)	S_{11}	S_{21}	S_{12}	S_{22}	S_{21} (dB)	K
6	$0.804\angle -106°$	$2.21\angle 97°$	$0.031\angle 50°$	$0.742\angle -29°$	6.89	1.00
8	$0.765\angle -125°$	$1.79\angle 80°$	$0.031\angle 49°$	$0.744\angle -40°$	5.06	1.16
10	$0.758\angle -135°$	$1.42\angle 66°$	$0.036\angle 55°$	$0.764\angle -50°$	3.03	1.42
12	$0.733\angle -141°$	$1.19\angle 57°$	$0.038\angle 64°$	$0.787\angle -54°$	1.52	1.56

Table 5.10 The input and output reflection coefficients of the HFET-1000 unilateral model with $R_i = 8.44\ \Omega$, $C_i = 0.77$ pF, $R_o = 320\ \Omega$, and $C_o = 0.138$ pF

f (GHz)	Γ_i	Γ_o
6	$0.800\angle -110°$	$0.745\angle -30°$
8	$0.766\angle -124°$	$0.756\angle -39°$
10	$0.750\angle -134°$	$0.768\angle -48°$
12	$0.738\angle -141°$	$0.782\angle -56°$

Table 5.11 The conjugate of the HFET-1101 optimum noise source reflection coefficient modeled by an $R_iL_iC_i$ series circuit with $R_i = 14.35\ \Omega$, $L_i = 0.63$ nH, and $C_i = 0.723$ pF

f (GHz)	Γ_m^*	Γ_i
6	$0.575\angle -140°$	$0.578\angle -148°$
8	$0.560\angle 170°$	$0.617\angle 170°$
10	$0.610\angle 135°$	$0.596\angle 139°$

Table 5.12 The conjugate of the HFET-2201 optimum noise source reflection coefficient modeled by an $R_iL_iC_i$ series circuit with $R_i = 5.93\ \Omega$, $L_i = 0.39$ nH, and $C_i = 0.8$ pF

f (GHz)	Γ_m^*	Γ_i
6	$0.78\angle -142°$	$0.81\angle -140°$
8	$0.80\angle -166°$	$0.79\angle -168°$
10	$0.80\angle 170°$	$0.79\angle 170°$
12	$0.78\angle 150°$	$0.80\angle 151.3°$

amplifiers, since a matching network must be design to equalize Γ_m^* to the source impedance R_s (50 Ω in practice) to obtain a minimum noise figure. Some examples of the modeling of Γ_m^* are given in Tables 5.11 and 5.12 (Γ_i here refers to the reflection coefficient of the series $R_iL_iC_i$ impedance).

In summary, the unilateral model shown in Fig. 5.1 can approximate the performance of microwave transistors fairly well over a broad bandwidth. The degradation in the transducer power gain by using this model can be predicted by (5.3). If the unilateral model is used at the frequency band at which the transistor is potentially unstable, the designer must make sure that the source and load reflection coefficients looking toward the generator and the load from the transistor's input and output fall outside the source and load stability circles.

5.2 GENERAL LUMPED BROADBAND MATCHING THEORY

Using the unilateral model of the transistor in Fig. 5.1, the design of broadband amplifiers is reduced to the design of the input matching network N_i and the output matching network N_o shown in Fig. 5.2 which can be designed to have a nonsloped transducer power gain characteristic (for minimum VSWR) or a sloped response (to compensate for the gain roll-off of the transistor, which is approximately 6 dB/octave). In summary, the design of N_i and N_o is a special case of the design of a lossless matching network N to equalize a passive complex load $Z_L(s)$ to a resistor source R_s as shown in Fig. 5.3.

Referring to Fig. 5.3, let

$$\mathbf{S}(j\omega) = \begin{bmatrix} S_{11}(j\omega) & S_{12}(j\omega) \\ S_{21}(j\omega) & S_{22}(j\omega) \end{bmatrix} \tag{5.4}$$

denote the scattering matrix of the lossless (reactive) matching two-port N normalized to the passive load $Z_L(j\omega)$ on the right and to the resistor R_s on the left as discussed in Section 2.2. According to (2.25), $\mathbf{S}(j\omega)$ is unitary, that is,

$$\mathbf{S}^{T*}(j\omega)\mathbf{S}(j\omega) = \mathbf{1}_2 \tag{5.5}$$

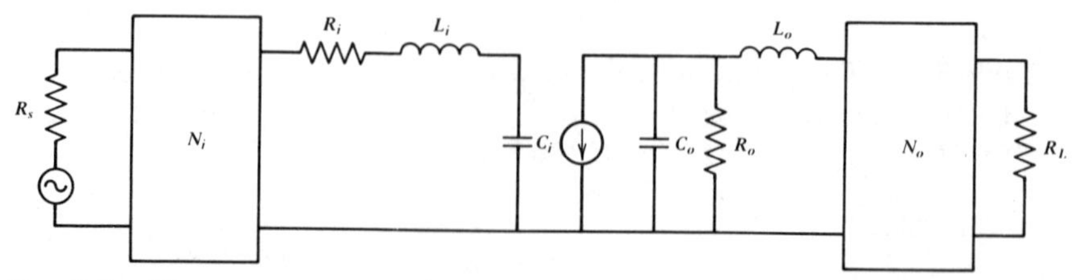

Fig. 5.2. Microwave transistor broadband matching configuration employing a unilateral model.

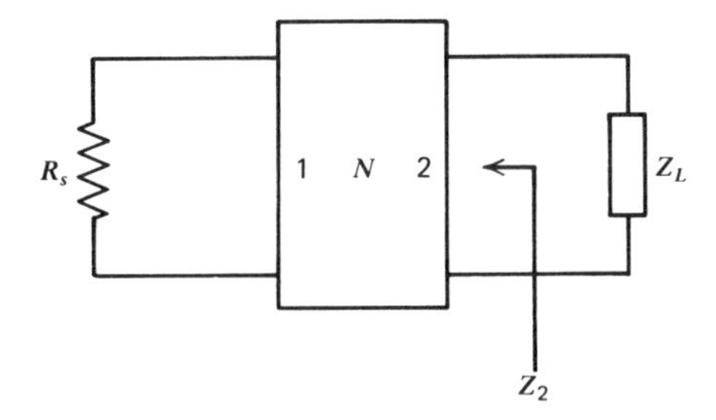

Fig. 5.3. Schematic of a broadband matching problem.

or in expanded form

$$S_{11}^*(j\omega)S_{11}(j\omega)+S_{21}^*(j\omega)S_{21}(j\omega)=1 \tag{5.6a}$$

$$S_{11}^*(j\omega)S_{12}(j\omega)+S_{21}^*(j\omega)S_{22}(j\omega)=0 \tag{5.6b}$$

$$S_{12}^*(j\omega)S_{12}(j\omega)+S_{22}^*(j\omega)S_{22}(j\omega)=1 \tag{5.6c}$$

From (5.6b) we obtain

$$S_{12}(j\omega)=-S_{21}^*(j\omega)S_{22}(j\omega)/S_{11}^*(j\omega) \tag{5.7}$$

Substituting (5.7) into (5.6c) yields

$$|S_{11}(j\omega)|=|S_{22}(j\omega)| \tag{5.8}$$

Using (5.6a), (5.6c), and (5.8) we have

$$|S_{12}(j\omega)|=|S_{21}(j\omega)| \tag{5.9}$$

Equation (2.40) tells us that the transducer power gain from port 1 to port 2 of N is given by

$$G(\omega^2)=|S_{21}(j\omega)|^2=1-|S_{22}(j\omega)|^2 \tag{5.10}$$

where $S_{22}(j\omega)$ is given by (2.30) as

$$S_{22}(j\omega)=\frac{Z_2(j\omega)-Z_L^*(j\omega)}{Z_2(j\omega)+Z_L(j\omega)}=\frac{Z_2(j\omega)-Z_L(-j\omega)}{Z_2(j\omega)+Z_L(j\omega)} \tag{5.11}$$

where Z_2 is the positive-real impedance at port 2 of N and $Z_L^*(j\omega)=Z_L(-j\omega)$ since Z_L is real and rational. Denote $S_{22}(s)$ as follows:

$$S_{22}(s)=\frac{Z_2(s)-Z_L(-s)}{Z_2(s)+Z_L(s)} \tag{5.12}$$

From (5.11) and (5.12) it is obvious that $S_{22}(j\omega)=S_{22}(s)|_{s=j\omega}$. We also see that the poles of $S_{22}(s)$ in $\text{Re}\, s>0$ are precisely those of $Z_L(-s)$ since $Z_2(s)$ and $Z_L(s)$ are positive-real and cannot have poles in $\text{Re}\, s>0$. Let s_i $(i=1,2,\ldots,m)$ denote the poles of $Z_L(-s)$ in $\text{Re}\, s>0$ and set

$$B(s)=\prod_{i=1}^{m}\frac{s-s_i}{s+s_i} \tag{5.13}$$

Then $B(s)$ is an all-pass function, that is, $B(s)$ is real for Re s, analytic in $\text{Re}\, s>0$ and $B(s)B(-s)=1$. Moreover

$$\rho(s)=B(s)S_{22}(s) \tag{5.14}$$

is analytic in $\text{Re}\, s>0$ and $|\rho(j\omega)|=|S_{22}(j\omega)|\leqslant 1$. Thus $\rho(s)$ is a bounded-real reflection coefficient. Substituting (5.14) into (5.12) and (5.10) yield

$$Z_2(s)=\frac{2r_L(s)B(s)}{B(s)-\rho(s)}-Z_L(s) \tag{5.15}$$

where

$$r_L(s)=\tfrac{1}{2}\left[Z_L(s)+Z_L(-s)\right] \tag{5.16}$$

and

$$|\rho(j\omega)|^2=1-G(\omega^2) \tag{5.17}$$

Since $\rho(s)$ is bounded-real, it is necessary that $G(\omega^2)$ be rational in ω and $0\leqslant G(\omega^2)\leqslant 1$. Also from (5.17) we notice that if $G(\omega^2)$ is given, then $\rho(s)$ can be determined by a spectral factorization due to the analytic continuation property, that is,

$$\rho(-s)\rho(s)=1-G(-s^2)=\frac{N(s^2)}{M(s^2)} \tag{5.18}$$

by substituting $\omega=-js$ into (5.17), where $N(s^2)$ and $M(s^2)$ denote the numerator and denominator of $1-G(-s^2)$, respectively. Let $N(s^2)$ and $M(s^2)$ be factored into a product of two polynomials, that is, $N(s^2)=n(-s)n(s)$ and $M(s^2)=m(-s)m(s)$, where $n(s)$ and $m(s)$ are Hurwitz polynomials formed by the zeros in $\text{Re}\, s<0$ of $N(s^2)$ and $M(s^2)$, respectively. Then from (5.18) $\rho(s)$ can be selected to be

$$\rho(s)=\pm\frac{n(s)}{m(s)} \tag{5.19}$$

The above procedure to obtain $\rho(s)$ is called spectral factorization. We note that

$\rho(s)$ in (5.19) has no zeros or poles in $\text{Re}\, s>0$. Such a function is termed minimum-phase and is uniquely obtained from $G(-s^2)$ up to within a plus or minus sign. The most general bounded-real reflection coefficient $\hat{\rho}(s)$ satisfying (5.18) is given by

$$\hat{\rho}(s)=\pm\eta(s)\rho(s) \tag{5.20}$$

where $\eta(s)$ is an arbitrary, real, rational, regular all-pass function of the form

$$\eta(s)=\prod_{r=1}^{k}\frac{s-\mu_r}{s+\mu_r^*}, \qquad \text{Re}\,\mu_r>0 \tag{5.21}$$

for $r=1,2,\ldots,k$. For our purpose, we will use $\rho(s)$ in (5.19) throughout the discussion.

In summary, given $Z_L(s)$ and $G(\omega^2)$, we can obtain $Z_2(s)$ from (5.15). If $Z_2(s)$ is positive-real, we can synthesize the lossless network N that is terminated at port 1 by a resistor. The remaining problem is to find constraints on $\rho(s)$ to make $Z_2(s)$ positive-real.

From (5.15) we have

$$B(s)-\rho(s)=\frac{2r_L(s)B(s)}{Z_2(s)+Z_L(s)}=\frac{2\lambda(s)B(s)}{Z_2(s)/Z_L(s)+1} \tag{5.22}$$

where

$$\lambda(s)=r_L(s)/Z_L(s) \tag{5.23}$$

and conclude that every zero of $\lambda(s)$ in $\text{Re}\,s\geqslant 0$ must also be a zero of $B(s)-\rho(s)$. Stated differently, regardless of the choice of reactive equalizer N, there exist points s_0 in $\text{Re}\,s\geqslant 0$, dictated solely by the choice of load $Z_L(s)$ such that $B(s_0)=\rho(s_0)$ where $B(s)$ is given by (5.13). A zero s_0 of $\lambda(s)$ in $\text{Re}\,s\geqslant 0$ of multiplicity k is said to be a zero of transmission of the load of order k. The constraints for $Z_2(s)$ to be positive-real will now be formulated based upon s_0. For convenience all s_0's are divided into four mutually exclusive classes.

Class 1: all s_0 such that $\text{Re}\,s_0>0$.
Class 2: all $s_0=j\omega_0$ such that $Z_L(j\omega_0)=0$.
Class 3: all $s_0=j\omega_0$ such that $0<Z_L(j\omega_0)<\infty$.
Class 4: all $s_0=j\omega_0$ such that $|Z_L(j\omega_0)|=\infty$.

Also we let $\rho(s)$, $B(s)$, and $F(s)=2r_L(s)B(s)$ be represented by power series expansions at s_0, that is,

$$\rho(s)=\sum_{i=0}^{\infty}\rho_i(s-s_0)^i \tag{5.24a}$$

$$B(s)=\sum_{i=0}^{\infty}B_i(s-s_0)^i \tag{5.24b}$$

$$F(s)=2r_L(s)B(s)=\sum_{i=0}^{\infty}F_i(s-s_0)^i \tag{5.24c}$$

Furthermore, let

$$Z_2'(j\omega)=\frac{d}{d\omega}Z_2(j\omega) \tag{5.24d}$$

$$Z_L'(j\omega)=\frac{d}{d\omega}Z_L(j\omega) \tag{5.24e}$$

and the residue of $Z_L(s)$ at a pole $s=j\omega_0$ be

$$a_{-1}=\lim_{s\to j\omega_0}(s-j\omega_0)Z_L(s) \tag{5.25a}$$

$$=\lim_{s\to\infty}\frac{Z_L(s)}{s} \tag{5.25b}$$

depending on whether ω_0 is finite or infinite, respectively. We then have the following theorem.

Broadband Matching Theorem

Let $Z_L(s)$ be a rational positive-real function and $\rho(s)$ be a rational bounded-real function. Then $Z_2(s)$ given in (5.15) is rational and positive-real if and only if

(a) $B_i=\rho_i$ $(i=0,1,\ldots,k-1)$ for Class 1 zero of transmission s_0 of order k.

(b) $B_i=\rho_i$ $(i=0,1,\ldots,k-1)$ and $(B_k-\rho_k)/F_{k+1}\geqslant 0$ for Class 2 zero of transmission s_0 of order k.

(c) $B_i=\rho_i$ $(i=0,1,\ldots,k-2)$ and $(B_{k-1}-\rho_{k-1})/F_k\geqslant 0$ for Class 3 zero of transmission s_0 of order k.

(d) $B_i=\rho_i$ $(i=0,1,\ldots,k-1)$ and $F_{k-1}/(B_k-\rho_k)\geqslant a_{-1}$ for Class 4 zero of transmission s_0 of order $k-1$.

Proof.

"If" part.

(a) At any Class 1 zero of transmission s_0 of order k of $\lambda(s)$, the difference $B(s)-\rho(s)$ must also have a zero of order k as seen from (5.22). Thus the first k coefficients in power series expansions of $B(s)$ and $\rho(s)$ in (5.24a) and (5.24b) must coincide.

(b) At any Class 2 zero of transmission $s_0=j\omega_0$ of order k of $\lambda(s)$, we also have $Z_L(j\omega_0)=0$. Thus $F(s)=2r_L(s)B(s)$ must have a zero at $s_0=j\omega_0$ of multiplicity $k+1$ since $Z_L(s)$ can only have simple zeros on the $j\omega$-axis by positive-real property. If $Z_2(j\omega_0)=0$, then $Z_2(j\omega_0)+Z_L(j\omega_0)=0$ and as is seen from (5.22), $s_0=j\omega_0$ must be a zero of $B(s)-\rho(s)$ of order k. Thus as in case a, the first k coefficients in the power series expansions of $B(s)$ and $\rho(s)$ must coincide. From the right side of (5.22) we see that the coefficient of $(s-j\omega_0)^k$ is $F_{k+1}/[Z_2'(j\omega_0)+Z_L'(j\omega_0)]$. Comparing this to the left side of (5.22) yields the equality

$$B_k-\rho_k=\frac{F_{k+1}}{Z_2'(j\omega_0)+Z_L'(j\omega_0)} \tag{5.26}$$

It can be shown that the positive-real property of Z_2 and Z_L implies that $Z_2'(j\omega_0)>0$ and $Z_L'(j\omega_0)>0$, that is, $(B_k-\rho_k)/F_{k+1}>0$ since $F_{k+1}\neq 0$ [$j\omega_0$ is a zero of multiplicity $k+1$ of $F(s)$]. In the case $Z_2(j\omega_0)\neq 0$, then $s_0=j\omega_0$ is a zero of multiplicity $k+1$ of $B(s)-\rho(s)$, that is, $B_i=\rho_i$ for $i=0,1,\ldots,k$. The two cases combine to give condition b.

(c) At any Class 3 zero of transmission $s_0=j\omega_0$ of $\lambda(s)$ of order k such that $0<|Z_L(j\omega_0)|<\infty$, $F(s)$ must have $s_0=j\omega_0$ as a zero of order k. Furthermore, $B(s)-\rho(s)$ must have a zero of order $k-1$ at $j\omega_0$. Therefore, condition c can be proved in similar way as condition b.

(d) If $s_0=j\omega_0$ is a zero of transmission of order k such that $|Z_L(j\omega_0)|=\infty$, then it is a zero of order k of $B(s)-\rho(s)$ and thus as in condition b we get $B_i=\rho_i$ $(i=0,1,\ldots,k-1)$. From (5.22) we see that the coefficient of $(s-j\omega_0)^k$ on both sides using power series expansions yields

$$A_k-\rho_k=\left.\frac{F_{k-1}}{(s-j\omega_0)Z_2(s)+(s-j\omega_0)Z_L(s)}\right|_{s\to j\omega_0} \tag{5.27}$$

If $|Z_2(j\omega_0)|\neq\infty$, then $A_k-\rho_k=F_{k-1}/a_{-1}(\omega_0)$, where $a_{-1}(\omega_0)$ is the residue of $Z_L(s)$ evaluated at the pole $j\omega_0$. If $|Z_2(j\omega_0)|=\infty$ then $A_k-\rho_k=F_{k-1}/[\hat{a}_{-1}(\omega_0)+a_{-1}(\omega_0)]>F_{k-1}/a_{-1}(\omega_0)$ since the residue of $Z_2(s)$ at the pole $j\omega_0$, which is $\hat{a}_{-1}(\omega_0)$, is positive and real. This completes the proof of the "if" part.

"Only if" part.

We now show that $Z_2(s)$ given by (5.15) is positive-real only if conditions a–d are satisfied. From (5.15) we obtain

$$\operatorname{Re} Z_2(j\omega)=\tfrac{1}{2}\left[Z_2(j\omega)+Z_2(-j\omega)\right]=\frac{r_L(j\omega)\left[1-|\rho(j\omega)|^2\right]}{|B(j\omega)-\rho(j\omega)|^2}\geqslant 0 \quad (5.28)$$

since $r_L(j\omega)\geqslant 0$ and $|\rho(j\omega)|<1$. It remains to be shown that $Z_2(s)$ is analytic in $\operatorname{Re} s>0$. Set

$$z(s)=\frac{2r_L(s)B(s)}{B(s)-\rho(s)}=\frac{F(s)}{B(s)-\rho(s)} \quad (5.29)$$

It is obvious that $\operatorname{Re} z(j\omega)\geqslant \operatorname{Re} Z_L(j\omega)\geqslant 0$ since $\operatorname{Re} Z_2(j\omega)\geqslant 0$. Our purpose is to prove $z(s)$ is positive-real, which is equivalent to showing that

$$y(s)=z^{-1}(s)=\frac{B(s)-\rho(s)}{F(s)} \quad (5.30)$$

is positive-real. We note that $\operatorname{Re} y(j\omega)\geqslant 0$ since $\operatorname{Re} y^{-1}(j\omega)=\operatorname{Re} z(j\omega)\geqslant 0$. Furthermore, the poles in $\operatorname{Re} s>0$ of $y(s)$ can only be the zeros in $\operatorname{Re} s>0$ of $F(s)$ since $B(s)$ and $\rho(s)$ are bounded-real and cannot have poles in $\operatorname{Re} s>0$. These zeros of $F(s)$ are precisely the Class 1 zeros of $\lambda(s)=r_L(s)/Z_L(s)$ with the same order. At such a zero, condition a guarantees that $B(s)-\rho(s)$ vanishes to at least the same order. Therefore (5.30) implies that $y(s)$ cannot have zeros in $\operatorname{Re} s>0$ and hence it is analytic in $\operatorname{Re} s>0$. It remains to show that poles of $y(s)$ on the $j\omega$-axis must be simple and possess real positive residues. Consider a Class 2 zero of transmission $s_0=j\omega_0$ of order k of $Z_L(s)$ such that $|Z_L(j\omega_0)|=0$. We know that it is also a zero of $F(s)$ but of order $k+1$. Condition b implies that $s_0=j\omega_0$ is also a zero of $B(s)-\rho(s)$ of at least order k, and hence $s_0=j\omega_0$ can be a pole of $y(s)$ of at most order one. The associated residue is $(B_k-\rho_k)/F_{k+1}\geqslant 0$. Now consider a Class 3 zero of transmission $s_0=j\omega_0$ of order k of $Z_L(s)$ such that $0<|Z_L(j\omega_0)|<\infty$. We know that s_0 is also a zero of $F(s)$ of order k. Condition c implies that $s_0=j\omega_0$ is also a zero of $B(s)-\rho(s)$ of at least order $k-1$, and therefore is a pole of $y(s)$ of at most order one with the associated residue $(B_{k-1}-\rho_{k-1})/F_k\geqslant 0$. If $s_0=j\omega_0$ is now a Class 4 zero of transmission of order k, then it is also a zero of $F(s)$ of order $k-1$ and is a zero of $B(s)-\rho(s)$ of order k according to condition d. Thus $y(j\omega_0)=0$. We can conclude from the above arguments that $y(s)$ and hence $z(s)$ is positive-real. Since $Z_2(s)=z(s)-Z_L(s)$ is also analytic in $\operatorname{Re} s>0$ it has simple poles on the $j\omega$-axis. It remains to show that the residues at these poles of $Z_2(s)$ are real and positive. Let $j\omega_0$ be a pole of $Z_2(s)$, if $|Z_L(j\omega_0)|\neq\infty$ then the residue of $Z_2(s)$ at $j\omega_0$ is the same as that of $z(s)$ at $j\omega_0$ and hence must be real and positive. If $|Z_L(j\omega_0)|=\infty$, then $j\omega_0$ is a Class 4 zero of transmission and according to the above argument, it is a zero of $y(s)$ and hence a pole of $z(s)$. The residue of $z(s)$ at this pole is $F_{k-1}/(B_k - \rho_k)$ and the corresponding residue of $Z_2(s)$ is therefore $F_{k-1}/(B_k - \rho_k) - a_{-1}(\omega_0) \geqslant 0$ by condition d. This completes the "only if" proof.

The broadband matching problem based upon this theorem can be stated in the following manner. A prescribed lumped passive impedance $Z_L(s)$ is to be equalized to a resistor R_s by means of a lumped lossless matching network to achieve a preassigned transducer power gain $G(\omega^2)$. The design of such equalizer can be outlined as follows:

Step 1. Verify that a given $G(\omega^2)$ is rational in ω^2 and satisfies $0 \leqslant G(\omega^2) \leqslant 1$ for all ω.

Step 2. Find the minimum-phase bounded-real reflection coefficient $\rho(s)$ such that $\rho(-s)\rho(s) = 1 - G(-s^2)$.

Step 3. From $Z_L(s)$, calculate $\lambda(s)$ in (5.23), $B(s)$ in (5.13), and $F(s)$ in (5.24c). Classify the zeros of transmissions of $Z_L(s)$ that are zeros of $\lambda(s)$ and obtain the power series expansion coefficients ρ_i, B_i, and F_i according to each class of zeros of transmission. Check the restrictions imposed on them by conditions a–d of the broadband matching theorem.

Step 4. If restrictions are all satisfied, determine $Z_2(s)$ according to (5.15). The theorem guarantees $Z_2(s)$ to be positive-real.

Step 5. Synthesize $Z_2(s)$ as the input impedance of a lossless network N terminated in a resistor R. The selection of $\rho(s)$ as a minimum-phase bounded-real function guarantees a ladder network N that contains no mutual coupled coils. This is of practical importance at microwave frequencies since lumped ladder networks are realizable either by inductors and capacitors described in Chapter 3 or by approximated transmission lines on microstrip substrates. In general an ideal transformer is necessary to scale R to R_s. The ideal transformer can usually be eliminated by equivalent inductive or capacitive T or pi networks at the expense of introducing one extra reactive element (detailed discussion of this will be postponed to later sections). If no equivalent transformed T or pi networks exist, the ideal transformer still can be realized by a conventional filter network between R and R_s at the expense of introducing more reactive elements.

The results developed here will now be used to design matching networks for microwave transistors in the next sections. We will approach the design using the analytical method, which utilizes the closed form transducer power gain responses $G(\omega^2)$ such as the Chebyshev and Butterworth characteristics, whose properties are well established so that explicit results in gain-bandwidth limitations can be obtained easily. In general any transducer power gain $G(\omega^2)$, which approximates an ideal response can be used. The coefficients of $G(\omega^2)$ can be determined by the least square approximation method. This approximation approach is by nature more general than the analytical approach, but is quite involved.

5.3 MICROWAVE TRANSISTOR AMPLIFIER DESIGN: ANALYTICAL APPROACH

The purpose of this section is to apply the result of the previous section to find the constraints on the gain-bandwidth limitations of microwave amplifiers and to present a procedure for designing matching networks that can compensate for the gain roll-off of microwave transistors, which is approximately 6 dB/octave using analytical responses.

5.3.1 Input Lossless Matching Network

The problem here is to derive the gain-bandwidth limitations and to synthesize the input matching network N_i between the source R_s and the input impedance $R_iL_iC_i$ of the transistor to achieve a desired transducer power gain $G(\omega^2)$ as shown in Fig. 5.2. The response $G(\omega^2)$ can be selected to have a nonsloped (or flat) response within the passband or to have a sloped response with roll-up of 6 dB/octave to compensate for the roll-off of the transistor. We note that the series $R_iL_iC_i$ impedance is of the bandpass type; hence it is necessary to select $G(\omega^2)$ as a bandpass response to maximize the gain-bandwidth product. For the nonsloped response, the following bandpass characteristic is selected:

$$G(\omega^2)=\frac{K_n}{1+P_n^2(\omega')}, \qquad 0\leqslant K_n\leqslant 1 \tag{5.31}$$

where the constant K_n is the passband gain of $G(\omega^2)$ and

$$\omega'=\omega/B-\omega_0^2/B\omega, \qquad B=\omega_2-\omega_1, \qquad \omega_0^2=\omega_1\omega_2 \tag{5.32}$$

Here ω_1 and ω_2 are the lower and upper passband frequencies. $P_n(\omega')=\epsilon C_n(\omega')$ for the Chebyshev response where $C_n(\omega')$ is the nth order Chebyshev polynomial of the first kind in ω' and ϵ is the ripple factor; $P_n(\omega')=\omega'^n$ for the Butterworth response. For the sloped response we let

$$G(\omega^2)=\frac{K_n(\omega/\omega_2)^2}{1+P_n^2(\omega')} \tag{5.33}$$

We note that $G(\omega^2)$ in (5.33) has a gain roll-up of 6 dB/octave over the band (ω_1, ω_2) and for it to be physically realizable as the transducer power gain of a lossless network, it is necessary that it is bounded by unity for all ω. To obtain such $G(\omega^2)$, it is necessary to find the frequency ω where $G(\omega^2)$ attains its maximum. One way to achieve this is to set $d\{G(\omega^2)\}/d\omega=0$ to find a real ω_m such that $\omega_m\geqslant\omega_2$; then if $G(\omega_m^2)\leqslant 1$, it is physically realizable. For example $G(\omega^2)\leqslant 1$ for $n=2$, $K_2=1$, $\epsilon^2=0.26$, $\omega_1=0.5\omega_2$ and for $n=2$, $K_2=0.9$, $\epsilon^2=0.01$, $\omega_1=0.8\omega_2$.

After selecting the transducer power gain $G(\omega^2)$, the minimum-phase bounded-real reflection coefficient $\rho(s)$ can be obtained from it as in (5.18). In general $\rho(s)$ can be put in the form

$$\rho(s)=\frac{s^{2n}+a'_{2n-1}s^{2n-1}+\cdots+a'_1 s+a'_0}{s^{2n}+a_{2n-1}s^{2n-1}+\cdots+a_1 s+a_0} \tag{5.34}$$

where $a_0=a'_0$ and $a_i>a'_i$ $(i=1,2,\ldots,2n-1)$ by the bounded-real property. The load impedance is now the input impedance $Z_i(s)$ of the transistor, which is $Z_i(s)=R_i+sL_i+1/sC_i$. Therefore from (5.23) we have

$$\lambda(s)=\frac{\frac{1}{2}[Z_i(s)+Z_i(-s)]}{Z_i(s)}$$

$$=\frac{R_i}{R_i+sL_i+1/sC_i}$$

$$=\frac{sR_iC_i}{s^2L_iC_i+sR_iC_i+1} \tag{5.35}$$

It is obvious that the zeros of $\lambda(s)$ are zero and infinity of order one. Also $|Z_i(0)|=\infty$ and $|Z_i(\infty)|=\infty$. Thus $s_0=0$ and $s_0=\infty$ are Class 4 zeros of transmission of $Z_i(s)$ of order one. The residues of $Z_i(s)$ at zero and infinity are

$$a_{-1}(0)=\lim_{s\to 0} sZ_i(s)=\frac{1}{C_i} \tag{5.36a}$$

$$a_{-1}(\infty)=\lim_{s\to\infty}\frac{Z_i(s)}{s}=L_i \tag{5.36b}$$

Also $Z_i(-s)=R_i-sL_i-1/sC_i$ has no poles in Re $s>0$, $B(s)=1$, and $F(s)=2R_i$. According to (5.24) the power series expansion of $\rho(s)$, $B(s)$, and $F(s)$ at zero and infinity can be written as

$$\rho(s)=\sum_{i=0}^{\infty}\rho_i s^i=\sum_{i=0}^{\infty}\rho'_i s^{-i} \tag{5.37}$$

$$B(s)=\sum_{i=0}^{\infty}B_i s^i=\sum_{i=0}^{\infty}B'_i s^{-i}=1 \tag{5.38}$$

$$F(s)=\sum_{i=0}^{\infty}F_i s^i=\sum_{i=0}^{\infty}F'_i s^{-i}=2R_i \tag{5.39}$$

Condition d of the broadband matching theorem dictates that

$$B_0 = \rho_0, \qquad \frac{F_0}{B_1 - \rho_1} \geqslant a_{-1}(0) \tag{5.40a}$$

$$B_0' = \rho_0', \qquad \frac{F_0'}{B_1' - \rho_1'} \geqslant a_{-1}(\infty) \tag{5.40b}$$

From (5.34) and (5.37)–(5.39) it is obvious that $B_0 = B_0' = 1$, $B_1 = B_1' = 0$, $F_0 = F_0' = 2R_i$, $\rho_0 = a_0'/a_0 = 1$, $\rho_0' = \lim_{s\to\infty} \rho(s) = 1$, $\rho_1 = (a_1' - a_1)/a_0$, and $\rho_1' = a_{2n-1}' - a_{2n-1}$. Thus (5.40a)–(5.40b) are satisfied if and only if

$$R_i C_i \geqslant \tfrac{1}{2}(a_1 - a_1')/a_0 \tag{5.41a}$$

$$R_i / L_i \geqslant \tfrac{1}{2}(a_{2n-1} - a_{2n-1}') \tag{5.41b}$$

Equations (5.41a) and (5.41b) determine the gain-bandwidth limitations of the input matching network. In the case of the nonsloped response, these constraints can be found upon using the closed form of the coefficient a_i and a_i' for both Chebyshev and Butterworth transducer power gain responses.

For the Chebyshev response, it can be computed that (Appendix A6)

$$a_0 = a_0' = \omega_0^{2n} \tag{5.42a}$$

$$a_1 = \frac{\omega_0^{2n-2} B \sinh x}{\sin u_1}, \qquad a_1' = \frac{\omega_0^{2n-2} B \sinh x'}{\sin u_1} \tag{5.42b}$$

$$a_{2n-1} = \frac{B \sinh x}{\sin u_1}, \qquad a_{2n-1}' = \frac{B \sinh x'}{\sin u_1} \tag{5.42c}$$

where

$$x = (1/n)\sinh^{-1}(1/\epsilon), \qquad x' = (1/n)\sinh^{-1}\left(\sqrt{1-K_n}/\epsilon\right), \qquad u_m = m\pi/2n \tag{5.42d}$$

Using (5.42a)–(5.42d) in (5.41) we obtain

$$(1-K_n)^{1/2} \geqslant \max\left\{ \epsilon \sinh\left[n \sinh^{-1}\left(\sinh x - \frac{2R_i C_i \omega_0^2 \sin u_1}{B} \right)\right], \right.$$
$$\left. \epsilon \sinh\left[n \sinh^{-1}\left(\sinh x - \frac{2R_i \sin u_1}{B L_i} \right)\right]\right\} \tag{5.43}$$

For the Butterworth response (Appendix A6)

$$a_0 = a'_0 = \omega_0^{2n} \tag{5.44a}$$

$$a_1 = \frac{\omega_0^{2n-2}B}{\sin u_1}, \qquad a'_1 = \frac{\omega_0^{2n-2}B(1-K_n)^{1/2n}}{\sin u_1} \tag{5.44b}$$

$$a_{2n-1} = \frac{B}{\sin u_1}, \qquad a'_{2n-1} = \frac{B(1-K_n)^{1/2n}}{\sin u_1} \tag{5.44c}$$

Substituting (5.44a)–(5.44c) into (5.41) yields

$$(1-K_n)^{1/2n} \geqslant \max\left\{1 - \frac{2R_iC_i\omega_0^2 \sin u_1}{B}, 1 - \frac{2R_i \sin u_1}{BL_i}\right\} \tag{5.45}$$

In general, the Chebyshev response is preferable to the Butterworth response in the design of matching networks. Therefore we will drop the Butterworth response in subsequent discussions. The impedance $Z_{2i}(s)$ needed to realize the input matching network N_i is given by (5.15), where $Z_L(s)$ is replaced by $Z_i(s)$, as follows

$$Z_{2i}(s) = R_i \frac{1+\rho(s)}{1-\rho(s)} - L_i s - \frac{1}{sC_i} \tag{5.46}$$

The impedance $Z_{2i}(s)$ in (5.46) can be realized as a bandpass ladder network with an ideal transformer that can be replaced by an equivalent T or pi inductive or capacitive network (Appendix A7) with a constraint in addition to (5.43). The synthesis of $Z_{2i}(s)$ can be carried out using the pole and zero extraction method.

In practice most microwave transistors can be matched with $n \leqslant 3$ (that is, six reactive elements at most in N_i) for a typical bandwidth of an octave.

Example 5.1

In this example we present closed form element values for the 6 dB/octave sloped response characteristic $G(\omega^2)$ in (5.33) with $n=2$ when (5.41a)–(5.41b) are satisfied. For $n=2$ it can be shown that

$$\frac{1+\rho(s)}{1-\rho(s)} = sL_1 + \frac{1}{sC_1} + \frac{1}{\dfrac{1}{sL_2} + \dfrac{1}{(1/sC_2 + R)}} \tag{5.47}$$

where

$$L_1 = 2/(a_3 - a_3') \tag{5.48a}$$

$$C_1 = (a_1 - a_1')/2a_0$$

$$R = [(a_3 + a_3') - 2(a_2 - a_2')/(a_3 - a_3')]/(a_3 - a_3') \tag{5.48b}$$

$$L_2 = [(a_2 + a_2') - 2(a_1 - a_1')/(a_3 - a_3') - 2a_0(a_3 - a_3')/(a_1 - a_1')]/(a_1 - a_1') \tag{5.48c}$$

$$C_2 = (a_3 - a_3')/[(a_2 + a_2') - 2(a_1 - a_1')/(a_3 - a_3') - 2a_0(a_3 - a_3')/(a_1 - a_1')] \tag{5.48d}$$

Since $\rho(s)$ is bounded-real, all quantities in (5.48) are positive. It can be seen from (5.41) and (5.48a)–(5.48b) that $R_i L_1 - L_i \geqslant 0$ and $R_i/C_1 - 1/C_i \geqslant 0$. The equality holds if (5.41) is satisfied with the equal sign. Therefore $Z_{2i}(s)$ in (5.46) has the form

$$Z_{2i}(s) = s(R_i L_1 - L_i) + \frac{1}{s(R_i/C_1 - 1/C_i)^{-1}} + \frac{1}{\dfrac{1}{sR_i L_2} + \dfrac{1}{(1/s(C_2/R_i) + RR_i)}} \tag{5.49}$$

which can be realized by the network shown in Fig. 5.4.

In many cases the designer wants to eliminate the series capacitor shown in the second term on the right side of (5.49). This is equivalent to absorbing the input capacitor C_i into the matching network. For a given bandwidth B and ripple factor ϵ, this can be done by adjusting the passband gain K_n. The following procedure is presented for $n=2$. In order to eliminate the second term in (5.49) we must have, using (5.48b)

$$C_1 = R_i C_i = \tfrac{1}{2}(a_1 - a_1')/a_0 \tag{5.50}$$

For a given B, ϵ and n, a_0 and a_1 are fixed. Thus in order to have $C_1 = R_i C_i$ we

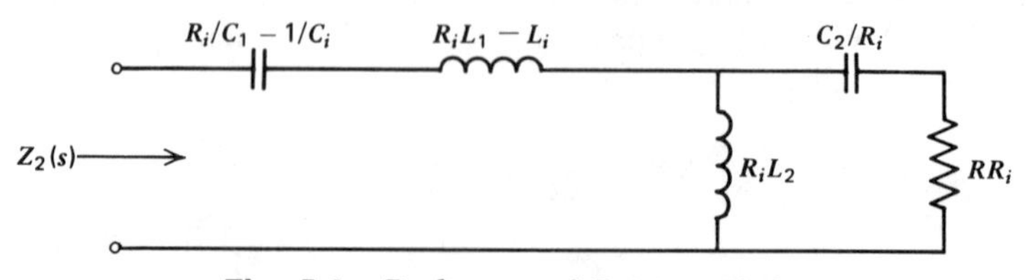

Fig. 5.4. Realization of $Z_2(s)$ in (5.49).

must adjust $a_1' = a_1 - 2a_0 R_i C_i$. For $n=2$, we get

$$\rho(-s)\rho(s) = 1 - G(-s^2) = \frac{s^8 + b_6' s^6 + b_4' s^4 + b_2' s^2 + b_0'}{s^8 + b_6 s^6 + b_4 s^4 + b_2 s^2 + b_0} \tag{5.51}$$

and

$$\rho(s) = \frac{s^4 + a_3' s^3 + a_2' s^2 + a_1' s + a_0'}{s^4 + a_3 s^3 + a_2 s^2 + a_1 s + a_0} \tag{5.52}$$

where

$$b_6' = 4\omega_0^2 + B^2 + K_2 B^4/4\epsilon^2 = 2a_2' - a_3'^2 \tag{5.53a}$$

$$b_6 = 4\omega_0^2 + B^2 = 2a_2 - a_3^2 \tag{5.53b}$$

$$b_4' = b_4 = B^4/4\epsilon^2 + 6\omega_0^4 + 2\omega_0^2 B + B^4/4 \tag{5.53c}$$

$$= 2a_0' + a_2'^2 - 2a_1' a_3' = 2a_0 + a_2^2 - 2a_1 a_3$$

$$b_2' = b_2 = 4\omega_0^6 + \omega_0^4 B^2 = 2a_0' a_2' - a_1'^2 = 2a_0 a_2 - a_1^2 \tag{5.53d}$$

$$b_0 = b_0' = a_0^2 = a_0'^2 = \omega_0^8 \tag{5.53e}$$

From (5.53) we see that

$$K_2 = (p_1 - p_2 - p_3 - p_4) p_5 \tag{5.54}$$

where

$$p_1 = \left(a_1'^2 + b_2\right)/a_0 - 4\omega_0^2 - B^2 \tag{5.55a}$$

$$p_2 = \left(a_1'^2 + b_2\right)^4 / 64 a_0^4 a_1'^2 \tag{5.55b}$$

$$p_3 = (2a_0 - b_4)\left(a_1'^2 + b_2\right)^2 / 8 a_0^2 a_1'^2 \tag{5.55c}$$

$$p_4 = (2a_0 - b_4)^2 / 4a_1^2 \tag{5.55d}$$

$$p_5 = 4\epsilon^2 / B^4 \tag{5.55e}$$

Example 5.2

A medium power GaAs FET has an equivalent input impedance composed of a series $R_i C_i$ circuit with $R_i = 5\ \Omega$ and $C_i = 0.68$ pF over the frequency band 4.8–6 GHz and a gain roll-off of about 6 dB/octave. We wish to design an input matching network N_i to compensate for this roll-off with less than 0.1 dB ripple in the passband.

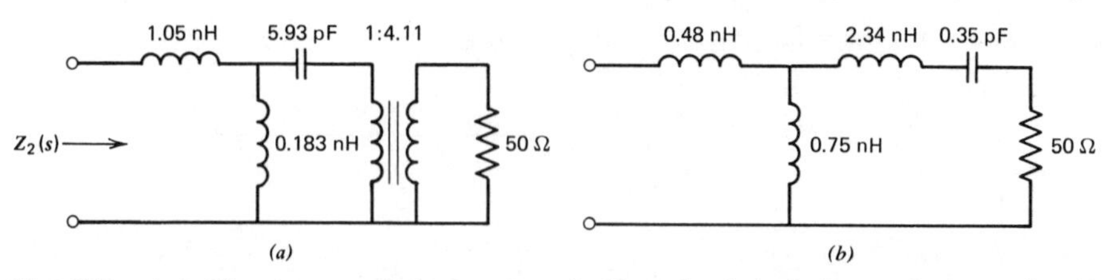

Fig. 5.5. (*a*) The input matching network N_i with the ideal transformer. (*b*) N_i without the ideal transformer.

First we normalize the frequency to $\omega_2 = 2\pi \times 6$ Grad/sec. Hence $\omega_{2N} = 1$ rad/sec, $\omega_{1N} = 0.8$ rad/sec, $B_N = 0.2$ rad/sec and $\omega_{0N}^2 = 0.8$ rad^2/sec^2, and $C_{iN} = 0.0256$ F are our normalized quantities. If we let $\epsilon = 0.1$, then $b_6 = 3.24$, $b_4 = 3.944$, $b_2 = 2.0736$ and $b_0 = 0.41$. The Hurwitz polynomial $M(s) = s^8 + b_6 s^6 + b_4 s^4 + b_2 s^2 + b_0$ can be found to be $s^4 + a_3 s^3 + a_2 s^2 + a_1 s + a_0$ by factoring the roots of $M(s)$ in Re $s > 0$, where $a_3 = 0.602$, $a_2 = 1.8$, $a_1 = 0.481$, $a_0 = 0.64$. In order to absorb the series capacitor C_{iN} into the matching network, it is necessary that $a_1' = a_1 - 2a_0 R_i C_{iN} = 0.317$. From (5.55) we have $p_1 = 0.157$, $p_2 = 20.67$, $p_3 = -38.18$, $p_4 = 17.63$ and $p_5 = 25$. Therefore $K_2 = 0.9$, which also guarantees by a simple check that $G(\omega^2)$ is bounded by unit. Also we get $b_6' = 3.276$ and the Hurwitz polynomial of $N(s) = s^8 + b_6' s^6 + b_4' s^4 + b_2' s^2 + b_0'$ can be found to be $s^4 + a_3' s^3 + a_2' s^2 + a_1' s + a_0' = s^4 + 0.348 s^3 + 1.7 s^2 + 0.317 s + 0.64$. Thus from (5.52) we have

$$\rho(s) = \frac{s^4 + 0.348 s^3 + 1.7 s^2 + 0.317 s + 0.64}{s^4 + 0.6 s^3 + 1.8 s^2 + 0.481 s + 0.64} \tag{5.56}$$

Substituting the values of a_i and a_i' into (5.48) yields $L_{1N} = 7.893$ H, $C_{1N} = 0.128$ F, $L_{2N} = 1.382$ H, $C_{2N} = 1.118$ F and $R = 0.591$ Ω. After renormalizing the frequency and substituting into (5.49) we get $R_i L_1 - L_i = 1.047$ nH, $R_i / C_1 - 1/C_i = 0$, $R_i L_2 = 0.183$ nH, $C_2 / R_i = 5.93$ pF and $RR_i = 2.96$. The final matching network N_i is shown in Fig. 5.5*a* with an ideal transformer and in Fig. 5.5*b* with the ideal transformer replaced by using an inductive T network (Appendix A7).

Before going further into real design problems we will discuss the output matching network.

5.3.2 Output Lossless Matching Network

In this section we discuss the gain-bandwidth limitation and the synthesis of lossless matching networks to operate between the load R_L and the output impedance of the transistor. For the microwave transistor contained in the package form, the output impedance $R_o C_o$-L_o shown in Fig. 5.1*a* is of the lowpass type, that is, $Z_o(s) = R_o/(sR_o C_o + 1) + sL_o$ is finite when $s = j\omega = 0$ ($Z_o(0) = R_o$), and $Z_o(\infty) = \infty$. Therefore in order to use the result of the broadband matching theorem with the Chebyshev or Butterworth transducer power gain, it is necessary to select $G(\omega^2)$ of the lowpass type as in (5.57a) for the nonsloped response and (5.57b) for the sloped response (ω_c is called the cutoff frequency of the lowpass response)

$$G(\omega^2)=\frac{K_n}{1+P_n^2(\omega)} \tag{5.57a}$$

$$G(\omega^2)=\frac{K_n(\omega/\omega_c)^2}{1+P_n^2(\omega)} \tag{5.57b}$$

where $P_n(\omega)=\epsilon C_n(\omega/\omega_c)$ for the Chebyshev case and $P_n(\omega)=(\omega/\omega_c)^n$ for the Butterworth case. In the case of the chip microwave transistor whose output impedance is a parallel R_oC_o circuit, the bandpass type responses in (5.33) are appropriate. The reader should note that the selection of a particular $G(\omega^2)$ to be compatible with a given load is important because it will dictate the realizability of $Z_2(s)$ in (5.15). For example, in the design of the input matching network N_i, the load $Z_i(s)$ is of the bandpass type; hence it is necessary to select a bandpass type $G(\omega^2)$ such that the continued expansion of $[1+\rho(s)]/[1-\rho(s)]$ in (5.46) contains two terms sL_1 and $1/sC_1$ such that $R_iL_1-L_i\geqslant 0$ and $R_i/C_1-1/C_i\geqslant 0$, that is, to guarantee the realizability of $Z_2(s)$. The Chebyshev and Butterworth bandpass $G(\omega^2)$ fulfill this purpose. There are of course other types of transducer power gains that can be employed for matching but they are not in analytical form such as the Chebyshev or Butterworth responses and can only be obtained by least-square approximation. Furthermore, we also notice that the lowpass R_oC_o-L_o load can indeed be matched by a nonsloped Chebyshev bandpass type $G(\omega^2)$ but with additional constraints to those that exist when $G(\omega^2)$ is a nonsloped lowpass response, and with appropriate transformation. This will be discussed after we study the matching of $Z_o(s)$ with $G(\omega^2)$ given in (5.57). After selecting $G(\omega^2)$, the minimum-phase bounded-real reflection coefficient $\rho(s)$ must be formed by the previously mentioned procedure. For the lowpass $G(\omega^2)$, $\rho(s)$ can be put in the form

$$\rho(s)=\frac{s^n+a'_{n-1}s^{n-1}+\cdots+a'_1s+a'_0}{s^n+a_{n-1}s^{n-1}+\cdots+a_1s+a_0} \tag{5.58}$$

where $a_i>a'_i$ $(i=0,1,2,\ldots,n-1)$. From (5.23) we get

$$\begin{aligned}\lambda(s)&=\tfrac{1}{2}\left[Z_o(s)+Z_o(-s)\right]/Z_o(s)\\&=\frac{R_o}{(1-sR_oC_o)\left[R_o+sL_o(1+sR_oC_o)\right]}\end{aligned} \tag{5.59}$$

It is seen that the zeros of $\lambda(s)$ are infinity and of order 3. Also since the residue of $Z_o(s)$ at infinity is

$$a_{-1}(\infty)=\lim_{s\to\infty}\frac{Z_o(s)}{s}=L_o \tag{5.60}$$

Furthermore, since $Z_o(-s)=R_o/(1-sR_oC_o)-sL_o$ possesses one pole in Re $s>0$ that is $s=1/R_oC_o$, we get

$$B(s)=\frac{sR_oC_o-1}{sR_oC_o+1} \tag{5.61}$$

and

$$F(s)=B(s)\left[Z_o(s)+Z_o(-s)\right]=\frac{-R_o}{(1+sR_oC_o)^2} \tag{5.62}$$

According to (5.24) the power series expansions of $\rho(s)$, $B(s)$, and $F(s)$ at infinity can be written as

$$\rho(s)=\sum_{i=0}^{\infty}\rho_i s^{-i}=\frac{s^n+a'_{n-1}s^{n-1}+\cdots+a'_1 s+a'_0}{s^n+a_{n-1}s^{n-1}+\cdots+a_1 s+a_0} \tag{5.63a}$$

$$B(s)=\sum_{i=0}^{\infty}B_i s^{-i}=\frac{sR_oC_o-1}{sR_oC_o+1} \tag{5.63b}$$

$$F(s)=\sum_{i=0}^{\infty}F_i s^{-i}=\frac{-R_o}{(1+sR_oC_o)^2} \tag{5.63c}$$

Condition d of the broadband matching theorem dictates that

$$B_i=\rho_i, \qquad i=0,1,2 \tag{5.64a}$$

and

$$\frac{F_2}{B_3-\rho_3}\geqslant a_{-1}(\infty)=L_o \tag{5.64b}$$

It can be easily found from (5.63b) and (5.63c) that $B_0=1$, $B_1=-2/R_oC_o$, $B_2=2/R_o^2C_o^2$ and $B_3=-2/R_o^3C_o^3$ and $F_2=-2/R_oC_o^2$. Since $\rho_0=\lim_{s\to\infty}\rho(s)=1$, $B_0=\rho_0$ and for $B_i=\rho_i$ $(i=1,2)$, we must have

$$-2/R_oC_o=a'_{n-1}-a_{n-1}=\rho_1 \tag{5.65a}$$

$$2/R_o^2C_o^2=a'_{n-2}-a_{n-2}-a_{n-1}(a'_{n-1}-a_{n-1})=\rho_2 \tag{5.65b}$$

and from (5.64b),

$$L_1 = \frac{-2/R_oC_o^2}{-2/R_o^3C_o^3 - a'_{n-3} + a_{n-3} + a_{n-1}(a'_{n-2} - a_{n-2}) - (a'_{n-1} - a_{n-1})(a_{n-1}^2 - a_{n-2})} \geqslant L_o \qquad (5.65c)$$

In the case of the nonsloped response, it can be found that for the lowpass Chebyshev case (Appendix A6)

$$a_{n-1} = \frac{\sinh x}{\sin u_1} \qquad (5.66a)$$

$$a_{n-2} = \tfrac{1}{4}n + \tfrac{1}{2}a_{n-1}^2 \qquad (5.66b)$$

$$a_{n-3} = \frac{\sinh x}{\sin u_1}\left(\tfrac{1}{4}n - \frac{\cos^2 u_1 \sin u_1}{\sin u_3} + \frac{\sinh^2 x \cos u_2}{2\sin u_1 \sin u_3}\right) \qquad (5.66c)$$

$$a_0 = 2^{1-n}\sinh nx \quad (n \text{ odd}), \qquad a_0 = 2^{1-n}\cosh nx \quad (n \text{ even}) \qquad (5.66d)$$

The a'_is are obtained simply by replacing x by x' where x and x' are given by (5.42d). Upon using (5.66) in (5.65) we get

$$\theta_1 \leqslant 1/\epsilon \leqslant \theta_2 \qquad (5.67a)$$

$$\alpha_2 \leqslant \sqrt{1-K_n} = \alpha_1 \leqslant \alpha_3 \qquad (5.67b)$$

$$\frac{\omega_c L_o}{R_o} \leqslant \frac{\left[2(\sin^2 u_3)(1-\cos u_2)\right]^{1/2}}{\sin^2 u_2} \qquad (5.67c)$$

where

$$\theta_1 = \sinh\left\{n \sinh^{-1}\left[-p/2 - (p^2 - 4q)^{1/2}/2\right]\right\} \qquad (5.68a)$$

$$\theta_2 = \sinh\left\{n \sinh^{-1}\left[-p/2 + (p^2 - 4q)^{1/2}/2\right]\right\} \qquad (5.68b)$$

$$p = \frac{2R_o(\sin u_3)(\cos u_2 - 1)}{\omega_c L_o \sin^2 u_2} \tag{5.68c}$$

$$q = 1 - \frac{R_o^2 \sin^2 u_3}{\omega_c^2 L_o^2 \sin^2 u_2} \tag{5.68d}$$

$$\alpha_1 = \epsilon \sinh\left[n \sinh^{-1}\left(\sinh x - \frac{2 \sin u_1}{R_o C_o \omega_c} \right) \right] \tag{5.68e}$$

$$\alpha_2 = \epsilon \sinh\left\{ n \sinh^{-1}\left[-b/2 - (b^2 - 4c)^{1/2}/2 \right] \right\} \tag{5.68f}$$

$$\alpha_3 = \epsilon \sinh\left\{ n \sinh^{-1}\left[-b/2 + (b^2 - 4c)^{1/2}/2 \right] \right\} \tag{5.68g}$$

$$b = \frac{2R_o \sin u_3}{\omega_c L_o} - 2 \sinh x \cos u_2 \tag{5.68h}$$

$$c = \sinh^2 x + \sin^2 u_2 - \frac{2R_o \sinh x \sin u_3}{\omega_c L_o} \tag{5.68i}$$

Equations (5.67a)–(5.67c) constitute the gain-bandwidth constraints for the output matching network N_o with the nonsloped Chebyshev lowpass response. The impedance $Z_{2o}(s)$ in (5.15) can be evaluated as

$$Z_{2o}(s) = \frac{F(s)}{B(s) - \rho(s)} - Z_o(s)$$

$$= \frac{-R_o[1 - \rho(s)]}{sR_o C_o[1 - \rho(s)] - [1 + \rho(s)]} - sL_o$$

$$= \frac{R_o}{[1 + \rho(s)]/[1 - \rho(s)] - sR_o C_o} - sL_o$$

$$= \frac{1}{R_o^{-1}[1 + \rho(s)]/[1 - \rho(s)] - sC_o} - sL_o \tag{5.69}$$

At this point the reader would probably ask, "Can $Z_o(s)$ be matched to R_L using the Chebyshev bandpass response $G(\omega^2)$ in (5.31) and (5.33) instead of

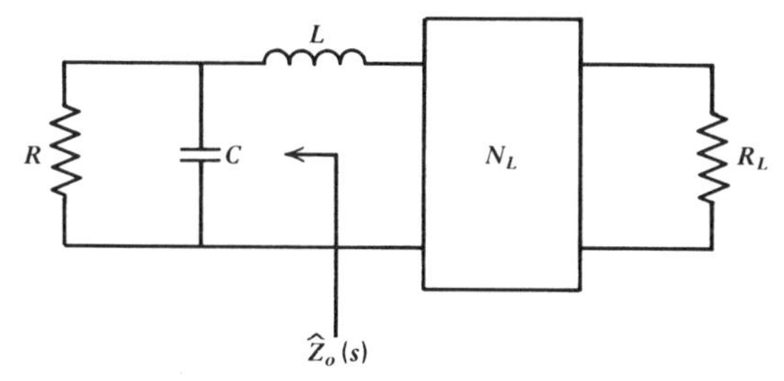

Fig. 5.6. Matching of the RC-L load using lowpass network N_L.

the lowpass response in (5.57a) and (5.57b)?". The answer is negative for the sloped response since the lowpass-to-bandpass transformation in (5.32) cannot transform (5.57b) to (5.33) while it transforms (5.57a) to (5.31). This does not mean that no sloped bandpass response exists that can match $Z_o(s)$ to R_L. Indeed these responses can only be obtained through least-square approximation.

Now in order to study the nonsloped bandpass matching of $Z_o(s)$, assume that the impedance $\hat{Z}_o(s)=R/(1+sRC)+sL$ can be matched to R_L by a lowpass network N_L, for example, using the procedure described above as shown in Fig. 5.6, where the cutoff frequency ω_c is set to 1 rad/sec. If we apply the lowpass to bandpass transformation in (5.32) we obtain the bandpass network shown in Fig. 5.7, in which the lowpass matching network N_L is transformed into the bandpass matching network N_B (Appendix A6) with

$$C_p=C/B, \qquad L_p=B/C\omega_0^2 \tag{5.70a}$$

$$C_s=B/L\omega_0^2, \qquad L_s=L/B \tag{5.70b}$$

The network in Fig. 5.7 is then transformed into that in Fig. 5.8 using an inductive T network with the following realizable constraint

$$1\leqslant t\leqslant 1+\frac{L_s}{L_p} \tag{5.71}$$

We note that this transformation scales R to $R_o=t^2R$. By observing Fig. 5.8 we can immediately see that the R_oC_o-L_o load can be matched to R_L by a bandpass

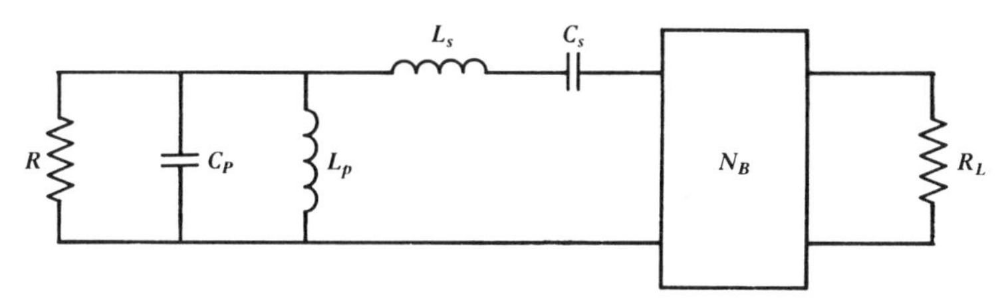

Fig. 5.7. The bandpass network of Fig. 5.6.

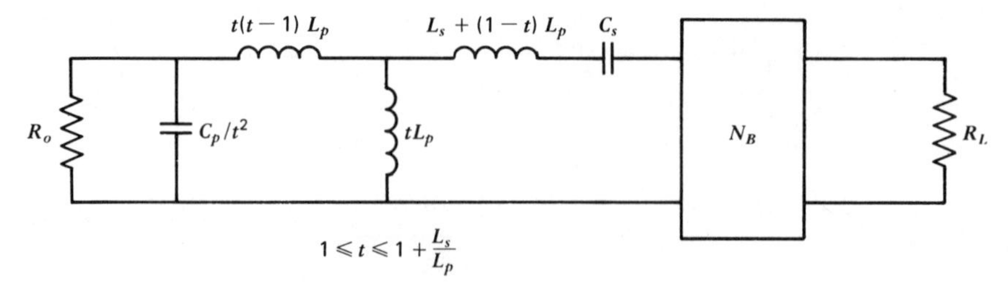

Fig. 5.8. The equivalent network of Fig. 5.7.

matching network if

$$R_o = t^2 R \tag{5.72a}$$

$$C_o = C_p / t^2 \tag{5.72b}$$

$$L_o \leq t(t-1) L_p \tag{5.72c}$$

From (5.70)–(5.72) it can be easily derived that

$$t \geq \max\left\{1, \frac{1}{1-\omega_0^2 L_o C_o}\right\} \tag{5.73a}$$

$$R = \frac{R_o}{t^2}, \qquad C = t^2 B C_o, \qquad L \geq \frac{B(t-1)}{t^2 \omega_0^2 C_o} = L_2 \tag{5.73b}$$

Thus for a given R_oC_o-L_o load we can select an arbitrary t such that (5.73a) holds and obtain the RC-L load from (5.73b) that can be matched to R_L using the lowpass response (5.57a) with $\omega_c = 1$ rad/sec if the gain-bandwidth constraints (5.65)–(5.68) are satisfied. After the matching network N_L for the RC-L load has been obtained, apply the bandpass transformation in (5.32) to obtain N_B and other elements as shown in Fig. 5.8. The complete bandpass matching network N_o for the R_oC_o-L_o load is shown in Fig. 5.9. The reader should note that this procedure can apply to any lowpass response $G(\omega^2)$ and not necessarily the one

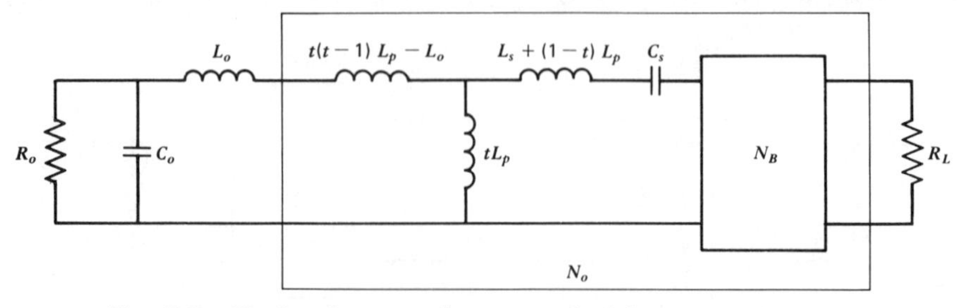

Fig. 5.9. The bandpass matching network of the transistor's output.

in (5.57a). Also the selection of t in (5.73a) will dictate the turn ratio of the ideal transformer used at the end of the network N_B in Fig. 5.9 to scale the appropriate resistance termination to R_L. From (5.69) the terminating resistance of the ladder can be determined for the RC-L load by setting $s=0$, giving

$$R_2 = Z_{2o}(0) = R\frac{1-\rho(0)}{1+\rho(0)} = R\frac{1-(a_0'/a_0)}{1+(a_0'/a_0)} = R\frac{a_0 - a_0'}{a_0 + a_0'} \tag{5.74}$$

Substituting (5.66d) and (5.72a) into (5.74) yields

$$R_2 = Z_{2o}(0) = \frac{R_o}{t^2}\cdot\frac{\sinh nx - \sinh nx'}{\sinh nx + \sinh nx'} = \frac{R_o}{t^2}\cdot\frac{1-\sqrt{1-K_n}}{1+\sqrt{1-K_n}}, \qquad n \text{ odd} \tag{5.75a}$$

$$= \frac{R_o}{t^2}\cdot\frac{\cosh nx - \cosh nx'}{\cosh nx + \cosh nx'} = \frac{R_o}{t^2}\cdot\frac{\sqrt{1+\epsilon^2} - \sqrt{1+\epsilon^2 - K_n}}{\sqrt{1+\epsilon^2} + \sqrt{1+\epsilon^2 - K_n}}, \qquad n \text{ even} \tag{5.75b}$$

In practice, it is always desirable to eliminate the ideal transformer that scales R_2 to R_L. This can be done if $R_2 = R_L$, which yields

$$t^2 = \frac{R_o}{R_L}\cdot\frac{1-\sqrt{1-K_n}}{1+\sqrt{1-K_n}}, \qquad n \text{ odd} \tag{5.76a}$$

$$= \frac{R_o}{R_L}\cdot\frac{\sqrt{1+\epsilon^2} - \sqrt{1+\epsilon^2 - K_n}}{\sqrt{1+\epsilon^2} + \sqrt{1+\epsilon^2 - K_n}}, \qquad n \text{ even} \tag{5.76b}$$

Equations (5.76a) and (5.76b) indicate that t would be fixed once ϵ and K_n are determined. Also from (5.67c), (5.68), and (5.72b) it is seen that the gain-bandwidth constraints do not depend on t once L is selected as $L = L_1$ in (5.65c). Thus the designer can determine ϵ and K_n first using (5.67)–(5.68) and fix t by (5.76).

For the microwave transistors in chip form shown in Fig. 5.1*b*, the output impedance is a simple parallel R_oC_o circuit and in this case $Z_o(s)$ possesses a Class 2 zero of transmission of order 1 at infinity, as can be easily verified. Since this particular load is of both bandpass and lowpass types, any result derived for the bandpass response will hold for the lowpass responses by letting the lower passband frequency ω_1 in (5.32) be zero. Using the bandpass response $G(\omega^2)$ in (5.31) an (5.33) and the broadband matching condition b, it can be shown that

$Z_o(s)$ can be matched to R_L if and only if

$$B_0 = \rho_0 \tag{5.77a}$$

$$\frac{B_1 - \rho_1}{F_2} \geqslant 0 \tag{5.77b}$$

where

$$\rho(s) = \sum_{i=0}^{\infty} \rho_i s^{-i} = \frac{s^{2n} + a'_{2n-1}s^{2n-1} + \cdots + a'_1 s + a'_0}{s^{2n} + a_{2n-1}s^{2n-1} + \cdots + a_1 s + a_0} \tag{5.78a}$$

$$B(s) = \sum_{i=0}^{\infty} B_i s^{-i} = \frac{sR_oC_o - 1}{sR_oC_o + 1} \tag{5.78b}$$

$$F(s) = \sum_{i=0}^{\infty} F_i s^{-i} = \frac{-R_o}{(1 + sR_oC_o)^2} \tag{5.78c}$$

As in the case of the R_oC_o-L_o load, it can be shown that $B_0 = 1$, $B_1 = -2/R_oC_o$, $\rho_0 = 1$, $F_2 = -2/R_oC_o^2$, $\rho_1 = a'_{2n-1} - a_{2n-1}$. Equation (5.77a) is automatically satisfied and inequality (5.77b) is equivalent to

$$\frac{2}{R_oC_o} \geqslant a_{2n-1} - a'_{2n-1} \tag{5.79}$$

which is the gain-bandwidth constraint of the R_oC_o load. In the case of the nonsloped Chebyshev response, (5.79) is equivalent to (5.80) using (5.42c)

$$(1 - K_n)^{1/2} \geqslant \epsilon \sinh\left[n \sinh^{-1}\left(\sinh x - \frac{2 \sin u_1}{R_oC_oB}\right)\right] \tag{5.80}$$

The impedance $Z_{2o}(s)$ in (5.15) can be evaluated as

$$Z_{2o}(s) = \frac{F(s)}{B(s) - \rho(s)} - Z_o(s)$$

$$= \frac{R_o}{[1 + \rho(s)]/[1 - \rho(s)] - sR_oC_o}$$

$$= \frac{1}{R_o^{-1}[1 + \rho(s)]/[1 - \rho(s)] - sC_o} \tag{5.81}$$

In summary:

(a) For the input matching network, both nonsloped and sloped bandpass Chebyshev (or Butterworth) responses can be employed.

(b) For the output matching network, if the transistor's output impedance is the parallel-series R_oC_o-L_o circuit, then the low pass Chebyshev response must be used for sloped compensation while both lowpass and bandpass Chebyshev responses can be employed for the nonsloped design. If the transistor's output impedance is the parallel R_oC_o circuit, then the bandpass Chebyshev response can be used for both sloped and nonsloped design.

5.3.3 Design Examples

To illustrate the matching network design procedure using the analytical transducer power gain response such as Chebyshev's, practical examples using GaAs FETs shown in Section 5.1 are presented here.

Example 5.3

The problem considered in this example is to design an amplifier to operate in the frequency band from 4 to 8 GHz. The GaAs FET used for this example is the NE-38806 whose scattering parameters are shown in Table 5.3. Its input impedance can be approximated by a series $R_iL_iC_i$ impedance where $R_i = 10.5\ \Omega$, $L_i = 0.65$ nH, and $C_i = 0.64$ pF. Its output impedance can be modeled by a parallel-series R_oC_o-L_o impedance where $R_o = 164\ \Omega$, $C_o = 0.428$ pF, and $L_o = 0.474$ nH. The problem is to match the input to a 50 Ω source to compensate for the gain roll-off and to match the output to a 50 Ω load for a flat response. From Table 5.3, the maximum unilateral gain $G_{u,\max}$ varies from 15.5 to 10.5 dB over the band 4–8 GHz and presents a roll-off of 5 dB/octave, which is close to the ideal 6 dB/octave. Let $\omega_1 = 2\pi \times 4 \times 10^9$ rad/sec and $\omega_2 = 2\pi \times 8 \times 10^9$ rad/sec. Then $B = 2\pi \times 4 \times 10^9$ rad/sec. For convenience of computation we normalize all the frequencies to ω_2; this yields $\omega_{1N} = 0.5$ rad/sec, $\omega_{2N} = 1$ rad/sec and $B_N = 0.5$ rad/sec.

For the input matching network, we select the Chebyshev sloped response in (5.33) with $n = 2$, $\epsilon = 0.5$ and $K_2 = 1$. The minimum phase reflection coefficient $\rho(s)$ can be obtained from $G(\omega^2)$ as

$$\rho(s) = \frac{s^4 + 0.318s^3 + 1.206s^2 + 0.2s + 0.25}{s^4 + 0.548s^3 + 1.275s^2 + 0.274s + 0.25}$$

The normalized input capacitor and inductor are given as $C_{iN} = \omega_2 C_i = 2\pi \times 8 \times 10^9 \times 0.64 \times 10^{-12} = 3.22 \times 10^{-2}$ F and $L_{iN} = \omega_2 L_i = 2\pi \times 8 \times 10^9 \times 0.65 \times 10^{-9} = 32.67$ H. The gain-bandwidth constraints in (5.41a) and (5.41b) are

$$R_i C_{iN} = 0.338 > \tfrac{1}{2}(a_1 - a_1')/a_0 = 0.148$$

$$R_i / L_{iN} = 0.32 > \tfrac{1}{2}(a_3 - a_3') = 0.115$$

Thus the impedance $Z_{2i}(s)$ is positive-real and can be realized by (5.48) and

(5.49) as

$$Z_{2i}=58.68s+\frac{1}{0.025s}+\frac{1}{\dfrac{1}{40.22s}+\dfrac{1}{1/0.077s+12.15}}$$

The realization on $Z_{2i}(s)$ is given in Fig. 5.10*a* after denormalization of the element values. An inductive equivalent T network is then employed to replace the ideal transformer. The final input matching network is shown in Fig. 5.10*b*. For the output matching network it is desired that $Z_o(s)$ be matched to R_L by a bandpass network. In order to achieve this, a parallel-series load RC-L must first be obtained from the parallel-series load R_oC_o-L_o. To begin we select $n=3$ and $\epsilon=0.1$. Substituting (5.73b) into (5.68e) we get

$$\alpha_1=\epsilon\sinh\left[n\sinh^{-1}\left(\sinh x-\frac{2\sin u_1}{R_oC_oB}\right)\right]$$

where $u_1=\pi/6$ as seen from (5.42d). With $R_o=164\ \Omega$, $C_o=0.428$ pF, and $B=2\pi\times4\times10^9$ rad/sec, we obtain $\alpha_1=0.2718$, which yields $K_3=0.926$ from $\sqrt{1-K_3}=\alpha_1$ in (5.67b). Substituting K_3 into (5.76a) with $R_L=50\ \Omega$ yields $t=1.37>\max\{1,1.3446\}$ as seen from (5.73). From (5.73b) we get

$$R=87.33\ \Omega,\qquad C=2.02\times10^{-2}\ \text{F},\qquad L\geqslant9.16\ \text{H}$$

From (5.66) the a_i and a_i' are given as $a_0=2.5$, $a_0'=0.68$, $a_1=3.51$, $a_1'=1.49$,

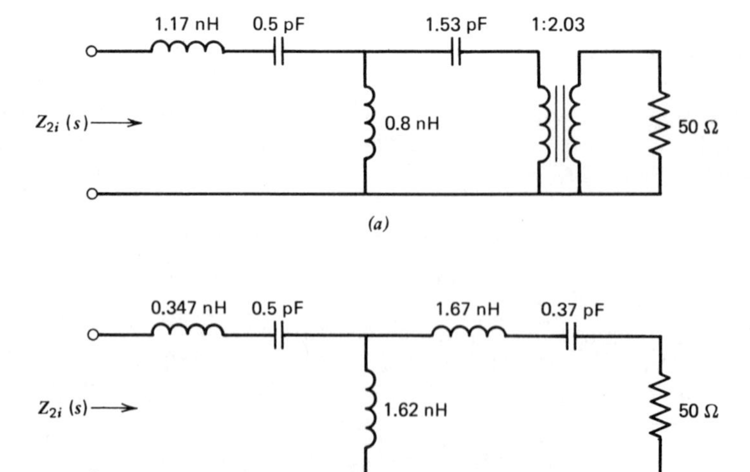

Fig. 5.10. (*a*) Realization of $Z_{2i}(s)$ with the ideal transformer. (*b*) Realization of $Z_{2i}(s)$ without the ideal transformer.

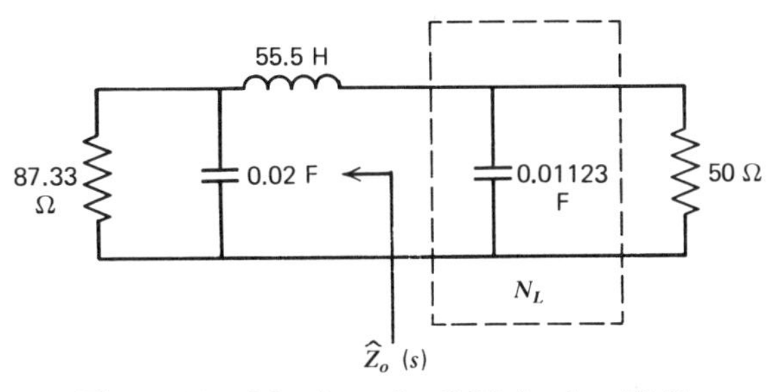

Fig. 5.11. Matching the *RC-L* load to 50 Ω.

$a_2=2.349$, and $a_2'=1.215$. The equations (5.65a) and (5.65b) are automatically satisfied, that is, $2/RC=1.134$ and $2/R^2C^2=0.643$, and from (5.65c) we obtain

$$L \leq L_1 = 55.5 \text{ H}$$

The reader can check that (5.67)–(5.68) are all satisfied with $R=87.33\ \Omega$, $C=2.02\times10^{-2}$ F, and $L=55.5$ H. Substituting a_i and a_i' into (5.63a) yields

$$\rho(s)=\frac{s^3+1.215s^2+1.49s+0.68}{s^3+2.349s^2+3.51s+2.5}$$

and

$$R^{-1}\frac{1+\rho(s)}{1-\rho(s)}=(2.02\times10^{-2})s+\cfrac{1}{55.5s+\cfrac{1}{(1.123\times10^{-2})s+0.02}}$$

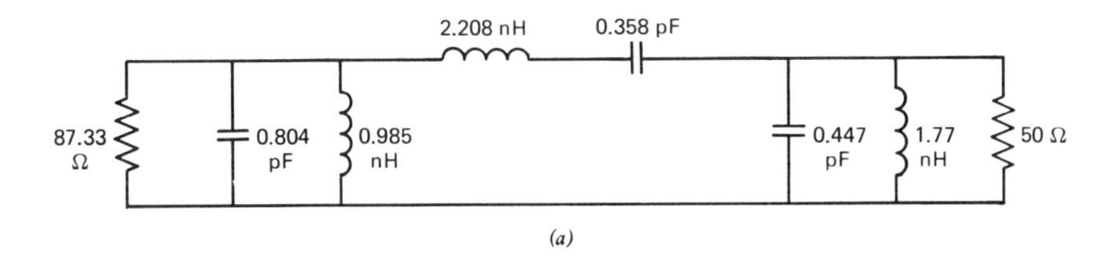

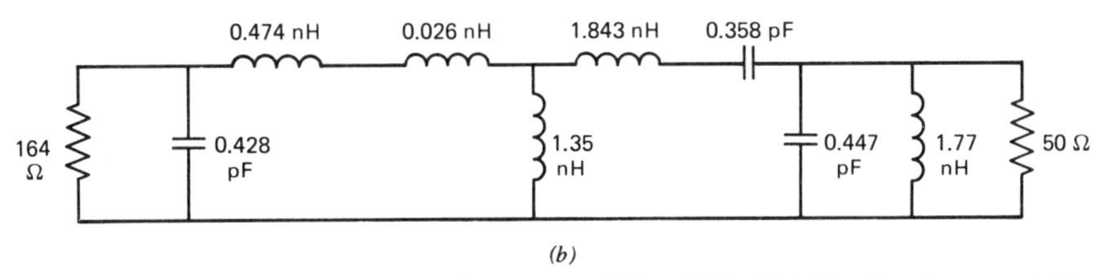

Fig. 5.12. (*a*) The bandpass transformation of Fig. 5.11. (*b*) The bandpass matching network of the R_oC_o-L_o load.

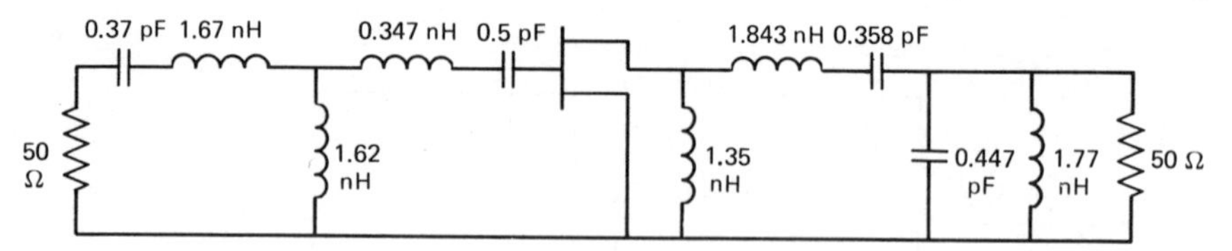

Fig. 5.13. The 4–8 GHz NE-38806 GaAs FET lumped amplifier.

Thus the impedance $Z_{2o}(s)$ can be found from (5.69) as

$$Z_{2o} = \frac{1}{(1.123 \times 10^{-2})s + \frac{1}{50}}$$

The *RC-L* load with the lowpass matching network N_L is shown in Fig. 5.11. Applying the lowpass-to-bandpass transformation in (5.32) to Fig. 5.11 we obtain the bandpass network of Fig. 5.12*a*. Using the inductive T network transformation of Fig. 5.8 we arrive at the bandpass matching network N_o of the R_oC_o-L_o load and $R_L = 50\ \Omega$ in Fig. 5.12*b*. The complete amplifier is shown in Fig. 5.13 and its performance is presented in Table 5.13.

Example 5.4

In this example the design of a broadband amplifier to operate in the frequency band from 6 to 12 GHz is considered. The GaAs FET is the HFET-1000 with its scattering parameters given in Table 5.9. Its input impedance can be approximated by a series R_iC_i impedance where $R_i = 8.44\ \Omega$ and $C_i = 0.77$ pF, and its output impedance can be modeled by a parallel R_oC_o impedance where $R_o = 320\ \Omega$ and $C_o = 0.138$ pF. The problem is to match the input to a 50 Ω source and the output to a 50 Ω load to achieve a gain of 8 dB with less than 1 dB ripple. From Table 5.9 we can compute that the transistor's maximum unilateral gain $G_{u,\max}$ varies from 14.9 dB at 6 GHz to 9.1 dB at 12 GHz. This presents a gain roll off of about 5.8 dB/octave, which is close to the ideal 6 dB/octave. Let $\omega_1 = 2\pi \times 6 \times 10^9$ rad/sec, $\omega_2 = 2\pi \times 12 \times 10^9$ rad/sec, then $B = 2\pi \times 6 \times 10^9$ rads/sec. For convenience of computation we normalize all the frequencies to ω_2; this yields $\omega_{1N} = 0.5$ rad/sec, $\omega_{2N} = 1$ rad/sec, and $B_N = 0.5$ rad/sec. We select to design the input matching network with non-sloped response and the output network with sloped response.

Table 5.13 Performance of the lumped 4–8 GHz NE-38806 GaAs FET amplifier

f (GHz)	S_{11}	S_{21}	S_{12}	S_{22}	S_{21} (dB)	K
4	$0.91\angle -175°$	$2.61\angle -24°$	$0.050\angle -75°$	$0.40\angle -132°$	8.34	0.92
5	$0.81\angle 50°$	$2.78\angle -154°$	$0.054\angle 162°$	$0.29\angle 86°$	8.89	1.27
6	$0.75\angle -7°$	$2.76\angle 132°$	$0.051\angle 108°$	$0.29\angle -20°$	8.81	1.51
7	$0.59\angle -74°$	$3.07\angle 59°$	$0.066\angle 73°$	$0.35\angle -113°$	9.74	1.36
8	$0.62\angle 167°$	$2.75\angle -32°$	$0.084\angle 22°$	$0.40\angle 140°$	8.78	1.05

For the input matching network, we select $n=2$, $\epsilon=0.5$. Thus $\sinh x=0.786$. Using (5.45) with $L_i=0$ we get $(1-K_2)^{1/2}\geqslant 0.937$ or $K_2\leqslant 0.9912$. Choosing $K_2=0.9912$, which is the maximum passband gain with such given ϵ. The minimum phase reflection coefficient $\rho(s)$ can be found through the spectral factorization of $1-G(-s^2)$. Using (5.46) the impedance $Z_{2i}(s)$ can be computed, after the frequency renormalization (by zero and pole extraction method)

$$Z_{2i}(s)=(0.457\times 10^{-9})s+\cfrac{1}{(1.893\times 10^{-12})s+\cfrac{1}{(0.186\times 10^{-9})s}+\cfrac{1}{22.53}}$$

The realization of $Z_{2i}(s)$ with an ideal transformer is shown in Fig. 5.14*a*. An inductive pi network is used to eliminate the transformer as is shown in Fig. 5.14*b*.

For the output matching network, we select $n=2$, $\epsilon=0.5$ and $K_2=1$ and obtain the minimum-phase reflection coefficient $\rho(s)$ through the spectral factorization of $1-G(-s^2)$ where $G(-s^2)$ is obtained from $G(\omega^2)$ in (5.33) with the purpose of compensating for the gain roll-off of the HFET-1000. The frequencies are as above and all are normalized to ω_2. As in the input matching network of Example 5.3, we get

$$\rho(s)=\frac{s^4+0.318s^3+1.206s^2+0.2s+0.25}{s^4+0.548s^3+1.275s^2+0.274s+0.25}$$

The normalized output capacitance C_{oN} is given as $C_{oN}=\omega_2 C_o=4.64\times 10^{-3}$ F;

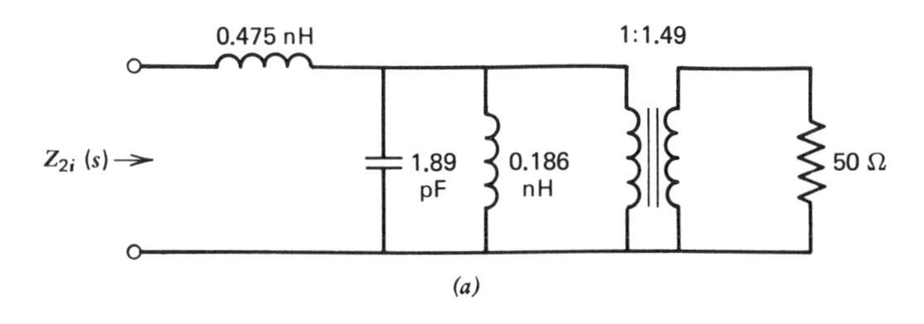

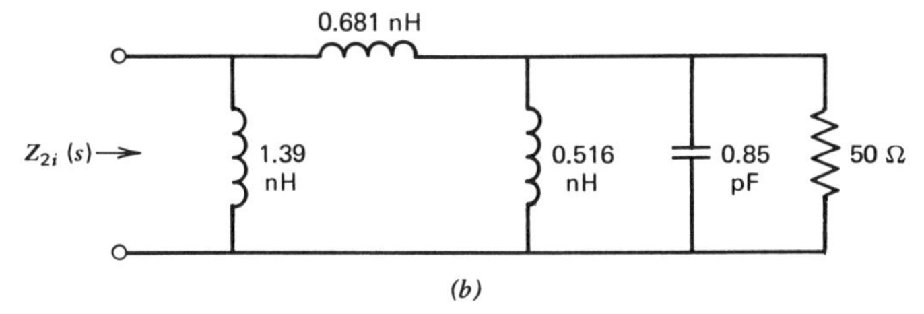

Fig. 5.14. Realization of $Z_{2i}(s)$. (*a*) With the ideal transformer. (*b*) With the inductive pi network.

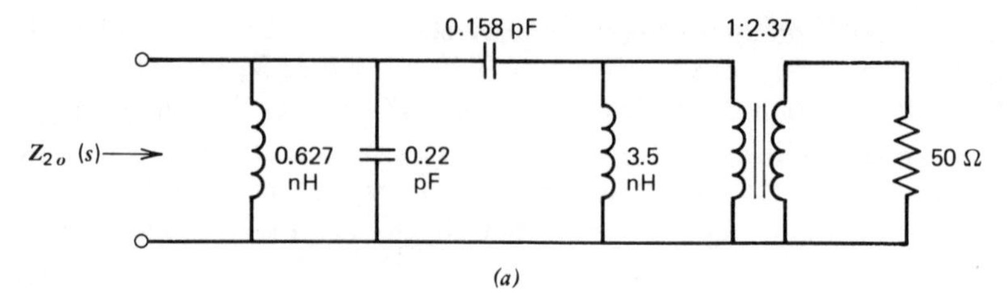

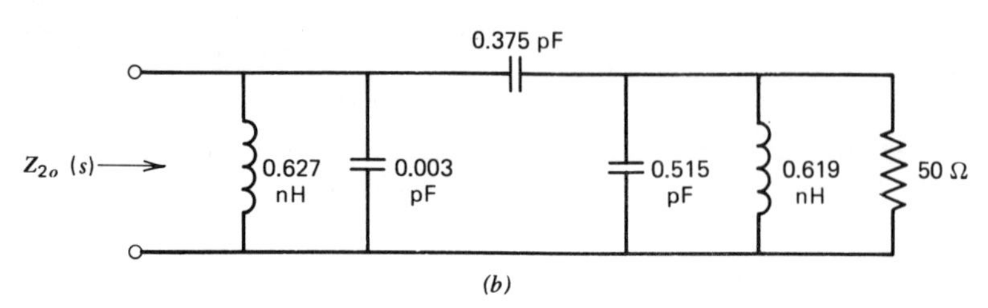

Fig. 5.15. Realization of $Z_{2o}(s)$. (a) With the ideal transformer. (b) With the capacitive pi network.

hence using (5.79) we get

$$\frac{2}{R_{oN}C_{oN}}=0.6\geqslant a_3-a_3'=0.548-0.318=0.23$$

The impedance $Z_{2o}(s)$ can be computed from (5.81), after the frequency renormalization

$$Z_{2o}(s)=\frac{1}{(0.22\times10^{-12})s+\dfrac{1}{(0.63\times10^{-9})s}}+\frac{1}{(0.16\times10^{-12})s+\dfrac{1}{\dfrac{1}{(3.5\times10^{-9})s}+\dfrac{1}{281.6}}}$$

The realization of $Z_{2o}(s)$ is shown in Fig. 5.15a with an ideal transformer and in

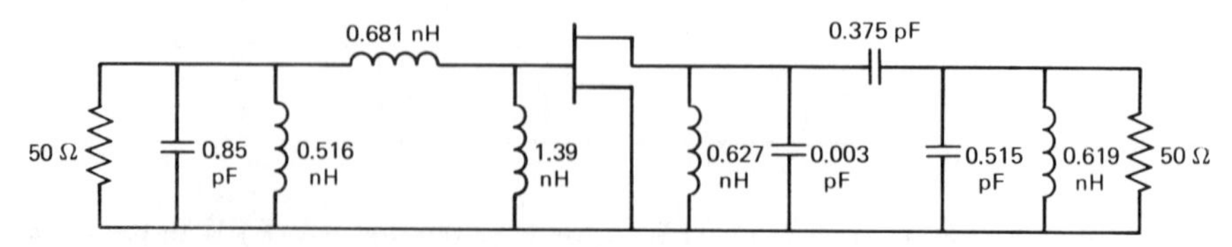

Fig. 5.16. The 6–12 GHz HFET-1000 GaAs FET lumped amplifier.

Table 5.14 Performance of the 6–12 GHz HFET-1000 GaAs FET amplifier

f (GHz)	S_{11}	S_{21}	S_{12}	S_{22}	S_{21} (dB)	K
6	$0.61\angle 65°$	$2.30\angle -13.4°$	$0.034\angle -60°$	$0.94\angle 1°$	7.3	0.96
8	$0.52\angle -162°$	$2.43\angle -177°$	$0.048\angle 152°$	$0.78\angle -151°$	7.7	1.16
10	$0.40\angle 129°$	$2.76\angle 95°$	$0.070\angle 84°$	$0.61\angle 128°$	8.8	1.42
12	$0.52\angle -67°$	$2.25\angle -16.3°$	$0.072\angle -9°$	$0.54\angle -23°$	7.1	1.56

Fig. 5.15*b* with the transformer replaced by a capacitive pi network. The complete lumped amplifier is shown in Fig. 5.16 with its performance given in Table 5.14.

5.4 EXPLICIT FORMULAS FOR INTERSTAGE LUMPED MATCHING NETWORKS

In the previous section we have presented a general procedure for designing matching networks between the input of the transistor and 50 Ω source and the output of the transistor and 50 Ω load. If a high gain amplifier is desired, many stages will be cascaded with an isolator (Section 3.5) in between stages. The purpose of the isolator is to present a close to 50 Ω input and output impedance to the input and output matching networks of the transistors in order to preserve the individual stage gain and ripple. A two-stage cascade amplifier is shown in Fig. 5.17. In many situations, it is necessary to eliminate isolators to reduce the size and cost of high gain amplifier (e.g., in integrated receivers) by designing interstage matching networks that can match the output of the first transistor to the input of the second transistor. In this section we will present an analytical technique for the design of such interstage matching networks using closed form expressions for gain-bandwidth limitations and element values. The procedure will use the unilateral model of microwave transistors as outlined in Fig. 5.18. In many cases in which the use of either interstage matching networks or circulators is not desirable, the cascade connection of balanced amplifiers employing hybrids (producing twice as much output power) can be attractive. This type of amplifier will be discussed in Section 6.10.4.

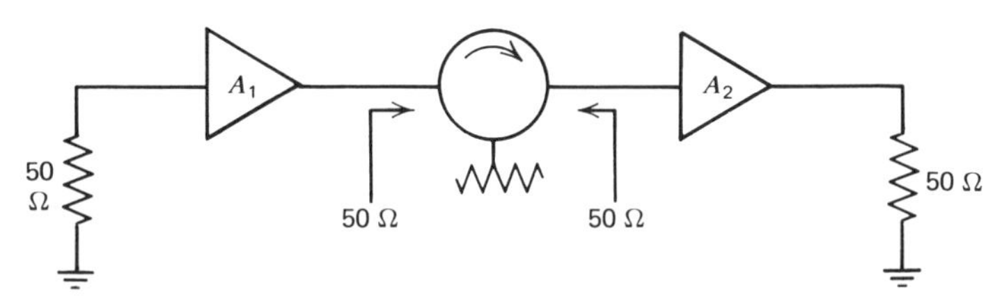

Fig. 5.17. A two-stage cascade amplifier using an isolator.

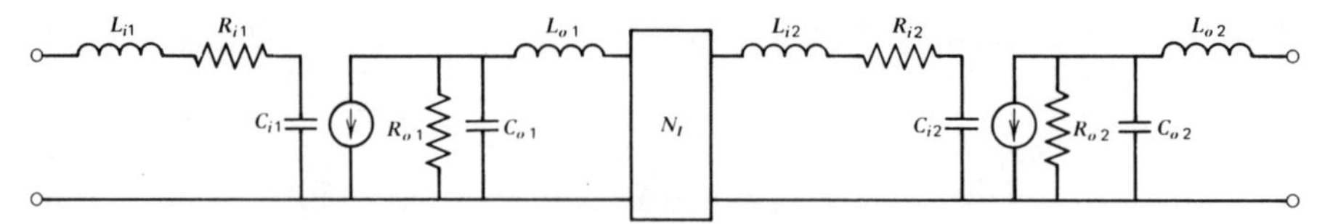

Fig. 5.18. An interstage matching network for cascade transistors.

5.4.1 LC Ladder Networks

Consider the LC bandpass ladder networks shown in Figs. 5.19*a–d* with the Chebyshev transducer power gain nonsloped response in (5.31). The explicit formulas for the element values of these networks are given as follows:

$$L_{si}=L_i/B, \qquad C_{si}=B/L_i\omega_0^2 \tag{5.82a}$$

$$L_{pi}=B/C_i\omega_0^2, \qquad C_{pi}=C_i/B \tag{5.82b}$$

where

$$R_1 \leqslant R_2: \quad L_i=R_1\alpha_i, \qquad C_i=\beta_i/R_1 \qquad (i=1,2,\ldots,n-1),$$

$$L_n=R_2\alpha_n, \qquad C_n=\beta_n/R_2$$

$$R_1 \geqslant R_2: \quad C_i=\alpha_i/R_1, \qquad L_i=R_1\beta_i \qquad (i=1,2,\ldots,n-1),$$

$$C_n=\alpha_n/R_2, \qquad L_n=R_2\beta_n$$

and

$$\alpha_1=\frac{2\sin u_1}{\sinh x-\sinh x'} \tag{5.83a}$$

$$\alpha_{2m-1}\beta_{2m}=\frac{16\sin u_{4m-3}\sin u_{4m-1}}{g_{2m-1}} \tag{5.83b}$$

$$\alpha_{2m+1}\beta_{2m}=\frac{16\sin u_{4m-1}\sin u_{4m+1}}{g_{2m}} \tag{5.83c}$$

$$\alpha_n=\frac{2\sin u_1}{\sinh x+\sinh x'}, \qquad n \text{ odd} \tag{5.83d}$$

$$\beta_n=\frac{2\sin u_1}{\sinh x+\sinh x'}, \qquad n \text{ even} \tag{5.83e}$$

$$u_m=m\pi/2n \tag{5.83f}$$

$$x=\frac{1}{n}\sinh^{-1}\frac{1}{\epsilon} \tag{5.83g}$$

$$x'=\frac{1}{n}\sinh^{-1}\frac{\sqrt{1-K_n}}{\epsilon} \tag{5.83h}$$

$$g_m=4\left(\sinh^2 x+\sinh^2 x'+\sin^2 u_{2m}-2\sinh x\sinh x'\cos u_{2m}\right) \tag{5.83i}$$

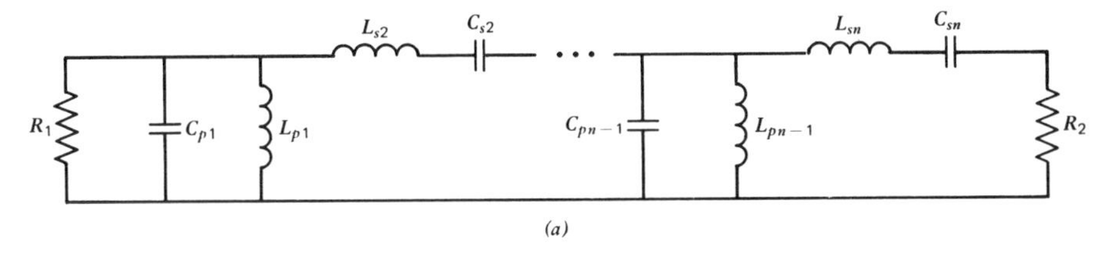

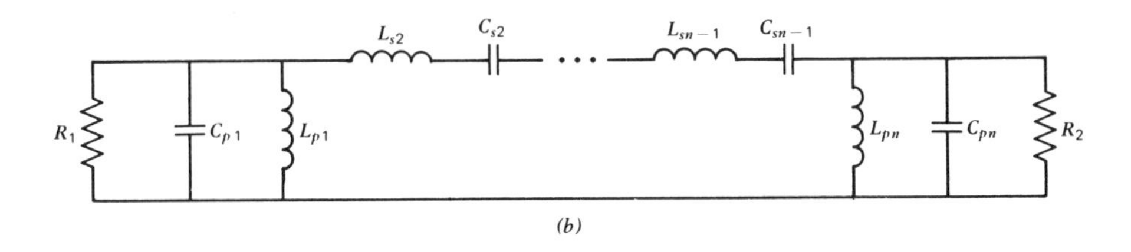

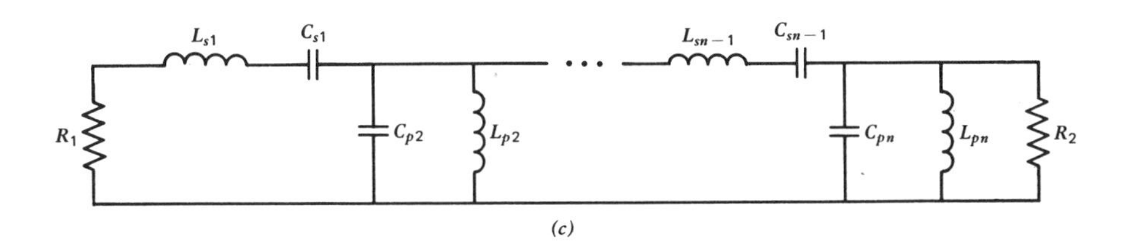

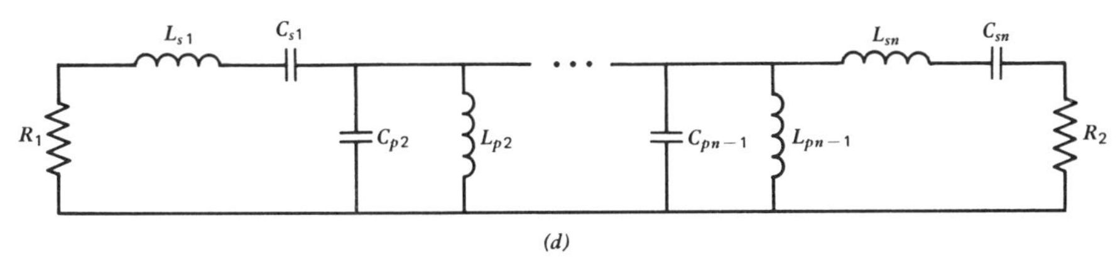

Fig. 5.19. The *LC* bandpass ladder networks. (*a*) $R_1 \geqslant R_2$, n: even. (*b*) $R_1 \geqslant R_2$, n: odd. (*c*) $R_1 \leqslant R_2$, n: even. (*d*) $R_1 \leqslant R_2$, n: odd.

The passband gain K_n and R_1 and R_2 are related and cannot be chosen arbitrarily for all three. Their relationship is given as

$$\frac{R_1}{R_2} = \left[\frac{1-\sqrt{1-K_n}}{1+\sqrt{1-K_n}}\right]^{\pm 1} = \delta^{\pm 1}, \qquad n \text{ odd} \tag{5.84a}$$

$$\frac{R_1}{R_2} = \left[\frac{\sqrt{1+\epsilon^2}-\sqrt{1+\epsilon^2-K_n}}{\sqrt{1+\epsilon^2}+\sqrt{1+\epsilon^2-K_n}}\right]^{\pm 1} = \gamma^{\pm 1}, \qquad n \text{ even} \tag{5.84b}$$

The minus sign is used when $R_1 \geqslant R_2$ and the plus sign is used when $R_1 \leqslant R_2$. For the Butterworth nonsloped response, $\sinh x \pm \sinh x'$ are replaced by $1 \pm k$ and g_m by $1-2k\cos u_{2m}+k^2$ where

$$k=(1-K_n)^{1/2n} \tag{5.85a}$$

and

$$\frac{R_1}{R_2}=\left[\frac{1-k^n}{1+k^n}\right]^{\pm 1} \tag{5.85b}$$

5.4.2 Interstage Matching Networks

The matching of two microwave transistors in direct cascade is shown in Fig. 5.18. The problem considered in this section is to derive the gain-bandwidth limitations and the element values for the interstage matching network N_I using the bandpass networks in Fig. 5.19 to match the $R_{o1}C_{o1}$-L_{o1} output impedance of the first transistor to the $R_{i2}L_{i2}C_{i2}$ input impedance of the second transistor. Since $R_{o1}C_{o1}$-L_{o1} is a lowpass impedance while $R_{i2}L_{i2}C_{i2}$ is a bandpass impedance, the matching is possible if appropriate transformation is used. Consider the network in Fig. 5.19*a* for even n and $R_1 \geqslant R_2$. According to (5.84b) we get $R_1=R_2/\gamma$ where $\gamma \leqslant 1$ since $R_1 \geqslant R_2$ for even n. Returning to the microwave transistor model in Fig. 5.1*a*, we note that $R_o \gg R_i$ in reality (more than 10 times in general); therefore if we let $R_{i2}=R_2$, $R_{o1}=R_1t^2$ where $t^2=R_{o1}\gamma/R_{i2}$, then the network in Fig. 5.19*a* is completely equivalent to the network in Fig. 5.20*a* by the use of an ideal transformer of turn ratio $t:1$. By using the inductive T network transformation, the network in Fig. 5.20*a* can be made equivalent to that in Fig. 5.20*b* if

$$1 \leqslant t \leqslant \frac{L_{s2}}{L_{p1}} \tag{5.86}$$

In a similar way, it can be shown that the network in Fig. 5.19*c* is equivalent to the network in Fig. 5.21 with the following realizable constraint

$$1 \leqslant t \leqslant 1+\frac{L_{sn-1}}{L_{pn}} \tag{5.87}$$

and with $R_{i2}=R_1$, $R_{o1}=R_2t^2$, $t^2=R_{o1}\gamma/R_{i2}$ where $R_2=R_1/\gamma$. Here $\gamma \leqslant 1$ since $R_1 \leqslant R_2$.

Using the networks in Figs. 5.20*b* and 5.21, the matching constraints for the interstage network N_I in Fig. 5.18 can be derived.

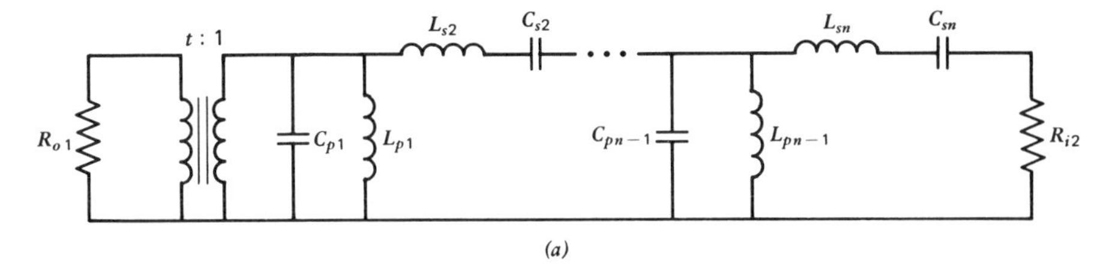

(a)

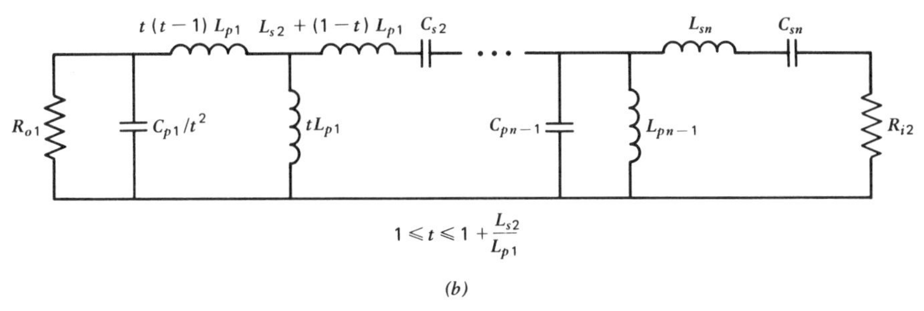

(b)

Fig. 5.20. The equivalent network of Fig. 5.19*a*. (*a*) With ideal transformer. (*b*) With inductive T network.

For Fig. 5.20*b*, the matching is possible if and only if

$$C_{i2} \geq C_{sn} \tag{5.88a}$$

$$L_{i2} \leq L_{sn} \tag{5.88b}$$

$$C_{o1} = C_{p1}/t^2 \tag{5.88c}$$

$$L_{o1} \leq t(t-1)L_{p1} \tag{5.88d}$$

$$1 \leq t \leq 1 + (L_{s2} - \mu L_{i2})L_{p1}^{-1} \tag{5.88e}$$

where $\mu = 1$ for $n=2$ and $\mu = 0$ for all even $n > 2$.

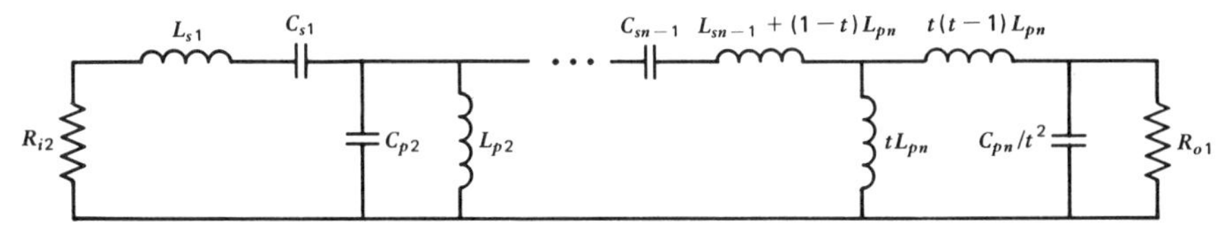

Fig. 5.21. The equivalent network of Fig. 5.19*c* with inductive T network transformation.

For Fig. 5.21, the matching is possible if and only if

$$C_{i2} \geqslant C_{s1} \tag{5.89a}$$

$$L_{i2} \leqslant L_{s1} \tag{5.89b}$$

$$C_{o1} = C_{pn}/t^2 \tag{5.89c}$$

$$L_{o1} \leqslant t(t-1)L_{pn} \tag{5.89d}$$

$$1 \leqslant t \leqslant 1 + (L_{s\bar{n}-1} - \mu L_{i2})L_{pn}^{-1} \tag{5.89e}$$

Substituting (5.82) and (5.83) into (5.88) and (5.89) yields the following gain-bandwidth limitations for the networks in Figs. 5.20*b* and 5.21, respectively.

Chebyshev response:
Fig. 5.20*b*:

$$\epsilon \leqslant \tfrac{1}{2}(\theta^{-1/2} - \theta^{1/2}) \tag{5.90a}$$

$$\sqrt{1-K_n} = a_1 = \min\{a_2, a_3, a_4\} \tag{5.90b}$$

$$b_1 \leqslant \frac{B^2}{\omega_0^2} \leqslant \begin{cases} b_2, & n=2 \\ b_3, & n>2\ (n \text{ even}) \end{cases} \tag{5.90c}$$

where

$$\theta = R_{i2}/R_{o1} \tag{5.90d}$$

$$a_1 = \epsilon \sinh\left[n \sinh^{-1}\left(\sinh x - \frac{2 \sin u_1}{R_{o1}C_{o1}B}\right)\right] \tag{5.90e}$$

$$a_2 = \frac{(1-\theta^2) - 4\epsilon^2\theta}{1+\theta} \tag{5.90f}$$

$$a_3 = \epsilon \sinh\left[n \sinh^{-1}\left(\frac{2R_{i2}\sin u_1}{BL_{i2}} - \sinh x\right)\right] \tag{5.90g}$$

$$a_4 = \epsilon \sinh\left[n \sinh^{-1}\left(\frac{2R_{i2}C_{i2}\omega_0^2 \sin u_1}{B} - \sinh x\right)\right] \tag{5.90h}$$

$$b_1 = \frac{2BL_{o1}\sin u_1}{R_{o1}(1-\sqrt{\theta/\gamma})(\sinh x - \sinh x')} \tag{5.90i}$$

$$b_2 = \frac{2\gamma \sin u_1[2R_{i2}\sin u_1 - BL_{i2}(\sinh x + \sinh x')]}{R_{i2}(\sqrt{\theta/\gamma} - 1)(\sinh^2 x - \sinh^2 x')} \tag{5.90j}$$

$$b_3 = \frac{16 \sin u_1 \sin u_3}{(\sqrt{\gamma/\theta} - 1)g_1} \tag{5.90k}$$

Fig. 5.21:

$$\epsilon \leq \tfrac{1}{2}(\theta^{-1/2} - \theta^{1/2}) \tag{5.91a}$$

$$\max\{-a_3, -a_4\} \leq \sqrt{1-K_n} = -a_1 \leq a_2 \tag{5.91b}$$

$$c_1 \leq \frac{B^2}{\omega_0^2} \leq \begin{cases} c_2, & n=2 \\ c_3, & n>2\ (n \text{ even}) \end{cases} \tag{5.91c}$$

where c_1 and c_2 are obtained from b_1 and b_2, respectively, by replacing $\sinh x'$ with $-\sinh x'$ and c_3 is obtained from b_3 by replacing u_1, u_2 and g_1 with u_{2n-1}, u_{2n-3}, and g_{n-1}, respectively.

Butterworth response:
Fig. 5.20*b*:

$$(1-K_n)^{1/2n} = a_1' = \min\{a_2', a_3', a_4'\} \tag{5.92a}$$

$$b_1' \leq \frac{B^2}{\omega_0^2} \leq \begin{cases} b_2', & n=2 \\ b_3', & n>2\ (n \text{ even}) \end{cases} \tag{5.92b}$$

where

$$a_1' = 1 - \frac{2\sin u_1}{R_{o1}C_{o1}B} \tag{5.92c}$$

$$a_2' = \frac{(1-\theta)^2}{(1+\theta)} \tag{5.92d}$$

$$a_3' = \frac{2R_{i2}\sin u_1}{BL_{i2}} \tag{5.92e}$$

$$a_4' = \frac{2R_{i2}C_{i2}\omega_0^2 \sin u_1}{B} - 1 \tag{5.92f}$$

where b_1', b_2', and b_3' are obtained from b_1, b_2, and b_3 in (5.90i)–(5.90k) by replacing $\sinh x \pm \sinh x'$ and g_m with $1 \pm k$ and $1-2k\cos u_{2m}+k^2$, respectively, where k is given by (5.85a).

Fig. 5.21:

$$\max\{-a_3', -a_4'\} \leq (1-K_n)^{1/2n} = -a_1' \leq a_2' \tag{5.93a}$$

$$c_1' \leq \frac{B^2}{\omega_0^2} \leq \begin{cases} c_2', & n=2 \\ c_3', & n>2\ (n \text{ even}) \end{cases} \tag{5.93b}$$

where c_1' and c_2' are obtained from b_1' and b_2' by replacing k with $-k$ and c_3' is obtained from b_3' by replacing u_1, u_2, and u_3 with u_{2n-1} and u_{2n-2} and u_{2n-3}, respectively. The element values of these matching networks are readily calculated from (5.82) and (5.83) when the above gain-bandwidth limitations are satisfied with t given as

$$t=(R_{o1}\gamma/R_{i2})^{1/2} \tag{5.94}$$

5.4.3 Input and Output Matching Networks

The input and output matching networks for microwave transistors to the source and the load have less restrictive constraints (50 Ω source and load in practice). The input matching network to the source R_s ($R_s > R_i$ in practice) is shown in Figs. 5.22*a–c*, which are obtained from Figs. 5.19*a*, 5.19*c*, and 5.19*d*, respectively, where we let $R_s=R_1$, $R_i=R_2$ in Fig. 5.19*a* and $R_s=R_2$, $R_i=R_1$ in Figs. 5.19*c* and 5.19*d*. Since the passband gain K_n and the ripple ϵ are fixed by the ratio $\gamma=R_i/R_s$ in (5.84b) for Figs. 5.19*a* and 5.19*c* and K_n is fixed by $\delta=R_i/R_s$ in (5.84*a*) for Fig. 5.19*d*, the transformation network N_T that can be either an inductive or capacitive T or pi network can be

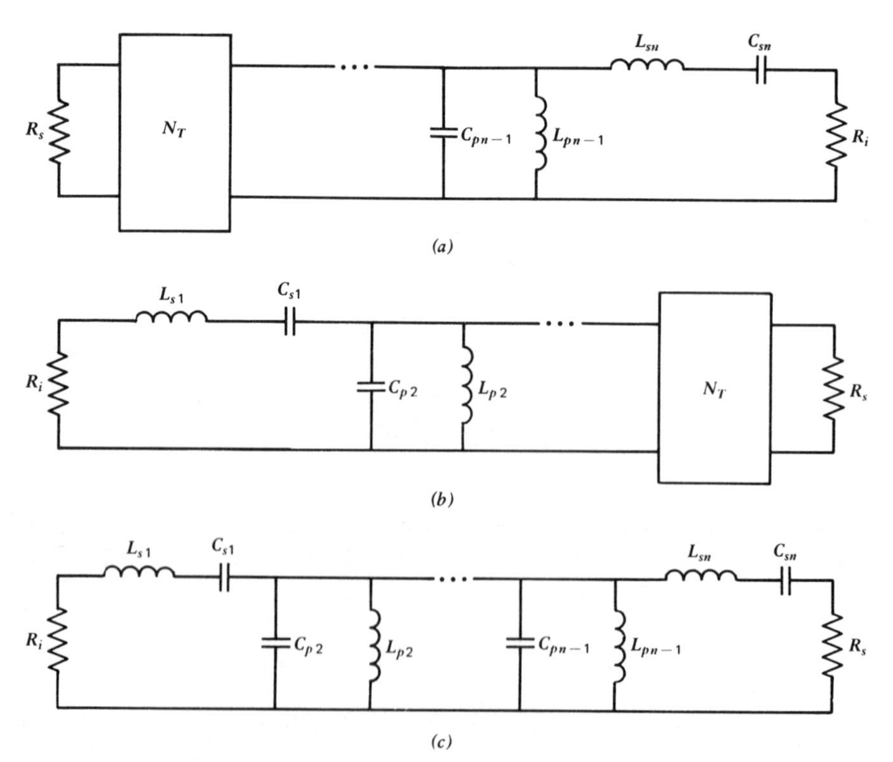

Fig. 5.22. Input matching networks. (*a*)–(*b*) With transformation network N_T for maximum gain-bandwidth product. (*c*) N_T is not applicable.

employed in Figs. 5.19*a* and 5.19*c* to obtain the maximum possible gain-bandwidth product. Configurations of N_T are shown in Figs. 5.23*a–b*. We note that such transformation is not applicable to the networks of Fig. 5.19*d* since in practice $R_s > R_i/\delta$ (in practice $R_i < 20\ \Omega$ and $R_s = 50\ \Omega$. This yields $K_{n,\max} = 0.816$) and since R_s is in series with L_{sn} and C_{sn}, it is not possible to scale R_s down to a smaller value as in Figs. 5.19*a* and 5.19*c*.

The gain-bandwidth constraints for the input matching network can be obtained from (5.90)–(5.93) by setting $R_s = R_{o1}$, $R_i = R_{i2}$, $L_i = L_{i2}$, $C_i = C_{i2}$, $L_{o1} = 0$ and $C_{o1} = 0$ for the inductive T or pi networks and $R_s = R_{o1}$, $(\omega_0^2 C_i)^{-1} = L_{i2}$, $(\omega_0^2 L_i)^{-1} = C_{i2}$, $L_{o1} = 0$ and $C_{o1} = 0$ for the capacitive T or pi networks. The element values of the transformation T network in Fig. 5.23*a* are given by its impedance $Z_i(s)$ $(i = 1, \ldots, 4)$ as

$$Z_1(s) = \left(\frac{T^2}{sM_{pi}} \right)^{\pm 1} \tag{5.95a}$$

$$Z_2(s) = \left[sN_{pi}(T^2 - T) \right]^{\pm 1} \tag{5.95b}$$

$$Z_3(s) = (sN_{pi}T)^{\pm 1} \tag{5.95c}$$

$$Z_4(s) = \left[sN_{sj} + s(1 - T)N_{pi} \right]^{\pm 1} \tag{5.95d}$$

where $i = 1$, $j = 2$ for Fig. 5.22*a* and $i = n$, $j = n - 1$ for Fig. 5.22*b*; also $M = C$,

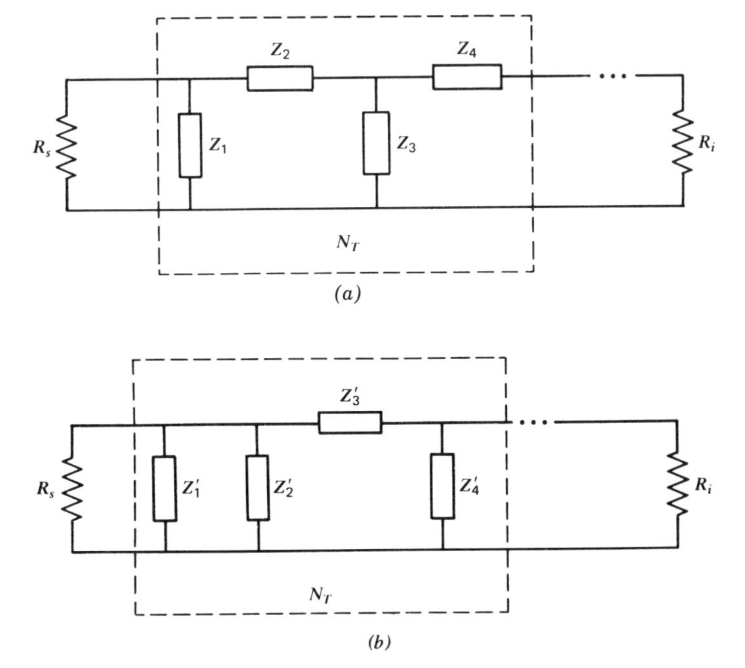

Fig. 5.23. (*a*) Transformation T network. (*b*) Transformation pi network.

$N=L$, $T=t$ and plus sign is used for inductive N_T and $M=L$, $N=C$, $T=1/t$, and minus sign is used for capacitive N_T.

The element values of the transformation pi network in Fig. 5.23*b* are given by its impedance $Z_i'(s)$ $(i=1,\ldots,4)$ as

$$Z_1'(s)=\left(\frac{T^2}{sM_{pi}}\right)^{\pm 1} \tag{5.96a}$$

$$Z_2'(s)=\left(\frac{s^2T^2N_{pi}N_{sj}}{N_{sj}+(1-T)N_{pi}}\right)^{\pm 1} \tag{5.96b}$$

$$Z_3'(s)=(sN_{sj}T)^{\pm 1} \tag{5.96c}$$

$$Z_4'(s)=\left(\frac{sTN_{sj}}{T-1}\right)^{\pm 1} \tag{5.96d}$$

where $i=1$, $j=2$ for Fig. 5.22*a* and $i=n$, $j=n-1$ for Fig. 5.22*b*; also $M=C$, $N=L$, $T=t$ and plus sign is used for inductive N_T and $M=L$, $N=C$, $T=1/t$, and minus sign is used for capacitive N_T.

For the output matching network, the gain-bandwidth constraints are obtained from (5.90)–(5.93) by setting $R_o=R_{o1}$, $R_L=R_{i2}$, $L_o=L_{o1}$, $C_o=C_{o1}$, $L_{i2}=0$ and $C_{i2}=\infty$. The output matching networks are obtained in an identical fashion as the input matching networks shown in Figs. 5.22*a*–*b* by setting $R_o=R_s$, $R_L=R_i$ with N_T as the inductive T network shown in Fig. 5.23*a*. In addition to these two networks, the output of the transistor can be matched using Fig. 5.19*b* with $R_o=R_1$, $R_L=R_2$ in conjunction with the inductive T network N_T for transformation shown in Fig. 5.23*a* with $R_o=R_s$ and $R_L=R_i$.

5.4.4 Gain Roll-Off Compensation Using Nonsloped Response

Because of the gain roll-off of microwave transistors of approximately 6 dB/octave, the matching networks must be designed to compensate for them in order to have an overall flat response. This can be achieved by using a sloped response for the transducer gain as described in detail in Section 5.3. Since explicit formulas for sloped Chebyshev and Butterworth responses do not exist, the previous procedure relies on network synthesis rather than closed form expressions. In this section, we attempt to use the asymptotic gain roll-up of about $6n$ dB/octave for $\omega\leqslant\omega_1$ of the Chebyshev bandpass matching network discussed in this section to compensate for the gain roll-off of the transistor. This enables us to use the closed form expressions derived above. We observe that when two transistors are in direct cascade, the overall gain roll-off is about 12 dB/octave. Thus if either one of the input, interstage, or output matching networks can be designed with $n=2$ with appropriate frequency shifting as discussed below, this gain roll-off can be compensated by the 12 dB/octave gain roll-up of the bandpass response for $\omega \leqslant \omega_1$. Suppose the roll-off of the two transistors in direct cascade is r dB over the operating frequency band (ω_L, ω_H) where ω_L and ω_H are the low and high cutoff frequencies. Let $\omega_H = \omega_1$ and select an arbitrary $\omega_2 > \omega_1$ where ω_1 and ω_2 are the lower and upper cutoff frequencies of the matching network response defined previously. The scheme is shown in Fig. 5.24. We note that for the Chebyshev response $G(\omega_1^2 = \omega_H^2) = K_n/(1+\epsilon^2)$ and hence at $\omega_L < \omega_1 = \omega_H$ we must have $10\log G(\omega_L^2) = 10\log G(\omega_H^2) - r$ or $G(\omega_H^2) = G(\omega_L^2)\log^{-1}(r/10)$. Using (5.31), we have

$$\epsilon^2 = \frac{\log^{-1}(r/10) - 1}{C_n^2(\omega_L/B - \omega_H\omega_2/B\omega_L) - \log^{-1}(r/10)} \tag{5.97}$$

where $n=2$ for 12 dB/octave compensation.

Equation (5.97) imposes an additional constraint on the gain-bandwidth limitations in (5.90)–(5.93). If ϵ is given, ω_2 can be calculated from (5.97) as

$$\omega_2 = \frac{\omega_L(\omega_L + \varphi\omega_H)}{\omega_H + \varphi\omega_L} \tag{5.98}$$

where

$$\varphi = \pm\frac{1}{\sqrt{2}}\sqrt{1+\sqrt{v}} \tag{5.99a}$$

$$v = \frac{(\epsilon^2+1)\log^{-1}(r/10) - 1}{\epsilon^2} \tag{5.99b}$$

There are two possible values of ω_2 in (5.98). The designer must select $\omega_2 > \omega_H$ such that the gain-bandwidth constraints are satisfied. Although this technique is

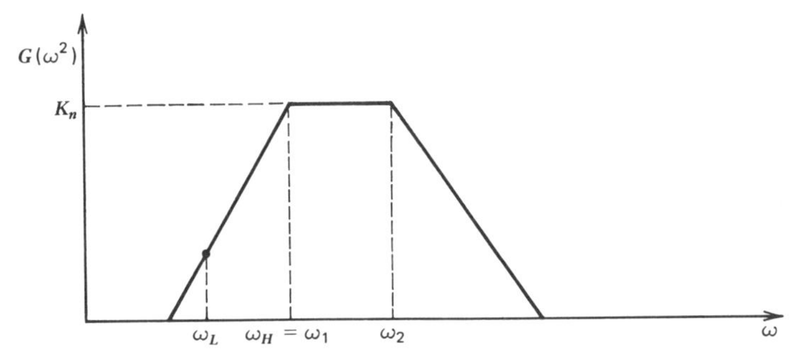

Fig. 5.24. Gain roll-off compensation using the roll-up slope of the flat transducer power gain response.

analytical, it is suggested that it be used in the case of moderate bandwidth amplifiers since a considerable amount of gain-bandwidth can be wasted if the operating bandwidth is too wide.

Example 5.5

In this example, the design of a cascade two-stage GaAs FET amplifier from 4 to 6 GHz is presented to illustrate the methods developed in this section. The source and load resistances are 50 Ω. The two GaAs FETs used are the NE-38806 FETs whose scattering parameters are shown in Table 5.15. The input impedance of this GaAs FET can be modeled by a series $R_iL_iC_i$ circuit where $R_i = 10.5$ Ω, $L_i = 0.65$ nH, $C_i = 0.64$ pF, and its output impedance can be modeled by a parallel-series R_oC_o-L_o circuit where $R_o = 164$ Ω, $C_o = 0.428$ pF, and $L_o = 0.474$ nH, as shown in Table 5.4. From 4 to 6 GHz the two FETs in cascade exhibit a gain roll-off of approximately 6 dB. An attempt to design the interstage matching network with $n=2$ employing the gain roll-up of the low side of the bandpass characteristic for high passband gain and small ripple has been unsuccessful; therefore it is decided to design the input matching network with $n=2$ to compensate for the gain roll-off and to design the output and the interstage matching networks for nonsloped responses.

(a) *Input matching network.* For 6 dB roll-off over the band 4–6 GHz, let $r=6$ dB. Thus from (5.99) we get $\varphi = \pm 5.412$ for $\epsilon = 0.03$. Substituting $\varphi = -5.412$ into (5.98) yields $\omega_2 = 2\pi \times 7.278 \times 10^9$ rad/sec with $\omega_H = \omega_1 = 2\pi \times 6 \times 10^9$ rad/sec and $\omega_L = 2\pi \times 4 \times 10^9$ rad/sec. Selecting $n=2$ yields $\sinh x = 4.02$ and using (5.91) for Fig. 5.22*b* with the appropriate replacement mentioned in Section 5.4.3 we have

$$\epsilon = 0.03 < 0.862$$

$$\max\{0.109, 0.263\} \leqslant \sqrt{1-K_2} \leqslant 0.515$$

This yields $K_{2,\max} = 0.931$. Using (5.83h) we get $\sinh x' = 1.976$ and from (5.84b) $\gamma = 0.582$. From (5.91c) we have

$$B^2/\omega_0^2 = B^2/\omega_H \omega_2 = 0.0374 < 0.04$$

Table 5.15 The NE-38806 S-parameters

f (GHz)	S_{11}	S_{21}	S_{12}	S_{22}
4.0	$0.78\angle -93°$	$3.03\angle 79°$	$0.058\angle 28°$	$0.55\angle -63°$
4.5	$0.74\angle -106°$	$2.92\angle 66°$	$0.058\angle 19.4°$	$0.52\angle -73°$
5.0	$0.71\angle -118°$	$2.80\angle 56°$	$0.054\angle 13.9°$	$0.52\angle -82°$
5.5	$0.69\angle -129°$	$2.67\angle 47°$	$0.050\angle 12°$	$0.54\angle -91°$
6.0	$0.68\angle -141°$	$2.51\angle 38°$	$0.046\angle 14°$	$0.56\angle -100°$

Using these calculated values in (5.82) and (5.83) we obtain

$$L_{s1}=0.905\ \text{nH},\qquad C_{s1}=0.64\ \text{pF}$$

$$L_{p2}=0.356\ \text{nH},\qquad C_{p2}=1.628\ \text{pF}$$

Using the transformed inductive T network shown in Fig. 5.22*b* with $t=1.665$, we obtain the input matching network N_i shown in Fig. 5.25.

(b) *Output matching network.* Since the output matching network is designed for a nonsloped response, let $\omega_1=2\pi\times4\times10^9$ rad/sec and $\omega_2=2\pi\times6\times10^9$ rad/sec. From (5.91) let

$$\sinh x=\frac{2\sin u_1}{R_oC_oB}=1.603$$

This yields for the selection of $n=2$

$$\epsilon=0.165$$

$$\sqrt{1-K_2}=0<0.345$$

Therefore $K_{2,\max}=1$ is possible. Using (5.83h) we get $\sinh x'=0$ and from (5.84b) $\gamma=0.72$. From (5.91c) we have

$$0.092<B^2/\omega_0^2=0.167<1.04$$

Using (5.82) and (5.83) we obtain

$$L_{s1}=3.51\ \text{nH},\qquad C_{s1}=0.3\ \text{pF}$$

$$L_{p2}=1.04\ \text{nH},\qquad C_{p2}=1.01\ \text{pF}$$

The output matching network N_o is obtained from Fig. 5.22*b* with $t=1.537$ (setting $R_L=R_i$ and $R_o=R_s$) and is shown in Fig. 5.25.

(c) *Interstage matching network.* The interstage matching network is also designed for a nonsloped response. Thus let $\omega_1=2\pi\times4\times10^9$ rad/sec and

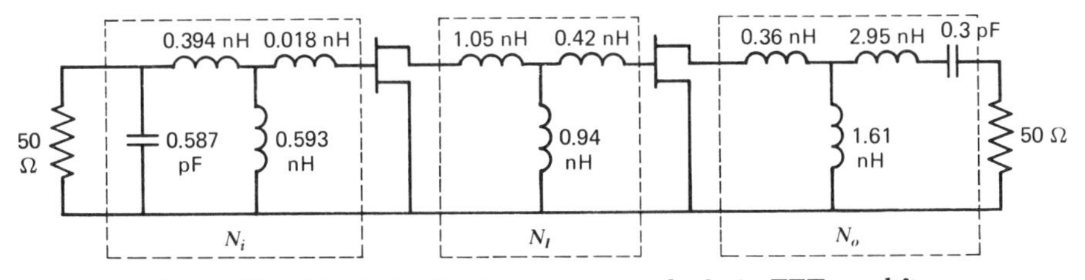

Fig. 5.25. The 4–6 GHz two-stage cascade GaAs FET amplifier.

Table 5.16 Performance of the 4–6 GHz two-stage cascade lumped GaAs FET amplifier

f (GHz)	S_{11}	S_{21}	S_{12}	S_{22}	S_{21} (dB)	K
4.0	$0.87\angle 79°$	$14.0\angle -42°$	$0.005\angle -140°$	$0.47\angle -75°$	23.0	1.72
4.5	$0.54\angle 15°$	$17.4\angle -158°$	$0.006\angle 124°$	$0.29\angle -66°$	24.8	2.86
5.0	$0.26\angle 173°$	$12.5\angle 159°$	$0.004\angle 84°$	$0.33\angle 28°$	22.0	7.53
5.5	$0.73\angle -25°$	$18.1\angle 166°$	$0.007\angle 77°$	$0.08\angle 59°$	25.1	1.93
6.0	$0.45\angle -133°$	$16.1\angle 53°$	$0.005\angle -5°$	$0.29\angle 133°$	24.1	4.07

$\omega_2 = 2\pi \times 6 \times 10^9$ rad/sec. Select $n=2$. For (5.91b) to be satisfied we set

$$\sinh x = \frac{1}{2}\left[\frac{2R_{i2}C_{i2}\omega_0^2 \sin u_1}{B} + \frac{2\sin u_1}{R_{o1}C_{o1}B}\right] = 1.16$$

This yields, when combined with (5.91a)

$$\epsilon = 0.281 < 1.85$$

From (5.91b) we have

$$\max\{-0.44, 0.273\} \leqslant \sqrt{1-K_2} = 0.273 < 0.842$$

which implies $K_{2,\max} = 0.925$, $\sinh x' = 0.443$, $\gamma = 0.439$, $t = 2.62$.
From (5.91c) we get

$$0.052 < B^2/\omega_0^2 = 0.167 < 0.286$$

Using (5.82) and (5.83) we obtain

$$L_{s1} = 1.65 \text{ nH}, \qquad C_{s1} = 0.64 \text{ pF}$$

$$L_{p2} = 0.359 \text{ nH}, \qquad C_{p2} = 2.937 \text{ pF}$$

Using the transformed inductive T network shown in Fig. 5.21 we obtain the interstage matching network N_I shown in Fig. 5.25. The performance of the lumped amplifier is shown in Table 5.16.

Table 5.17 Γ_m and R_n of the 6–8 GHz HFET-1101

f (GHz)	Γ_m	R_n	F (dB)
6.0	$0.575\angle 138°$	6.65	2.2
7.0	$0.595\angle 164°$	4.00	2.5
8.0	$0.617\angle 170°$	2.00	2.8

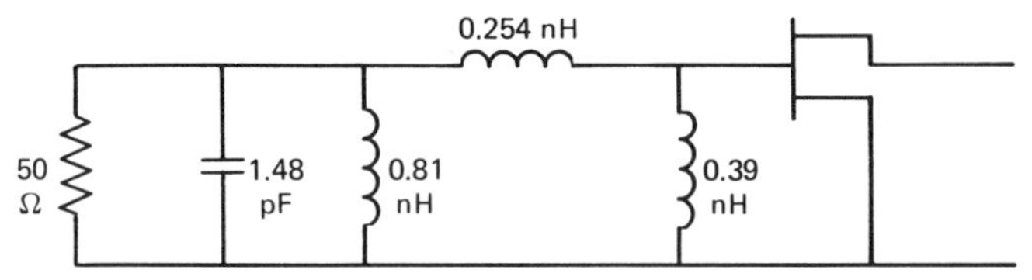

Fig. 5.26. The lumped noise matching network for the 6–8 GHz HFET-1101 GaAs FET amplifier.

Example 5.6

This example concerns the design of the noise matching network for the GaAs FET HFET-1101 to operate in the frequency band 6–8 GHz. The optimum noise source reflection coefficient Γ_m of the HFET-1101 over this frequency band is given in Table 5.17 together with the noise resistance R_n and its noise figure F.

The conjugate of Γ_m which is Γ_m^*, can be approximated by a series $R_i L_i C_i$ impedance where $R_i = 12.7\ \Omega$, $L_i = 0.786$ nH, and $C_i = 0.562$ pF. The reflection coefficient Γ_i of this series impedance is $0.63\angle -139°$ at 6 GHz, $0.60\angle -165°$ at 7 GHz and $0.597\angle 170°$ at 8 GHz. The problem is to match this impedance as closely as possible to a 50 Ω source over the band for good noise figure. Let $\omega_1 = 2\pi \times 6 \times 10^9$ rad/sec, $\omega_2 = 2\pi \times 8 \times 10^9$ rad/sec and $n=2$. Then from (5.91b) with the replacement $R_s = R_{o1} = 50\ \Omega$, $R_i = R_{i2}$, $L_i = L_{i2}$, $C_i = C_{i2}$, $L_{o1} = 0$ and $C_{o1} = 0$ we get $(2R_i \sin u_1)/BL_i = 1.82$, $(2R_i C_i \omega_0^2 \sin u_1)/B = 1.52$. Let $\sinh x = 1.52$ yield $\epsilon = 0.18 < 0.74$ and $0 \leq \sqrt{1-K_2} \leq 0.417$, which implies $K_{2,\max} = 1$. Then (5.91c) also yields $0.083 < B^2/\omega_0^2 < 0.149$. From these we obtain $\sinh x' = 0$, $\gamma = 0.699$, $t = 1.658$, $L_{s1} = 0.939$ nH, $C_{s1} = 0.562$ pF, $L_{p2} = 0.13$ nH, and $C_{p2} = 4.06$ pF. Using Fig. 5.22*b* and the inductive pi network in Fig. 5.23*b*, the desired matching network is shown in Fig. 5.26. The noise figure F (dB) of the amplifier is 2.32 dB, 2.57 dB, and 2.84 dB at 6, 7, and 8 GHz, respectively, as calculated from (2.59).

5.5 DISTRIBUTED APPROXIMATION AND KURODA TRANSFORMATION

Although capacitors and inductors can now be made to remain truly lumped at frequencies as high as 14 GHz, the disadvantage of lumped-element circuits at microwave frequencies is the design cost and time compared with conventional microstrip circuits. For frequencies below 2 GHz, distributed circuits usually occupy a large area. Therefore it is more appropriate to use lumped elements such as thick film capacitors or inductors that are commercially available or a mixture of lumped and distributed elements in a circuit. In this section we will discuss the procedure for approximation of certain types of lumped elements. From (1.29) the chain matrix of a two-port lossless transmission line is given as

$$\mathcal{C}=\begin{bmatrix} \cos\theta & jZ_0\sin\theta \\ \dfrac{j\sin\theta}{Z_0} & \cos\theta \end{bmatrix} \tag{5.100}$$

where $\theta=\beta l$ and Z_0 are the electrical length and the characteristic impedance of the line. From (1.30a) the input impedance of an open-circuited line is

$$Z_{oc}=\frac{-jZ_0}{\tan\theta} \tag{5.101a}$$

and from (1.30b) the input impedance of a short-circuited line is (also see Section 3.3)

$$Z_{sc}=jZ_0\tan\theta \tag{5.101b}$$

Defining the complex variable $\Omega=j\tan\theta$, then (5.101a) and (5.101b) become

$$Z_{oc}=\frac{Z_0}{\Omega} \tag{5.102a}$$

$$Z_{sc}=Z_0\Omega \tag{5.102b}$$

From (5.102a) and (5.102b) it is seen that an open-circuited line is a capacitor of capacitance $1/Z_0$ and a short-circuited line is an inductor of inductance Z_0 in terms of the new frequency Ω. We note that $\theta=\beta l=(l/v)\omega$ where l is the physical length of the line, v is the velocity of propagation of the signal, and ω is the frequency. The ratio $\tau=l/v$ is the time it takes a signal to travel across the line and is often called the delay time. Therefore $\Omega=j\tan\theta=j\tan\tau\omega$ is seen to be periodic in terms of ω with period π/τ. Thus in a lumped circuit, if all the capacitors and inductors are replaced by open- and short-circuited lines of equal lengths, the result will be the same frequency response in terms of ω but with periodicity, since Ω is periodic in ω, that is, the response is multiband, which for the matching networks should not pose any problems.

In microwave transistor amplifiers, all the matching circuits are usually fabricated on planar microstrip substrates. Therefore, series open- and short-circuited lines are not appropriate; series capacitors are usually metal-oxide-metal or interdigitated capacitors, and series inductors are usually approximated by a short length of transmission line of high characteristic impedance as in (3.16). In order to have a good approximation, the characteristic impedance of this line must be large compared to the impedances looking toward the right and the left of it over the entire bandwidth. For microstrip lines, the practical highest realizable characteristic impedance is about 100 Ω for a substrate with $\epsilon_r=10$ and about 170 Ω for $\epsilon_r=2.2$ with 0.025 in. thickness, that is, the width of this line is about 0.004 in. From experience, this type of approximation is usually good for a bandwidth of about 50% of the center frequency.

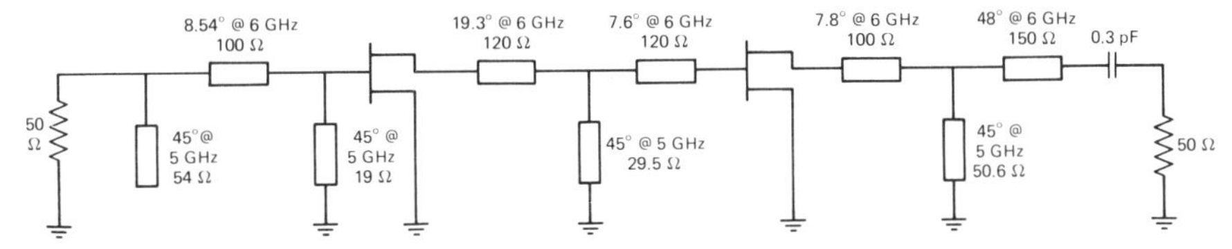

Fig. 5.27. The 4–6 GHz two-stage cascade distributed GaAs FET amplifier.

Example 5.7

The 4–6 GHz lumped amplifier in Example 5.5 (Fig. 5.25) is approximated by distributed circuits as follows:

(a) The series inductor L_s can be approximated by the transmission line with its electrical length in degrees given as $\theta=(180/\pi)\sin^{-1}(\omega_2 L_s/Z_{0H})$ or $\theta=180(\omega_2 L_s)/(\pi Z_{0H})$ (if $\omega_2 L_s/Z_{0H}>1$) at the frequency ω_2 and Z_{0H} is the highest realizable characteristic impedance.

(b) The shunt inductor L_p is converted to a short-circuited commensurate line of electrical length θ (usually 45°) at the center frequency $(\omega_1+\omega_2)/2$ (or $\omega_0=\sqrt{\omega_1\omega_2}$) with the characteristic impedance $Z_0=L_p(\omega_1+\omega_2)/2$ (or $Z_0=\omega_0 L$).

(c) The shunt capacitor C_p is converted to an open-circuited commensurate line of electrical length θ (usually 45°) at the frequency $(\omega_1+\omega_2)/2$ (or $\omega_0=\sqrt{\omega_1\omega_2}$) with the characteristic impedance $Z_0=2/[C_p(\omega_1+\omega_2)]$ (or $Z_0=1/(\omega_0 C_p)$.

The equivalent distributed amplifier is shown in Fig. 5.27 and its performance in Table 5.18.

Example 5.8

The noise matching network of the 6–8 GHz amplifier of Example 5.6. (Fig. 5.26) is approximated by the distributed circuit shown in Fig. 5.28. The noise figures F_1, F_2, and F_3 of the GaAs FET HFET-1101, the lumped amplifier in Fig. 5.26, and the distributed amplifier in Fig. 5.28 are shown in Table 5.19 for comparison. Figure 5.28 and Table 5.19 are shown on page 171.

Table 5.18 Performance of the 4–6 GHz two-stage cascade distributed GaAs FET amplifier

f (GHz)	S_{11}	S_{21}	S_{12}	S_{22}	S_{21} (dB)	K
4.0	$0.87\angle 89°$	$13.73\angle -54°$	$0.005\angle -152°$	$0.28\angle -58°$	22.8	1.71
4.5	$0.60\angle 31°$	$16.96\angle -176°$	$0.006\angle 106°$	$0.30\angle -30°$	24.6	2.80
5.0	$0.20\angle 176°$	$11.00\angle 132°$	$0.004\angle 57°$	$0.55\angle -4°$	20.8	7.89
5.5	$0.78\angle -46°$	$16.82\angle 126°$	$0.006\angle 37°$	$0.28\angle 6°$	24.5	1.92
6.0	$0.52\angle -141°$	$14.53\angle 7°$	$0.005\angle -51°$	$0.39\angle 65°$	23.3	4.15

In distributed circuits it is sometimes difficult to fabricate transmission lines on microstrip substrates because of a crowded junction. This can be facilitated by inserting a redundant lossless commensurate transmission line and performing the Kuroda transformation. This transformation is illustrated in Figs. 5.29*a–b*. The two-port networks N and N' are equivalent if their chain matrix $\mathcal{C}_N$ and $\mathcal{C}'_N$ are the same, that is, $\mathcal{C}_N = \mathcal{C}'_N$. We note that the chain matrix of a two-port lossless transmission line is given by (5.100) and in terms of the transformed variable $\Omega = j\tan\theta$, it can be written as

$$\begin{bmatrix} \cos\theta & jZ_0\sin\theta \\ \dfrac{j\sin\theta}{Z_0} & \cos\theta \end{bmatrix} = \frac{1}{\sqrt{\Omega^{-2}-1}} \begin{bmatrix} \dfrac{1}{\Omega} & Z_0 \\ \dfrac{1}{Z_0} & \dfrac{1}{\Omega} \end{bmatrix} \tag{5.103}$$

Using (5.103) we can write $\mathcal{C}_N$ and $\mathcal{C}'_N$ in terms of the chain matrix $\mathcal{C}$ and $\mathcal{C}'$ as

$$\mathcal{C}_N = \frac{1}{\sqrt{\Omega^{-2}-1}} \begin{bmatrix} \mathcal{C}_{11} & \mathcal{C}_{12} \\ \mathcal{C}_{21} & \mathcal{C}_{22} \end{bmatrix} \begin{bmatrix} \dfrac{1}{\Omega} & Z_0 \\ \dfrac{1}{Z_0} & \dfrac{1}{\Omega} \end{bmatrix}$$

$$= \frac{1}{\sqrt{\Omega^{-2}-1}} \begin{bmatrix} \dfrac{\mathcal{C}_{11}}{\Omega} + \dfrac{\mathcal{C}_{12}}{Z_0} & \mathcal{C}_{11}Z_0 + \dfrac{\mathcal{C}_{12}}{\Omega} \\ \dfrac{\mathcal{C}_{21}}{\Omega} + \dfrac{\mathcal{C}_{22}}{Z_0} & \mathcal{C}_{21}Z_0 + \dfrac{\mathcal{C}_{22}}{\Omega} \end{bmatrix} \tag{5.104a}$$

$$\mathcal{C}'_N = \frac{1}{\sqrt{\Omega^{-2}-1}} \begin{bmatrix} \dfrac{1}{\Omega} & Z'_0 \\ \dfrac{1}{Z'_0} & \dfrac{1}{\Omega} \end{bmatrix} \begin{bmatrix} \mathcal{C}'_{11} & \mathcal{C}'_{12} \\ \mathcal{C}'_{21} & \mathcal{C}'_{22} \end{bmatrix}$$

$$= \frac{1}{\sqrt{\Omega^{-2}-1}} \begin{bmatrix} \dfrac{\mathcal{C}'_{11}}{\Omega} + \mathcal{C}'_{21}Z'_0 & \dfrac{\mathcal{C}'_{12}}{\Omega} + \mathcal{C}'_{22}Z'_0 \\ \dfrac{\mathcal{C}'_{11}}{Z'_0} + \dfrac{\mathcal{C}'_{21}}{\Omega} & \dfrac{\mathcal{C}'_{12}}{Z'_0} + \dfrac{\mathcal{C}'_{22}}{\Omega} \end{bmatrix} \tag{5.104b}$$

From (1.5a), (5.101a), and (5.104) the open-circuit impedance Z_{11} and Z'_{11} of the two-ports N and N' can be computed as

$$Z_{11} = \frac{\mathcal{C}_{11}Z_0 + \mathcal{C}_{12}\Omega}{\mathcal{C}_{21}Z_0 + \mathcal{C}_{22}\Omega} \tag{5.105a}$$

$$Z'_{11} = \frac{\mathcal{C}'_{11}/\Omega + \mathcal{C}'_{21}Z'_0}{\mathcal{C}'_{11}/Z'_0 + \mathcal{C}'_{21}/\Omega} \tag{5.105b}$$

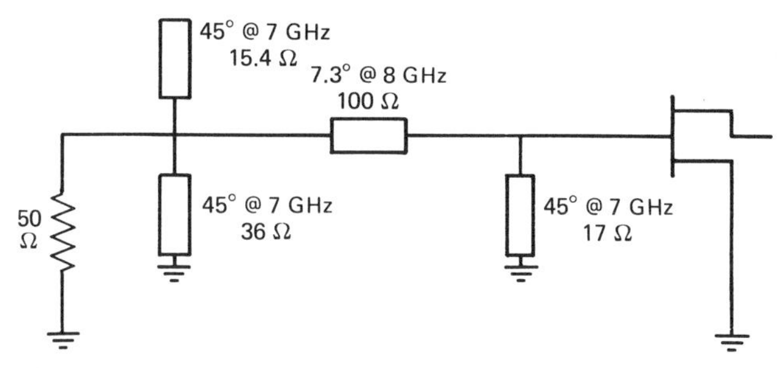

Fig. 5.28. The 6–8 GHz distributed noise matching network for the HFET-1101 GaAs FET amplifier.

From (5.105b) it is seen that $Z_0' = Z_{11}'$ $(\Omega = 1)$ for the two-ports N and N' to be equivalent. Thus we conclude that

$$Z_0' = \frac{\mathcal{C}_{11} Z_0 + \mathcal{C}_{12}}{\mathcal{C}_{21} Z_0 + \mathcal{C}_{22}} \tag{5.106}$$

Furthermore $\mathcal{C}_N = \mathcal{C}_N'$ yields, by equating (5.104a) and (5.104b),

$$\mathcal{C}_{11}' = \frac{\mathcal{C}_{11} + \mathcal{C}_{12}\Omega/Z_0 - \mathcal{C}_{21} Z_0' \Omega - \mathcal{C}_{22}\Omega^2 Z_0'/Z_0}{1-\Omega^2} \tag{5.107a}$$

$$\mathcal{C}_{12}' = \frac{\mathcal{C}_{12} + \mathcal{C}_{11} Z_0 \Omega - \mathcal{C}_{22} Z_0' \Omega - \mathcal{C}_{21} Z_0 Z_0' \Omega^2}{1-\Omega^2} \tag{5.107b}$$

$$\mathcal{C}_{21}' = \frac{\mathcal{C}_{21} + \mathcal{C}_{22}\Omega/Z_0 - \mathcal{C}_{11}\Omega/Z_0' - \mathcal{C}_{12}\Omega^2/Z_0 Z_0'}{1-\Omega^2} \tag{5.107c}$$

$$\mathcal{C}_{22}' = \frac{\mathcal{C}_{22} + \mathcal{C}_{21} Z_0 \Omega - \mathcal{C}_{12}\Omega/Z_0' - \mathcal{C}_{11}\Omega^2 Z_0/Z_0'}{1-\Omega^2} \tag{5.107d}$$

Table 5.19 Noise figures of the HFET-1101 and its lumped and distributed amplifiers

f (GHz)	F_1 (dB)	F_2 (dB)	F_3 (dB)
6	2.2	2.32	2.64
7	2.5	2.57	2.58
8	2.8	2.84	3.00

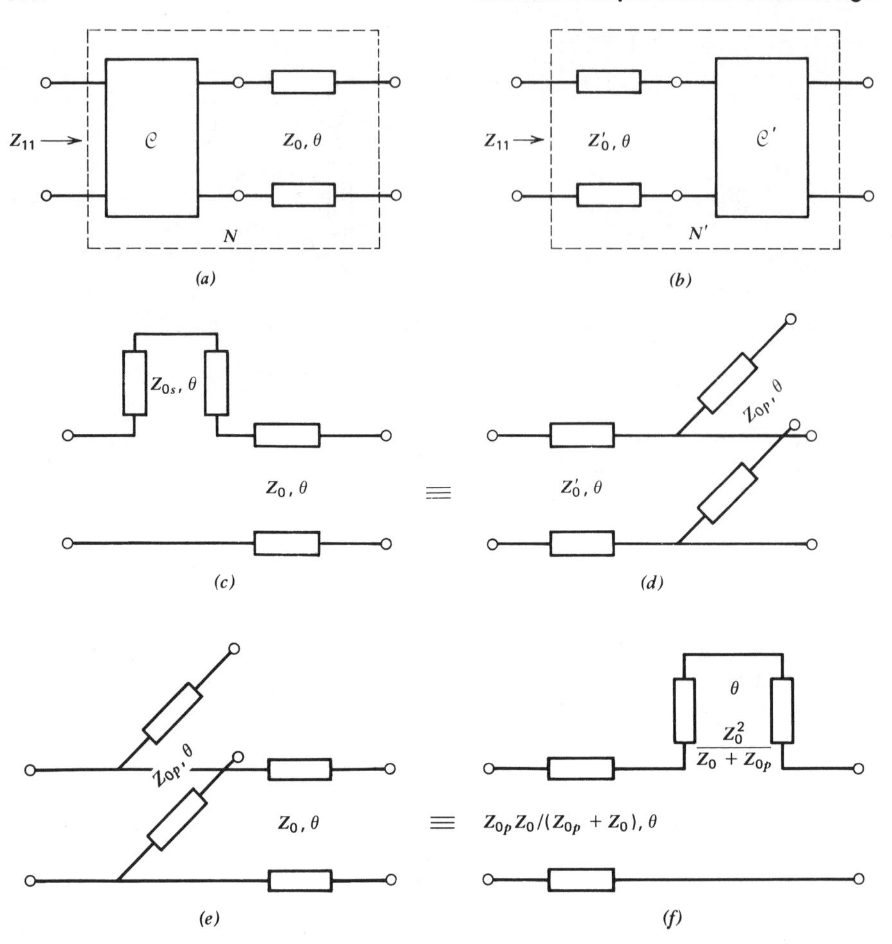

Fig. 5.29. Kuroda transformations.

Example 5.9

Consider the two-ports shown in Figs. 5.29*c*–*d*. The chain matrix of a two-port series short-circuited stub is given by (1.25a) and (5.101b) as

$$\mathcal{C}=\begin{bmatrix} 1 & Z_{0s}\Omega \\ 0 & 1 \end{bmatrix}$$

From (5.106) we get

$$Z_0' = Z_0 + Z_{0s}$$

and

$$\mathcal{C}'=\begin{bmatrix} 1 & 0 \\ \dfrac{Z_{0s}\Omega}{Z_0(Z_0+Z_{0s})} & 1 \end{bmatrix}$$

Comparing $\mathcal{C}'$ with (1.25b) using (5.101a), we conclude that $\mathcal{C}'$ is the chain matrix of a two-port shunt open-circuited stub with characteristic impedance $Z_{0p} = Z_0(Z_0 + Z_{0s})/Z_{0s}$. In a similar way, Figs. 5.29*e*–*f* are equivalent.

The Kuroda transformation remains the same if the two-ports are interchanged. In matching networks, the Kuroda transformation can be implemented by inserting commensurate lines in cascade with the source or the load provided that they have the same characteristic impedance as that of the source or the load (50 Ω in general).

5.6 FUNDAMENTALS OF OPTIMIZATION

Often in the design of a microwave amplifier, the initial responses—gain, noise figure, input or output VSWR, and so on—do not exactly match the specified values over the frequency band of interest for many reasons. For example, the model may not exactly represent the measured data; other causes of error include the unilateral assumption in the case of transistors, or the error introduced by distributed approximation of the lumped elements. Also encountered frequently in practice is the difficulty of fabricating distributed circuits on microstrip substrates because of the predetermined topology of the network, for example, a series lumped capacitor sandwiched between two parallel stubs. In such cases, it is necessary to insert a short length of line on each side of the capacitor for ease of fabrication, and this undoubtedly changes the response of the original circuit. In order to obtain the specified response, the elements of the original network can be adjusted by the cut-and-try method (tuning) in the lab; this usually limits a network to a few elements and a narrowband response. For a large broadband network, this type of tuning is a tremendous and often impossible task. Therefore, it is more appropriate to adjust the response of the initial network by varying its elements iteratively according to some numerical algorithm carried out on a digital computer. This process is called optimization. Since optimization is an iterative process, the procedure is laborious, and for a broadband amplifier the number of iterations could easily be a few hundred—even with the most-efficient currently available optimization methods—if the response of the initial network deviates too far from the specified one. Furthermore, the optimization process may fail or may converge to an inferior network with poor starting values. The optimization process is shown schematically in Fig. 5.30 and consists of the following steps:

Step 1. **Initial circuit**

The purpose of optimization is to improve the initial circuit. It is not meant to design a circuit to meet the specified response. Therefore the designer must use design procedures such as those described in the previous section to obtain a good initial circuit to ensure convergence within reasonable time and at reasonable cost.

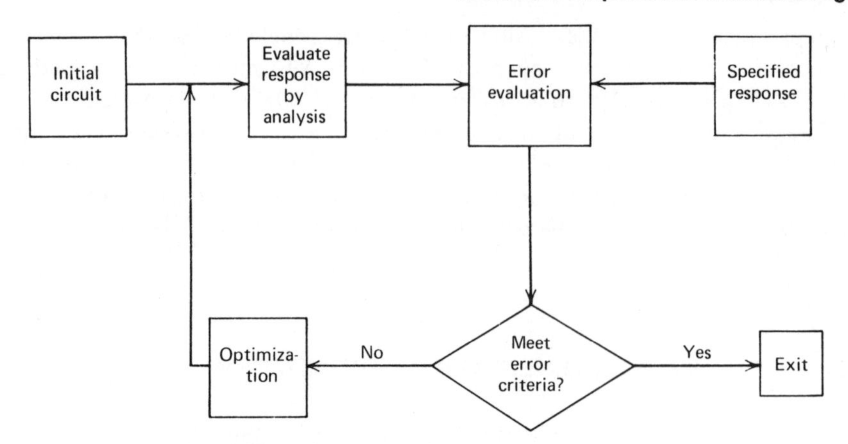

Fig. 5.30. Flow chart of an optimization process.

Step 2. **Analysis**

In any optimization process an analysis program to evaluate the desired response of the circuit must be included so that it can be compared with the specified values. For microwave amplifiers, such analysis can be carried out using the chain or chain scattering matrix methods described in Chapter 1.

Step 3. **Error evaluation**

There are several error functions capable of evaluating the deviation of the calculated response from the desired one. Let $F_c(\omega,\mathbf{p})$ be the calculated response and $F_d(\omega)$ be the desired response where $\mathbf{p}=(p_1, p_2,\ldots, p_n)^T$ is the parameter n-vector in which p_i represents the element value of the circuit, for example. One of the most popular form of error function is the least-square error criterion written

$$E_2 = \sum_{i=1}^{k} |F_d(\omega_i) - F_c(\omega_i,\mathbf{p})|^2 \tag{5.108}$$

where the error criterion E_2 is summed over k sampling frequencies $\omega_1,\ldots,\omega_k$ over the frequency band of interest. If there are more responses that need to be optimized simultaneously, for example, in a low noise amplifier, the designer may want to optimize the noise figure and the flat gain response simultaneously (although this is not necessary, and often optimization of individual responses separately leads to faster convergence), we can define the least-square error criterion as

$$E_2 = \sum_{i=1}^{k} W_1|F_{d1}(\omega_i) - F_{c1}(\omega_i,\mathbf{p})|^2 + \cdots + W_m|F_{dm}(\omega_i) - F_{cm}(\omega_i,\mathbf{p})|^2 \tag{5.109}$$

where $W_1, \ldots, W_m$ are weighing factors assigned to each type of response. If the designer wishes to emphasize one response over the others, then he should assign a larger value for the associated weighing factor. Another type of error evaluation is the min-max error criterion written

$$E_\infty = \max_{\omega} \sum_{i=1}^{k} W_1 |F_{d1}(\omega_i) - F_{c1}(\omega_i, \mathbf{p})| + \cdots + W_m |F_{dm}(\omega_i) - F_{cm}(\omega_i, \mathbf{p})| \tag{5.110}$$

A generalization of the least-square error criterion is the least-qth error criterion defined as

$$E_q = \sum_{i=1}^{k} W_1 |F_{d1}(\omega_i) - F_{c1}(\omega_i, \mathbf{p})|^q + \cdots + W_m |F_{dm}(\omega_i) - F_{cm}(\omega_i, \mathbf{p})|^q \tag{5.111}$$

Once the error criterion is established, the optimization scheme will adjust the circuit elements, that is, changing $\mathbf{p}$ to $\mathbf{p} + \boldsymbol{\Delta}\mathbf{p}$ for each iteration in such a way as to decrease the error E. The error criterion E actually defines an error surface in the n-dimensional coordinates $\mathbf{p} = (p_1, p_2, \ldots, p_n)^T$ in which E can possess a true minimum and many local minima. The purpose of the optimization process is to produce a local minimum E that is acceptable, that is, the error between the calculated and desired responses is within given specifications. It is known that the rate of convergence and the final accuracy of an optimization algorithm can be improved if the range of variation of the parameters $p_1, \ldots, p_n$ is similar. Therefore it is sometimes advantageous to compress the parameter range by using logarithmic scaling. Instead of updating the vector parameter $\mathbf{p}^{(j)} = (p_1^{(j)}, p_2^{(j)}, \ldots, p_n^{(j)})^T$ at the jth iteration as

$$p_i^{(j)} = p_i^{(j-1)} + \Delta p_i^{(j)}, \qquad i = 1, 2, \ldots, n \tag{5.112}$$

one can do it as

$$p_i^{(j)} = p_i^{(j-1)} \exp\left[\Delta p_i^{(j)}\right], \qquad i = 1, 2, \ldots, n \tag{5.113}$$

This change of variables can be interpreted as taking a linear step in the space of the logarithms of the variables, since

$$\ln p_i^{(j)} = \ln p_i^{(j-1)} + \Delta p_i^{(j)}, \qquad i = 1, 2, \ldots, n \tag{5.114}$$

Furthermore, logarithmic scaling constrains the sign of each variable to its original one during the entire optimization process.

In the following discussion we will present various known optimization algorithms that can be used to adjust $\mathbf{p}$ to decrease the error E. When the parameter vector $\mathbf{p}$ is updated to $\mathbf{p}+\mathbf{\Delta p}$ for $\Delta p_i \ll p_i$ for $i=1,2,\ldots,n$, then the updated error $E(\mathbf{p}+\mathbf{\Delta p})$ can be expanded in a Taylor series as

$$E(\mathbf{p}+\mathbf{\Delta p})=E(\mathbf{p})+\sum_{i=1}^{n}\frac{\partial E(\mathbf{p})}{\partial p_i}\Delta p_i+\frac{1}{2}\sum_{r=1}^{n}\sum_{s=1}^{n}\frac{\partial^2 E(\mathbf{p})}{\partial p_r\,\partial p_s}\Delta p_r \Delta p_s+\cdots$$

$$=E(\mathbf{p})+\left[\mathfrak{G}\mathbf{E}(\mathbf{p})\right]^T\mathbf{\Delta p}+\tfrac{1}{2}(\mathbf{\Delta p})^T\mathbf{H}\,\mathbf{\Delta p}+\cdots \tag{5.115}$$

where

$$\mathbf{\Delta p}=\begin{bmatrix}\Delta p_1\\ \Delta p_2\\ \vdots\\ \Delta p_n\end{bmatrix} \tag{5.116}$$

$$\mathfrak{G}\mathbf{E}(\mathbf{p})=\begin{bmatrix}\dfrac{\partial E(\mathbf{p})}{\partial p_1}\\ \dfrac{\partial E(\mathbf{p})}{\partial p_2}\\ \vdots\\ \dfrac{\partial E(\mathbf{p})}{\partial p_n}\end{bmatrix} \tag{5.117}$$

$$\mathbf{H}=\begin{bmatrix}\dfrac{\partial^2 E(\mathbf{p})}{\partial p_1^2} & \dfrac{\partial^2 E(\mathbf{p})}{\partial p_1\partial p_2} & \cdots & \dfrac{\partial^2 E(\mathbf{p})}{\partial p_1\partial p_n}\\ \dfrac{\partial^2 E(\mathbf{p})}{\partial p_2\partial p_1} & \dfrac{\partial^2 E(\mathbf{p})}{\partial p_2^2} & \cdots & \dfrac{\partial^2 E(\mathbf{p})}{\partial p_2\partial p_n}\\ \dfrac{\partial^2 E(\mathbf{p})}{\partial p_n\partial p_1} & \dfrac{\partial^2 E(\mathbf{p})}{\partial p_n\partial p_2} & \cdots & \dfrac{\partial^2 E(\mathbf{p})}{\partial p_n^2}\end{bmatrix} \tag{5.118}$$

The vector $\mathfrak{G}\mathbf{E}(\mathbf{p})$ is called the error gradient and $\mathbf{H}$ is called the Hessian matrix, which is symmetric. If (5.113)–(5.114) are used, then (5.115) becomes

$$E(\ln\mathbf{p}+\mathbf{\Delta p})=E(\ln\mathbf{p})+\sum_{i=1}^{n}\frac{\partial E(\ln\mathbf{p})}{\partial(\ln p_i)}\Delta p_i+\frac{1}{2}\sum_{r=1}^{n}\sum_{s=1}^{n}\frac{\partial^2 E(\ln\mathbf{p})}{\partial(\ln p_r)\,\partial(\ln p_s)}\Delta p_r\Delta p_s+\cdots \tag{5.119}$$

Since

$$\frac{\partial E(\ln \mathbf{p})}{\partial(\ln p_i)} = \frac{\partial E(\ln \mathbf{p})}{\partial p_i} \cdot \frac{\partial p_i}{\partial(\ln p_i)} = \frac{\partial E(\ln \mathbf{p})}{\partial p_i} p_i \tag{5.120}$$

$$\frac{\partial^2 E(\ln \mathbf{p})}{\partial(\ln p_r)\,\partial(\ln p_s)} = \frac{\partial^2 E(\ln \mathbf{p})}{\partial p_r\,\partial p_s} \cdot \frac{\partial p_r\,\partial p_s}{\partial(\ln p_r)\,\partial(\ln p_s)}$$

$$= \frac{\partial^2 E(\ln \mathbf{p})}{\partial p_r\,\partial p_s} p_r p_s \tag{5.121}$$

hence

$$E(\ln \mathbf{p}+\mathbf{\Delta p}) = E(\ln \mathbf{p}) + \left[\mathfrak{G}\mathbf{E}(\ln \mathbf{p})\right]^T \mathbf{A}(\mathbf{p})\,\mathbf{\Delta p} + \tfrac{1}{2}(\mathbf{\Delta p})^T \mathbf{A}^T(\mathbf{p})\mathbf{H}\mathbf{A}(\mathbf{p})\,\mathbf{\Delta p} + \cdots \tag{5.122}$$

where

$$\mathbf{A}(\mathbf{p}) = \begin{bmatrix} p_1 & 0\ldots 0 \\ 0 & p_2\ldots 0 \\ \vdots & \ddots \\ 0 & 0\ldots p_n \end{bmatrix} \tag{5.123}$$

The gradient method then utilizes (5.115) or (5.122) to evaluate $\mathbf{\Delta p}$ such that $E(\mathbf{p}+\mathbf{\Delta p})<E(\mathbf{p})$ or $E(\ln \mathbf{p}+\mathbf{\Delta p})<E(\ln \mathbf{p})$. We observe that the difference between (5.115) and (5.122) is the transformation of $\mathbf{\Delta p}$ to $\mathbf{A}(\mathbf{p})\mathbf{\Delta p}$. Therefore we discuss only (5.115) and referred the result to (5.122).

Steepest Descent Method

In this method only the second term of (5.115) is retained, that is,

$$E(\mathbf{p}+\mathbf{\Delta p}) \approx E(\mathbf{p}) + \left[\mathfrak{G}\mathbf{E}(\mathbf{p})\right]^T \mathbf{\Delta p} \tag{5.124}$$

We see that $E(\mathbf{p}+\mathbf{\Delta p})<E(\mathbf{p})$ if $\Delta E(\mathbf{p}) = [\mathfrak{G}\mathbf{E}(\mathbf{p})]^T \mathbf{\Delta p}<0$ and the more negative $\Delta E(\mathbf{p})$ the smaller $E(\mathbf{p}+\mathbf{\Delta p})$. It can easily be proved that the value of $\mathbf{\Delta p}$ that minimizes $\Delta E(\mathbf{p})$, and thus minimizes $E(\mathbf{p}+\mathbf{\Delta p})$, is given by

$$\mathbf{\Delta p} = -\frac{\|\mathbf{\Delta p}\|}{\|\mathfrak{G}\mathbf{E}(\mathbf{p})\|}\mathfrak{G}\mathbf{E}(\mathbf{p}) = \alpha\,\mathfrak{G}\mathbf{E}(\mathbf{p}) \tag{5.125}$$

where $\|\cdot\|$ is the Euclidean norm defined as $\|\mathbf{x}\| = (\mathbf{x}^T\mathbf{x})^{1/2}$ for any n-vector $\mathbf{x}$.

Thus using (5.125) in (5.124) we obtain

$$E(\mathbf{p}+\mathbf{\Delta p}) \approx E(\mathbf{p}) - \frac{\|\mathbf{\Delta p}\|}{\|\mathfrak{G}\mathbf{E}(\mathbf{p})\|}\|\mathfrak{G}\mathbf{E}(\mathbf{p})\|^2$$

$$\approx E(\mathbf{p}) - \|\mathbf{\Delta p}\|\,\|\mathfrak{G}\mathbf{E}(\mathbf{p})\| \qquad (5.126)$$

Thus the step $\mathbf{\Delta p}$ should be taken in the steepest descent direction. The success of this method depends on scaling, that is, the selection of $\|\mathbf{\Delta p}\|$, and thus if the circuit parameters are compressed as in (5.113) the convergence will be faster than if the parameter spreads are considerably different. Usually, the steepest descent method is efficient if $\|\mathfrak{G}\mathbf{E}(\mathbf{p})\|$ does not vary much over the range of $\mathbf{p}$; otherwise the convergence can be very slow near a local minimum. If (5.122) is used, $\|\mathbf{\Delta p}\|$ can be replaced by $\|\mathbf{A}(\mathbf{p})\mathbf{\Delta p}\|$.

In the steepest descent method, the calculation of $\mathbf{\Delta p}$ depends on the selection of the scalar α in (5.125). One way of doing this is to start with a small value α_1 and calculate

$$\alpha_k = \alpha_1 + l\alpha_1 + l^2\alpha_1 + \cdots + l^{k-1}\alpha_1, \qquad k=1,2,\ldots \qquad (5.127)$$

until $E(\mathbf{p}+\alpha_k\mathbf{\Delta p}) > E(\mathbf{p}+\alpha_{k-1}\mathbf{\Delta p})$. The value of l can be chosen in the range $1.5 \leqslant l \leqslant 2$ typically. This type of calculation will result in three values α_{k-2}, α_{k-1} and α_k such that

$$E(\mathbf{p}+\alpha_M\mathbf{\Delta p}) < E(\mathbf{p}+\alpha_{k-1}\mathbf{\Delta p}) < E(\mathbf{p}+\alpha_{k-2}\mathbf{\Delta p}) < E(\mathbf{p}+\alpha_k\mathbf{\Delta p}) \qquad (5.128)$$

or

$$E(\mathbf{p}+\alpha_M\mathbf{\Delta p}) < E(\mathbf{p}+\alpha_{k-1}\mathbf{\Delta p}) < E(\mathbf{p}+\alpha_k\mathbf{\Delta p}) < E(\mathbf{p}+\alpha_{k-2}\mathbf{\Delta p}) \qquad (5.129)$$

where α_M is the value of α that minimizes E. Here we assume that α_M is unique, that is, E is strictly increasing or both sides of α_M. Using (5.128) and (5.129) we can approximate E by a quadratic polynomial that matches E at α_{k-2}, α_{k-1}, and α_k whose minimum $\hat{\alpha}_M$ is in the interval (α_{k-2}, α_k), that is,

$$\hat{\alpha}_M = \frac{1}{2}\,\frac{\begin{array}{l}(\alpha_{k-1}^2-\alpha_k^2)E(\mathbf{p}+\alpha_{k-2}\mathbf{\Delta p})\\ \quad +(\alpha_k^2-\alpha_{k-2}^2)E(\mathbf{p}+\alpha_{k-1}\mathbf{\Delta p})+(\alpha_{k-2}^2-\alpha_{k-1}^2)E(\mathbf{p}+\alpha_k\mathbf{\Delta p})\end{array}}{\begin{array}{l}(\alpha_{k-1}-\alpha_k)E(\mathbf{p}+\alpha_{k-2}\mathbf{\Delta p})\\ \quad +(\alpha_k-\alpha_{k-2})E(\mathbf{p}+\alpha_{k-1}\mathbf{\Delta p})+(\alpha_{k-2}-\alpha_{k-1})E(\mathbf{p}+\alpha_k\mathbf{\Delta p})\end{array}}$$

(5.130)

By comparing $E(\mathbf{p}+\hat{\alpha}_M\mathbf{\Delta p})$ and $E(\mathbf{p}+\alpha_{k-1}\mathbf{\Delta p})$ we can narrow the bound on α_M as follows.

For $\alpha_{k-1} > \hat{\alpha}_M$:

$$\alpha_{k-2} < \alpha_M < \alpha_{k-1} \quad \text{if } E(\mathbf{p}+\alpha_{k-1}\boldsymbol{\Delta}\mathbf{p}) > E(\mathbf{p}+\hat{\alpha}_M\boldsymbol{\Delta}\mathbf{p})$$

$$\hat{\alpha}_M < \alpha_M < \alpha_k \quad \text{if } E(\mathbf{p}+\hat{\alpha}_M\boldsymbol{\Delta}\mathbf{p}) > E(\mathbf{p}+\alpha_{k-1}\boldsymbol{\Delta}\mathbf{p})$$

For $\hat{\alpha}_M > \alpha_{k-1}$:

$$\alpha_{k-1} < \alpha_M < \alpha_k \quad \text{if } E(\mathbf{p}+\alpha_{k-1}\boldsymbol{\Delta}\mathbf{p}) > E(\mathbf{p}+\hat{\alpha}_M\boldsymbol{\Delta}\mathbf{p})$$

$$\alpha_{k-2} < \alpha_M < \hat{\alpha}_M \quad \text{if } E(\mathbf{p}+\hat{\alpha}_M\boldsymbol{\Delta}\mathbf{p}) > E(\mathbf{p}+\alpha_{k-1}\boldsymbol{\Delta}\mathbf{p})$$

By continuing this way, the procedure will converge to α_M.

Newton Method

In this method the second and the third terms of (5.115) are retained, that is,

$$E(\mathbf{p}+\boldsymbol{\Delta}\mathbf{p}) \approx E(\mathbf{p}) + \left[\mathfrak{G}\mathbf{E}(\mathbf{p})\right]^T \boldsymbol{\Delta}\mathbf{p} + \tfrac{1}{2}(\boldsymbol{\Delta}\mathbf{p})^T \mathbf{H}\,\boldsymbol{\Delta}\mathbf{p} \tag{5.131}$$

We note that $E(\mathbf{p}+\boldsymbol{\Delta}\mathbf{p})$ is minimum when $\partial E(\mathbf{p}+\boldsymbol{\Delta}\mathbf{p})/\partial \Delta p_i = 0$ for $i = 1, 2, \ldots, n$. Using the matrix method, the following vector equation is obtained:

$$\mathfrak{G}\mathbf{E}(\mathbf{p}) + \mathbf{H}\,\boldsymbol{\Delta}\mathbf{p} = 0 \tag{5.132}$$

Equation (5.132) represents a system of n linear equations in n unknowns Δp_i and can be solved by standard techniques such as Gaussian elimination or Cholesky's method. Notice that the symmetric matrix $\mathbf{H}$ will approach a positive semidefinite matrix as $E(\mathbf{p}+\boldsymbol{\Delta}\mathbf{p})$ approaches its minimum $E(\mathbf{p}_M)$, since near this minimum, $\|\mathfrak{G}\mathbf{E}(\mathbf{p})\|$ approaches zero and thus $E(\mathbf{p}+\boldsymbol{\Delta}\mathbf{p}) \approx E(\mathbf{p}_M) + \frac{1}{2}(\boldsymbol{\Delta}\mathbf{p})^T \mathbf{H}\,\boldsymbol{\Delta}\mathbf{p} \geqslant E(\mathbf{p}_M)$, and we conclude that $\mathbf{H} \geqslant 0$.

The Newton method is effective if the error criterion is quadratic in nature, that is, its third and higher order derivatives are small such that (5.131) holds over the range of $\mathbf{p}$. In such a case convergence can be very fast although the calculation of $\mathbf{H}$ that involves the second derivative can be time consuming.

Since the steepest descent method is usually excellent for poor initial values—the region where the error surface is near planar—and can be very slow when it approaches the minimum, where the surface is likely quadratic and for which the Newton method can be very effective, a combination of the two methods can be used to obtain fast convergence as described below.

Davidson-Fletcher-Powell Method

Returning to (5.131) we see that the Newton method requires the calculation of the second order derivative matrix $\mathbf{H}$ and its inverse $\mathbf{H}^{-1}$ if it is nonsingular in order to find $\boldsymbol{\Delta}\mathbf{p}$ as in (5.132), that is, $\boldsymbol{\Delta}\mathbf{p} = -\mathbf{H}^{-1}\mathfrak{G}\mathbf{E}(\mathbf{p})$. The Davidson-Fletcher-Powell method instead uses the knowledge of $E(\mathbf{p})$ and $\mathfrak{G}\mathbf{E}(\mathbf{p})$

at the previous iteration to improve the current iteration and thus avoid finding the inverse matrix and the second-order derivatives. From (5.131) assume that the $n \times n$ Hessian matrix $\mathbf{H}$ is positive-definite. Then $\mathfrak{G}\mathbf{E}$ has a unique minimum. At any point $\mathbf{p}^{(j+1)} = \mathbf{p}^{(j)} + \Delta\mathbf{p}^{(j)}$ the gradient is

$$\mathfrak{G}\mathbf{E}(\mathbf{p}^{(j+1)}) = \mathfrak{G}\mathbf{E}(\mathbf{p}^{(j)}) + \mathbf{H}\,\Delta\mathbf{p}^{(j)} \tag{5.133}$$

Assume $\mathbf{H}$ is known, then

$$\Delta\mathbf{p}^{(j)} = \mathbf{H}^{-1}\left[\mathfrak{G}\mathbf{E}(\mathbf{p}^{(j+1)}) - \mathfrak{G}\mathbf{E}(\mathbf{p}^{(j)})\right] \tag{5.134}$$

At the minimum $\mathbf{p}^{(j+1)} = \mathbf{p}_M$ the gradient vanishes, that is, $\mathfrak{G}\mathbf{E}(\mathbf{p}^{(j+1)} = \mathbf{p}_M) = \mathbf{0}$ thus

$$\Delta\mathbf{p}_M = -\mathbf{H}^{-1}\mathfrak{G}\mathbf{E}(\mathbf{p}^{(j)}) \tag{5.135}$$

Subtracting (5.134) from (5.135) yields

$$\Delta\mathbf{p}_M = \Delta\mathbf{p}^{(j)} \tag{5.136}$$

From (5.136) we see that the optimum incremental size can be obtained in one step in the direction indicated by (5.135) if $\mathbf{H}^{-1}$ is known. Note that this is not along the direction of the gradient unless $\mathbf{H}^{-1} = \alpha\mathbf{1}_n$ where $\alpha > 0$ is a scalar. Fletcher and Powell implement (5.136) by using a matrix $\mathbf{K}_j > \mathbf{0}$ at the jth iteration such that

$$\mathbf{x}^{(j+1)} = \Delta\mathbf{p}^{(j+1)} - \Delta\mathbf{p}^{(j)} = -\alpha_j\mathbf{K}_j\,\mathfrak{G}\mathbf{E}(\mathbf{p}^{(j)}) \tag{5.137}$$

The scalar α_j is chosen to optimize E along the line of search. Usually the starting value for $\mathbf{K}_0$ is the unit matrix (this means that the first operation is the steepest descent method). The method generates m vectors $\Delta\mathbf{p}^{(1)}, \ldots, \Delta\mathbf{p}^{(m)}$ such that the gradient $\mathfrak{G}\mathbf{E}(\mathbf{p}^{(m)})$ is orthogonal to all the preceding $\mathbf{x}^{(1)}, \ldots, \mathbf{x}^{(m)}$, that is,

$$\left[\mathfrak{G}\mathbf{E}(\mathbf{p}^{(m)})\right]^T \mathbf{x}^{(j)} = \mathbf{0},\ j = 1, 2, \ldots, m;\ m = 1, 2, \ldots, n \tag{5.138}$$

This means that if $\mathbf{x}^{(1)}, \ldots, \mathbf{x}^{(n)}$ are linearly independent, the gradient $\mathfrak{G}\mathbf{E}(\mathbf{p}^{(n)})$ must vanish, that is, the optimum $\mathbf{p}^{(n)}$ is found. Furthermore the sequence of $\mathbf{K}_j > \mathbf{0}$, $j = 1, 2, \ldots, n$ is constructed such that

$$\mathbf{K}_n = \mathbf{H}^{-1} \tag{5.139}$$

In practice $\mathfrak{G}\mathbf{E}(\mathbf{p}^{(n)})$ may not vanish, for example, due to round-off error. Then the $(n+1)$-th step can be carried out as

$$\mathbf{x}^{(n+1)} = \Delta\mathbf{p}^{(n+1)} - \Delta\mathbf{p}^{(n)} = -\alpha_{n+1}\mathbf{K}_n \mathfrak{G}\mathbf{E}(\mathbf{p}^{(n)})$$

$$= -\alpha_{n+1}\mathbf{H}^{-1}\mathfrak{G}\mathbf{E}(\mathbf{p}^{(n)})$$

$$= \alpha_{n+1}(\Delta\mathbf{p}_M - \Delta\mathbf{p}^{(n)}) \tag{5.140}$$

Thus if $\alpha_{n+1}=1$ then $\Delta\mathbf{p}^{(n+1)}=\Delta\mathbf{p}_M$, which is the optimum step size after $n+1$ iterations. In practice $\mathfrak{G}\mathbf{E}$ is not usually quadratic but after n iterations the matrix $\mathbf{K}_n$ is usually a good approximation of the inverse of the Hessian matrix $\mathbf{H}$ that contains the second derivatives, and will converge rapidly as soon as $\mathfrak{G}\mathbf{E}$ becomes near quadratic. In the following section, the procedure for obtaining the matrix $\mathbf{K}_{j+1}$ at the $(j+1)$th iteration from the matrix $\mathbf{K}_j$ at the jth iteration is discussed. Let

$$\mathbf{K}_{j+1} = \mathbf{K}_j + \mathbf{A}_j + \mathbf{B}_j \tag{5.141}$$

The matrix $\mathbf{A}_j$ is used to generate $\mathbf{H}^{-1}$ in n steps and $\mathbf{B}_j$ is a correction term used to cancel out the initial starting value $\mathbf{K}_0$, that is,

$$\sum_{j=0}^{n-1} \mathbf{A}_j = \mathbf{H}^{-1} \tag{5.142}$$

$$\sum_{j=0}^{n-1} \mathbf{B}_j = -\mathbf{K}_0 \tag{5.143}$$

By the above definition it can be easily seen that $\mathbf{K}_n = \mathbf{H}^{-1}$, that is,

$$\mathbf{K}_n = \mathbf{K}_{n-1} + \mathbf{A}_{n-1} + \mathbf{B}_{n-1}$$

$$= \mathbf{K}_{n-2} + (\mathbf{A}_{n-1} + \mathbf{A}_{n-2}) + (\mathbf{B}_{n-1} + \mathbf{B}_{n-2})$$

$$= \mathbf{K}_0 + \sum_{j=0}^{n} \mathbf{A}_j + \sum_{j=0}^{n} \mathbf{B}_j = \mathbf{H}^{-1} \tag{5.144}$$

Now define

$$\mathbf{y}^{(j)} = \mathfrak{G}\mathbf{E}(\mathbf{p}^{(j+1)}) - \mathfrak{G}\mathbf{E}(\mathbf{p}^{(j)}) \tag{5.145}$$

By (5.134) the quadratic form E is minimized if $\mathbf{H}$ satisfies

$$\mathbf{H}^{-1}\mathbf{y}^{(j)} = \Delta\mathbf{p}^{(j)} \tag{5.146}$$

In other words, $\mathbf{K}_{j+1}$ must be found such that

$$\mathbf{K}_{j+1}\mathbf{y}^{(i)}=\mathbf{\Delta p}^{(i)} \tag{5.147}$$

where $\mathbf{K}_{j+1}$ is given in (5.141). The matrix $\mathbf{A}_j$ and $\mathbf{B}_j$ must be found such that $\mathbf{K}_{j+1}$ is positive-definite and symmetric to guarantee convergence. Substituting (5.141) into (5.147) yields

$$(\mathbf{K}_j+\mathbf{A}_j+\mathbf{B}_j)\mathbf{y}^{(i)}=\mathbf{\Delta p}^{(i)} \tag{5.148}$$

A simple choice is

$$\mathbf{A}_j\mathbf{y}^{(i)}=\mathbf{\Delta p}^{(i)} \tag{5.149a}$$

$$\mathbf{B}_j\mathbf{y}^{(i)}=-\mathbf{K}_j\mathbf{y}^{(i)} \tag{5.149b}$$

(5.149a) can be satisfied by selecting

$$\mathbf{A}_j=\frac{\mathbf{\Delta p}^{(i)}\left[\mathbf{\Delta p}^{(i)}\right]^T}{\left[\mathbf{\Delta p}^{(i)}\right]^T\mathbf{y}^{(i)}} \tag{5.150}$$

It is obvious that the solution of (5.149b) is $\mathbf{B}_j=-\mathbf{K}_j$, but this then would give $\mathbf{K}_{j+1}=\mathbf{A}_j$ which is not positive-definite. For this reason, the following selection is made

$$\mathbf{B}_j=\frac{-\mathbf{K}_j\mathbf{y}^{(i)}\left[\mathbf{y}^{(i)}\right]^T\mathbf{K}_j^T}{\left[\mathbf{y}^{(i)}\right]^T\mathbf{K}_j\mathbf{y}^{(i)}} \tag{5.151}$$

which satisfies (5.149b). Substitution of (5.150) and (5.151) into (5.141) gives

$$\mathbf{K}_{j+1}=\mathbf{K}_j+\frac{\mathbf{\Delta p}^{(i)}\left[\mathbf{\Delta p}^{(i)}\right]^T}{\left[\mathbf{\Delta p}^{(i)}\right]^T\mathbf{y}^{(i)}}-\frac{\mathbf{K}_j\mathbf{y}^{(i)}\left[\mathbf{y}^{(i)}\right]^T\mathbf{K}_j^T}{\left[\mathbf{y}^{(i)}\right]^T\mathbf{K}_j\mathbf{y}^{(i)}} \tag{5.152}$$

To show convergence, note that for $\mathbf{p}^{(i+1)}=\mathbf{p}^{(i)}-\alpha_j\mathbf{K}_j\,\mathfrak{G}\mathbf{E}(\mathbf{p}^{(i)})$, we have

$$\left.\frac{\partial}{\partial\alpha_j}\mathbf{E}(\mathbf{p}^{(i+1)})\right|_{\alpha_j=0}=-\left[\mathfrak{G}\mathbf{E}(\mathbf{p}^{(i)})\right]^T\mathbf{K}_j\,\mathfrak{G}\mathbf{E}(\mathbf{p}^{(i)}) \tag{5.153a}$$

$$\left.\frac{\partial}{\partial\alpha_j}\mathbf{E}(\mathbf{p}^{(i+1)})\right|_{\alpha_j=\alpha_M}=-\left[\mathfrak{G}\mathbf{E}(\mathbf{p}^{(i+1)}=\mathbf{p}_M)\right]^T\mathbf{K}_j\,\mathfrak{G}\mathbf{E}(\mathbf{p}^{(i)})=0 \tag{5.153b}$$

where α_M is the minimizing value of α_j. The derivative in (5.153a) will be negative if $\mathbf{K}_j > \mathbf{0}$. By using inductive proof we will show that $\mathbf{K}_{j+1} > \mathbf{0}$, that is, $\mathbf{g}^T\mathbf{K}_{j+1}\mathbf{g} > 0$ for all vector $\mathbf{g}$ if $\mathbf{g}^T\mathbf{K}_j\mathbf{g} > 0$. From (5.151) we note that $\mathbf{B}_j < \mathbf{0}$ if $\mathbf{K}_j > \mathbf{0}$, thus

$$\mathbf{g}^T\mathbf{K}_{j+1}\mathbf{g} \geqslant \mathbf{g}^T\mathbf{K}_j\mathbf{g} + \mathbf{g}^T\mathbf{A}_j\mathbf{g} \tag{5.154}$$

From (5.150), we see that $\mathbf{g}^T\mathbf{A}_j\mathbf{g} > 0$ if $[\Delta\mathbf{p}^{(i)}]^T\mathbf{y}^{(i)} > 0$, that is,

$$\left[\Delta\mathbf{p}^{(i)}\right]^T\mathbf{y}^{(i)} = \left[\Delta\mathbf{p}^{(i)}\right]^T\left[\nabla\mathbf{E}(\mathbf{p}^{(i+1)}) - \nabla\mathbf{E}(\mathbf{p}^{(i)})\right] \tag{5.155}$$

According to (5.137), if $\alpha_i = \alpha_M$, then $\nabla\mathbf{E}(\mathbf{p}^{(i+1)}) = \mathbf{0}$ and $\Delta\mathbf{p}^{(i)} = -\alpha_M\mathbf{K}_j \cdot \nabla\mathbf{E}(\mathbf{p}^{(i)})$. Thus since $\mathbf{K}_j = \mathbf{K}_j^T > \mathbf{0}$, we have

$$\left[\Delta\mathbf{p}^{(i)}\right]^T\mathbf{y}^{(i)} = \alpha_M\left[\nabla\mathbf{E}(\mathbf{p}^{(i)})\right]^T\mathbf{K}_j\,\nabla\mathbf{E}(\mathbf{p}^{(i)}) > 0 \tag{5.156}$$

Since $\mathbf{A}_j > \mathbf{0}$, $\mathbf{K}_{j+1} > \mathbf{0}$ according to (5.154).

In conclusion, if $\mathbf{K}_0$ is chosen as the unit matrix, then the first iteration is the same as the steepest descent method. If the initial value is far from the minimum, convergence will be fast. When the minimum is approached the error function behaves quadratically and convergence will be obtained in a finite number of iterations. Thus the Davidson-Fletcher-Powell method provides the gradual transition from the steepest descent method to the Newton method without calculating the Hessian matrix and its inverse. Indeed it approximates the inverse Hessian matrix in a continuous manner.

Least-Square Method

When the least-square error criterion is specified, then the quantity to be minimized by the optimization is $(k > n)$

$$E(\mathbf{p}) = \sum_{i=1}^{k} |F_d(\omega_i) - F_c(\omega_i, \mathbf{p})|^2$$

$$= \sum_{i=1}^{k} \{e_i(\mathbf{p})\}^2 \tag{5.157}$$

Defining the vector $\mathbf{e}(\mathbf{p}) = [e_1(\mathbf{p}), e_2(\mathbf{p}), \ldots, e_k(\mathbf{p})]^T$, then (5.157) can be written as

$$E(\mathbf{p}) = \mathbf{e}^T\mathbf{e} \tag{5.158}$$

and

$$\nabla\mathbf{E}(\mathbf{p}) = 2\mathbf{J}^T\mathbf{e} \tag{5.159}$$

where

$$\mathbf{J}=\begin{bmatrix} \frac{\partial e_1}{\partial p_1} & \frac{\partial e_1}{\partial p_2} & \cdots & \frac{\partial e_1}{\partial p_n} \\ \frac{\partial e_2}{\partial p_1} & \frac{\partial e_2}{\partial p_2} & \cdots & \frac{\partial e_2}{\partial p_n} \\ \cdots & \cdots & \cdots & \cdots \\ \frac{\partial e_k}{\partial p_1} & \frac{\partial e_k}{\partial p_2} & \cdots & \frac{\partial e_k}{\partial p_n} \end{bmatrix} \tag{5.160}$$

is the $k \times n$ Jacobian matrix. Using the first two terms of a Taylor series expansion we get

$$\mathbf{e}(\mathbf{p}+\Delta\mathbf{p}) \approx \mathbf{e}(\mathbf{p})+\mathbf{J}\Delta\mathbf{p} \tag{5.161}$$

Using (5.159) and (5.161) we arrive at

$$\nabla E(\mathbf{p}+\Delta\mathbf{p}) \approx 2\mathbf{J}^T[\mathbf{e}(\mathbf{p})+\mathbf{J}\Delta\mathbf{p}] \tag{5.162}$$

Thus for $\nabla E(\mathbf{p}+\Delta\mathbf{p})=\mathbf{0}$, we must have

$$\mathbf{J}^T\mathbf{e}(\mathbf{p})+\mathbf{J}^T\mathbf{J}\Delta\mathbf{p}=\mathbf{0} \tag{5.163}$$

Note that $\mathbf{J}^T\mathbf{J}$ is an $n \times n$ matrix of rank n. Thus its inverse exists and

$$\begin{aligned}\Delta\mathbf{p} &= -[\mathbf{J}^T\mathbf{J}]^{-1}\mathbf{J}^T\mathbf{e}(\mathbf{p}) \\ &= -\tfrac{1}{2}[\mathbf{J}^T\mathbf{J}]^{-1}\nabla E(\mathbf{p})\end{aligned} \tag{5.164}$$

Thus the term $\mathbf{J}^T\mathbf{J}$ corresponds to the Hessian matrix discussed previously. $E(\mathbf{p})$ is minimized when $[\mathbf{J}^T\mathbf{J}]^{-1}$ is positive-definite, which is generally true under the assumption shown in (5.161). In practice, to avoid divergence, the jth iteration is often taken as

$$\mathbf{p}^{(j+1)}=\mathbf{p}^{(j)}+\alpha_j\Delta\mathbf{p}^{(j)} \tag{5.165}$$

where α_j, as in the previous methods, is chosen to minimize $E(\mathbf{p}^{(j+1)})$. It is necessary to choose $\alpha_j<1$ as the minimum is approached to avoid divergence.

5.7 REAL FREQUENCY BROADBAND MATCHING TECHNIQUE

In section 5.2 the broadband matching of microwave transistors was treated as the matching of a series $R_iL_iC_i$ impedance to a resistive source and of a parallel-series R_oC_o-L_o impedance to a resistive load by assuming that the transistor is unilateral. Once the transducer power gain characteristic $G(\omega^2)$ is chosen that satisfies the load constraints; the topology of the matching network is fixed by $G(\omega^2)$. In this section, we present another procedure for designing the matching networks using the scattering parameter data directly without approximating by the unilateral model. The procedure is shown schematically in Figs. 5.31*a–b*. Denote the following quantities as

(a) $\mathbf{S}=[S_{ij}]$ is the transistor scattering matrix normalized to 50 Ω.

(b) $\mathbf{Z}_{2i}$ is the positive-real impedance at port 2 of the input matching network N_i ($Y_{2i}=1/Z_{2i}$ is the corresponding admittance).

(c) Z_1 is the transistor's input impedance when it is terminated by 50 Ω at the output ($Y_1=1/Z_1$ is the corresponding admittance). Z_1 is given as

$$Z_1=Z_0\frac{1+S_{11}}{1-S_{11}}, \qquad Z_0=50\ \Omega \tag{5.166}$$

(d) ρ_1 is the reflection coefficient at port 2 of N_i, normalized to Z_1 and is given

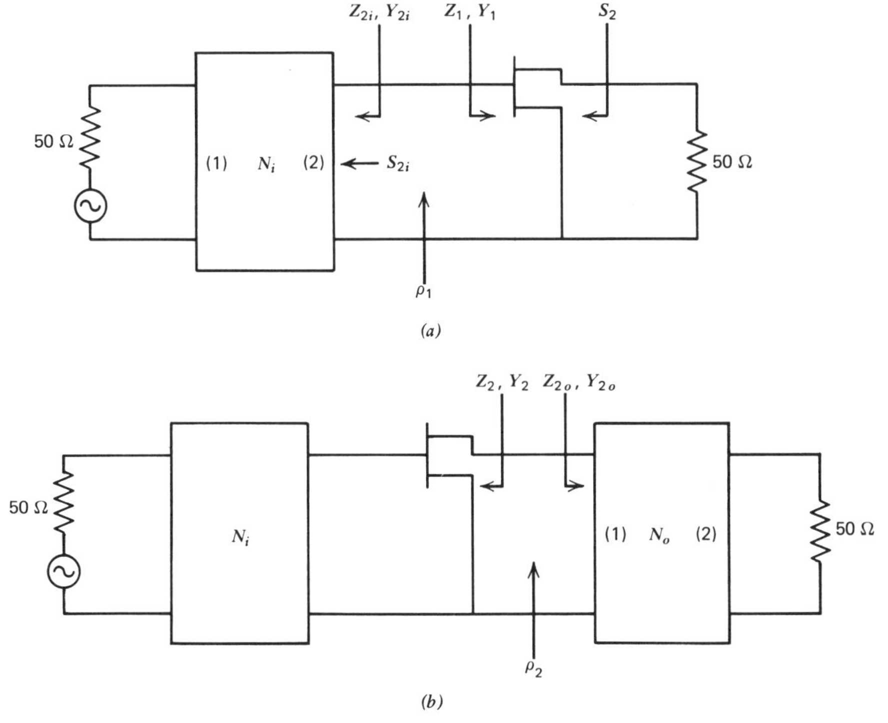

Fig. 5.31. (*a*) An amplifier with an input matching network. (*b*) An amplifier with input and output matching networks.

by (5.11) as

$$\rho_1 = \frac{Z_{2i} - Z_1^*}{Z_{2i} + Z_1} \tag{5.167}$$

(e) Z_{2o} is the positive-real impedance at port 1 of the output matching network N_o ($Y_{2o} = 1/Z_{2o}$ is the corresponding admittance).

(f) Z_2 is the transistor's output impedance when its input is terminated in N_i and 50 Ω (Y_2 is the corresponding admittance). Z_2 is given as

$$Z_2 = Z_0 \frac{1 + S_2}{1 - S_2}, \qquad Z_0 = 50\ \Omega \tag{5.168a}$$

where S_2 is given by the cascade-load formula in (1.21) as

$$S_2 = S_{22} + \frac{S_{21} S_{12} S_{2i}}{1 - S_{11} S_{2i}} \tag{5.168b}$$

and

$$S_{2i} = \frac{Z_{2i} - Z_0}{Z_{2i} + Z_0} \tag{5.168c}$$

(g) ρ_2 is the reflection coefficient at port 1 of N_o, normalized to Z_2 and is given by (5.11) as

$$\rho_2 = \frac{Z_{2o} - Z_2^*}{Z_{2o} + Z_2} \tag{5.169}$$

From Fig. 5.31*a* we see that the transducer power gain $G_1(\omega^2)$ of the network is the product of the transducer power gain of the input matching network N_i, which is $1 - |\rho_1|^2$ according to (5.10), and the power gain of the transistor, which is given by (4.4) as $|S_{21}|^2/(1 - |S_{11}|^2)$. Thus

$$\begin{aligned} G_1(\omega^2) &= \frac{\text{power delivered to 50 } \Omega \text{ load}}{\text{power available from 50 } \Omega \text{ source}} \\ &= \frac{\text{power delivered to transistor input}}{\text{power available from 50 } \Omega \text{ source}} \cdot \frac{\text{power delivered to 50 } \Omega \text{ load}}{\text{power delivered to transistor input}} \\ &= (1 - |\rho_1|^2) \frac{|S_{21}|^2}{1 - |S_{11}|^2} \end{aligned} \tag{5.170}$$

Similarly, from Fig. 5.31*b* we see that the transducer power gain $G(\omega^2)$ of the network is given as

$$G(\omega^2) = \frac{\text{power delivered to 50 }\Omega\text{ load}}{\text{power available from 50 }\Omega\text{ source}}$$

$$= \frac{\text{power delivered to transistor input}}{\text{power available from 50 }\Omega\text{ source}}$$

$$\cdot \frac{\text{power available at transistor output}}{\text{power delivered to transistor input}}$$

$$\cdot \frac{\text{power delivered to 50 }\Omega\text{ load}}{\text{power available at transistor output}}$$

But

$$\frac{\text{power available at transistor output}}{\text{power delivered to transistor input}} = \frac{\text{power delivered to 50 }\Omega\text{ reference}}{\text{power delivered to transistor input}}$$

$$\cdot \frac{1}{\dfrac{\text{power delivered to 50 }\Omega\text{ reference}}{\text{power available at transistor output}}}$$

$$= \frac{|S_{21}|^2}{1-|S_{11}|^2} \cdot \frac{1}{1-|S_2|^2}$$

since the power delivered to the 50 Ω reference termination is equal to the power available at the transistor output minus the reflected power from the termination, and since the ratio of the termination reflected power to its available power is $|S_2|^2$. Thus

$$G(\omega^2) = (1-|\rho_1|^2)\frac{|S_{21}|^2}{(1-|S_{11}|^2)(1-|S_2|^2)}(1-|\rho_2|^2)$$

$$= \frac{|S_{21}|^2}{1-|S_{11}|^2}(1-|\rho_1|^2)\frac{1-|\rho_2|^2}{1-|S_2|^2} \tag{5.171}$$

Equation (5.171) is the fundamental equation for the design. We note that $|S_{21}|^2/(1-|S_{11}|^2)$ is a fixed quantity since it is determined solely by the transistor scattering parameters. Thus once $G(\omega^2)$ is specified we can design N_i and N_o by determining Z_1 first from the assigned value of $1-|\rho_1|^2$ and then Z_2 from the assigned value of $1-|\rho_2|^2$ such that the product of the three terms on the right of (5.171) is $G(\omega^2)$. For example, the designer may choose to specify $1-|\rho_1|^2$ for flat response and then $1-|\rho_2|^2$ for gain slope compensation or vice versa. Furthermore, for noise matching networks, the designer can specify $1-|\rho_1|^2$ for minimum noise figure where Z_1 is now the transistor's noise impedance.

From (5.167) we have

$$1-|\rho_1(j\omega)|^2 = 1-\left|\frac{Z_{2i}-Z_1^*}{Z_{2i}+Z_1}\right|^2 = 1-\left|\frac{Y_{2i}^*-Y_1}{Y_{2i}+Y_1}\right|^2 \tag{5.172}$$

Once $1-|\rho_1|^2$ is specified, $|\rho_1|$ is also specified and the problem reduces to finding Z_{2i} such that the error $E(\omega)$ defined below is minimized

$$E(\omega) = \sum_{i=1}^{k}\left|\left|\frac{Z_{2i}(j\omega_i)-Z_1^*(j\omega_i)}{Z_{2i}(j\omega_i)+Z_1(j\omega_i)}\right|^2 - |\rho_1(j\omega_i)|^2\right|^2 \tag{5.173}$$

If the form of $Z_{2i}(j\omega)$ is specified with unknown coefficients, $E(\omega)$ can be minimized using the least-square method. One way to specify $Z_{2i}(j\omega)=R_{2i}(\omega)+jX_{2i}(\omega)$ is to represent $R_{2i}(\omega)$ and $X_{2i}(\omega)$ by a sum of semi-infinite constant slope characteristics to be used in (5.173) to minimize $E(\omega)$. Once $E(\omega)$ is minimized, the optimized $R_{2i}(\omega)$ will then be approximated by a rational function $\hat{R}_{2i}(\omega)$ using the least-square method, for example, and the rational function $\hat{Z}_{2i}$ that approximates Z_{2i} can now be found from $\hat{R}_{2i}(\omega)$ by a known method. To represent $R_{2i}(\omega)$, the designer can select a number of N break point frequency ω_m (N is usually much less than k, the number of sampling frequencies) from $\omega=0$ to ω_N to cover the frequency band of interest. Then $R_{2i}(\omega)$ can be represented by a sum of line segments each with unknown resistance excursion r_m as follows (note that R_{2i} must be positive for all ω for realizability)

$$R_{2i}(\omega) = r_0 + \sum_{m=1}^{N} r_m a_m(\omega) = r_0 + \mathbf{a}^T(\omega)\mathbf{r} \tag{5.174}$$

where $\mathbf{a}^T(\omega)=[a_1(\omega), a_2(\omega), \ldots, a_N(\omega)]$ and $\mathbf{r}^T=[r_1, r_2, \ldots, r_N]$ and

$$a_m(\omega) = \begin{cases} 0, & \omega<\omega_{m-1} \\ \dfrac{\omega-\omega_{m-1}}{\omega_m-\omega_{m-1}}, & \omega_{m-1}<\omega<\omega_m \\ 1, & \omega>\omega_m \end{cases} \tag{5.175}$$

It is noted that $r_0=R_{2i}(0)$ and since $R_{2i}(\omega)=0$ for $\omega>\omega_N$,

$$r_0 + \sum_{m=1}^{N} r_m = 0 \tag{5.176}$$

An example of such representation is shown in Fig. 5.32*a–c*. The representation

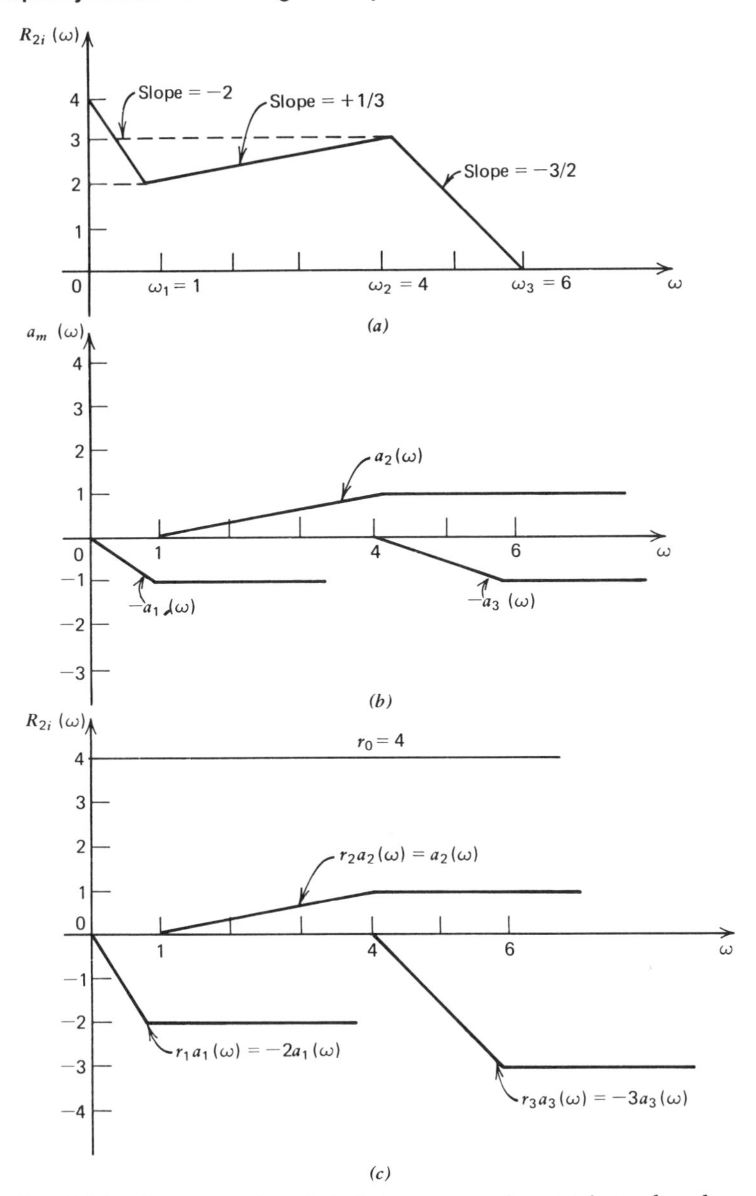

Fig. 5.32. Representation of $R_{2i}(\omega)$ as a sum of semi-infinite slope lines.

of the minimum reactance $X_{2i}(\omega)$ can be obtained from $R_{2i}(\omega)$ by Hilbert transform techniques as [1]

$$X_{2i}(\omega) = \frac{1}{\pi}\int_0^\infty \frac{dR_{2i}(\omega)}{d\omega} \ln\left|\frac{x+\omega}{x-\omega}\right| dx \tag{5.177}$$

$$X_{2i}(\omega) = \sum_{m=1}^{N} r_m b_m(\omega) = \mathbf{b}^T(\omega)\mathbf{r} \tag{5.178}$$

where

$$b_m(\omega)=\frac{1}{\pi(\omega_m-\omega_{m-1})}\int_{\omega_{m-1}}^{\omega_m}\ln\left|\frac{x+\omega}{x-\omega}\right|dx$$

$$=\frac{1}{\pi(\omega_m-\omega_{m-1})}\left[\int_0^{\omega_m}\ln\left|\frac{x+\omega}{x-\omega}\right|dx-\int_0^{\omega_{m-1}}\ln\left|\frac{x+\omega}{x-\omega}\right|d\omega\right] \tag{5.179}$$

A closed form solution of (5.179) is given as [1]

$$b_m(\omega)=\frac{K_{m1}-K_{m2}}{\pi(\omega_m-\omega_{m-1})} \tag{5.180}$$

where

$$K_{m1}=\int_0^{\omega_m}\ln\left|\frac{x+\omega}{x-\omega}\right|dx=\omega_m\left[\left(\frac{\omega}{\omega_m}+1\right)\ln\left(\frac{\omega}{\omega_m}+1\right)\right.$$
$$\left.+\left(\frac{\omega}{\omega_m}-1\right)\ln\left|\frac{\omega}{\omega_m}-1\right|-2\frac{\omega}{\omega_m}\ln\frac{\omega}{\omega_m}\right]$$

$$K_{m2}=\int_0^{\omega_{m-1}}\ln\left|\frac{x+\omega}{x-\omega}\right|dx=\omega_{m-1}\left[\left(\frac{\omega}{\omega_{m-1}}+1\right)\ln\left(\frac{\omega}{\omega_{m-1}}+1\right)\right.$$
$$\left.+\left(\frac{\omega}{\omega_{m-1}}-1\right)\ln\left|\frac{\omega}{\omega_{m-1}}-1\right|-2\frac{\omega}{\omega_{m-1}}\ln\frac{\omega}{\omega_{m-1}}\right]$$

As an example, consider $R_{2i}(\omega)$ in Fig. 5.32a; according to (5.180) we have

$$b_1(\omega)=\frac{1}{\pi}[(\omega+1)\ln(\omega+1)+(\omega-1)\ln|\omega-1|-2\omega\ln\omega]$$

$$b_2(\omega)=\frac{1}{3\pi}\left[4\left\{\left(\frac{\omega}{4}+1\right)\ln\left(\frac{\omega}{4}+1\right)+\left(\frac{\omega}{4}-1\right)\ln\left|\frac{\omega}{4}-1\right|-2\frac{\omega}{4}\ln\frac{\omega}{4}\right\}\right.$$
$$\left.-\{(\omega+1)\ln(\omega+1)+(\omega-1)\ln|\omega-1|-2\omega\ln\omega\}\right]$$

$$b_3(\omega)=\frac{1}{2\pi}\left[6\left\{\left(\frac{\omega}{6}+1\right)\ln\left(\frac{\omega}{6}+1\right)+\left(\frac{\omega}{6}-1\right)\ln\left|\frac{\omega}{6}-1\right|-2\frac{\omega}{6}\ln\frac{\omega}{6}\right\}\right.$$
$$\left.-4\left\{\left(\frac{\omega}{4}+1\right)\ln\left(\frac{\omega}{4}+1\right)+\left(\frac{\omega}{4}-1\right)\ln\left|\frac{\omega}{4}-1\right|-2\frac{\omega}{4}\ln\frac{\omega}{4}\right\}\right]$$

Using (5.174) and (5.178), (5.173) can be rewritten as

$$E(\omega)=\sum_{i=1}^{k}\left|\frac{4R_{2i}(\omega_i)R_1(\omega_i)}{\left[R_{2i}(\omega_i)+R_1(\omega_i)\right]^2+\left[X_{2i}(\omega_i)+X_2(\omega_i)\right]^2}-\left(1-|\rho_1(\omega_i)|^2\right)\right|^2 \tag{5.181a}$$

$$E(\omega)=\sum_{i=1}^{k}\left|\frac{4\left[r_0+\mathbf{a}^T(\omega_i)\mathbf{r}\right]R_1(\omega_i)}{\left[r_0+\mathbf{a}^T(\omega_i)\mathbf{r}+R_1(\omega_i)\right]^2+\left[\mathbf{b}^T(\omega_i)\mathbf{r}+X_2(\omega_i)\right]^2}-\left(1-|\rho_1(\omega_i)|^2\right)\right|^2 \tag{5.181b}$$

When r_m, $m=0,1,\ldots,N$ are determined, R_{2i} can be approximated by a rational function $\hat{R}_{2i}$. From $\hat{R}_{2i}$ the rational and positive-real impedance Z_{2i} can be obtained by a known method [2]. For example if the designer wishes to obtain a lowpass ladder network N_i, then $\hat{R}_{2i}$ can take the form

$$\hat{R}_{2i}=\frac{1}{a_n\omega^{2n}+a_{n-1}\omega^{2(n-1)}+\cdots+a_1\omega^2+a_0}$$

for a lumped network and

$$\hat{R}_{2i}=\frac{(1+\Omega^2)^n}{P_m(\Omega^2)}$$

for a cascade of n commensurate lines and $m-n$ stubs of electrical length θ where $\Omega=\tan\theta$ and $P_m(\Omega^2)$ is an even strictly positive polynomial of degree $2m\geqslant 2n$ [3].

The output matching network N_o can be designed in a similar fashion from $1-|\rho_2|^2$. For the input noise matching network, equation (2.59a) is used where $G_s=G_{2i}$ and $B_s=B_{2i}$ ($Y_{2i}=G_{2i}+jB_{2i}$).

5.8 BROADBAND MICROWAVE AMPLIFIER DESIGN USING POTENTIALLY UNSTABLE TRANSISTORS

The presence of internal feedback in microwave transistors such as the gate-to-drain capacitance in the FET and the base-to-collector capacitance in the bipolar transistor causes the devices to be potentially unstable at certain frequency ranges and they can go into oscillation with certain combinations of source and load impedances. In the case of the GaAs FET, the region of unstability can lie below 3 GHz, however its low noise figure (1 dB at 2 GHz for a 0.5 μ gate length device) and higher gain compared with the bipolar device make it attractive for a broadband amplifier from 1 to 3 GHz if the device can be made unconditionally stable in the frequency band of interest, by some form of lossless feedback to offset the internal feedback. One way to do this is to use source inductance series feedback for the GaAs FET or emitter inductance series feedback for the bipolar transistor as is shown in Fig. 5.33. The use of this type of feedback usually reduces the gain by a small amount and if L can be selected appropriately, the minimum noise figure of the device can be kept the same. An example of this type of series feedback is shown in Tables 5.20 and 5.21 for the GaAs FET HFET-1000 with $L=0.32$ nH, where the potentially unstable device is made unconditionally stable. Note that the transducer power gain G_{max} under the simultaneously matched condition given by (2.50b) is very high compared to that of bipolar transistors in the same frequency range. The value of L that makes the device unconditionally stable can be found easily by using the series connection. The scattering matrix $\mathbf{S}_1$ of the transistor can be converted to the impedance matrix $\mathbf{Z}_1$ by the relation (with Z_0 as the normalized characteristic impedance)

$$\mathbf{Z}_1 = Z_0(\mathbf{1}_2 + \mathbf{S}_1)(\mathbf{1}_2 - \mathbf{S}_1)^{-1} \tag{5.182}$$

The impedance matrix $\mathbf{Z}_2$ of the two-port that has L as a shunt inductance is

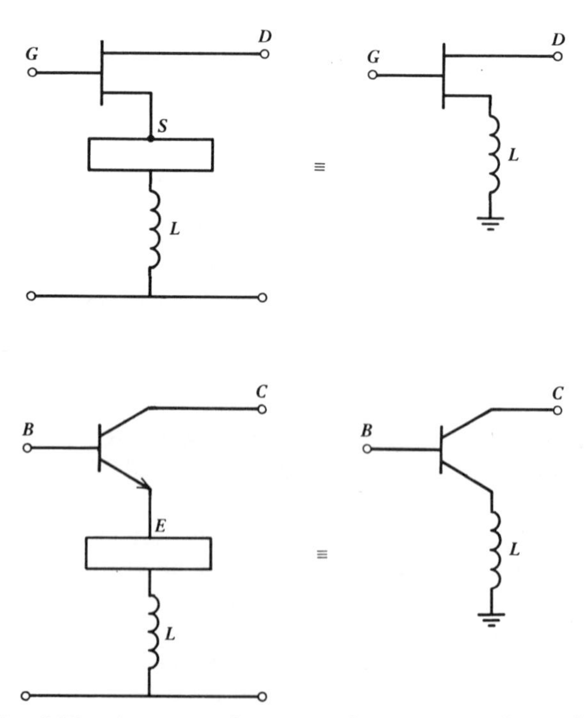

Fig. 5.33. A source and emitter inductance series feedback.

Table 5.20 The HFET-1000 S-parameters

f (GHz)	S_{11}	S_{21}	S_{12}	S_{22}	S_{21} (dB)	K
3	$0.90\angle-61°$	$3.18\angle 132°$	$0.028\angle 62°$	$0.76\angle-17°$	10	0.45
4	$0.85\angle-79°$	$2.80\angle 119°$	$0.031\angle 57°$	$0.76\angle-18°$	9	0.65
5	$0.82\angle-95°$	$2.50\angle 106°$	$0.032\angle 52°$	$0.74\angle-23°$	8	0.83

given as

$$\mathbf{Z}_2 = j\omega L\begin{bmatrix} 1 & 1 \\ 1 & 1 \end{bmatrix} \tag{5.183}$$

The scattering matrix $\mathbf{S}$ of the interconnected network is obtained from its corresponding impedance matrix $\mathbf{Z}=\mathbf{Z}_1+\mathbf{Z}_2$ as

$$\mathbf{S}=(\mathbf{Z}+Z_0\mathbf{1}_2)^{-1}(\mathbf{Z}-Z_0\mathbf{1}_2) \tag{5.184}$$

The elements of $\mathbf{S}$ in (5.184) can then be used in (2.51) to calculate K and $|\Delta|$.

The use of inductance series feedback at the source of the GaAs FET or the bipolar transistor usually do not alter the model of the device shown in Fig. 5.1*a*. As an example, the HFET-1000 described in Table 5.21 can be modeled very closely by the circuit in Fig. 5.1*a* with $R_i=27.7\ \Omega$, $L_i=0.307$ nH, $C_i=0.617$ pF and $R_o=381.6\ \Omega$, $C_o=0.1$ pF, $L_o=0.31$ nH. The use of series feedback will undoubtedly increase the minimum noise figure F_m of the transistor. How much increase depends on the value of the inductance L and can be predicted using the results derived in Appendix A8. By using (A8.33) and (A8.53) it can be proved that the minimum noise figure $\hat{F}_m$ of the interconnected feedback network remains approximately independent of L, and $\hat{F}_m \approx F_m$ if

$$L \ll \frac{1}{\omega}|Z_{21}| = \frac{2Z_0}{\omega}\cdot\frac{|S_{21}|}{|(1-S_{11})(1-S_{22})-S_{12}S_{21}|} \tag{5.185}$$

From (5.185) we note that L can be easily selected if the original two-port has a large $|S_{21}|$, which is usually the case with potentially unstable GaAs FETs at frequencies below 3 GHz.

In a similar way, one can show that it is also possible to use the parallel capacitance feedback C from the gate to the drain of the GaAs FET and from the base to the collector of the bipolar transistor. In this case, the minimum noise

Table 5.21 The HFET-1000 (with source inductance $L=0.32$ nH) S-parameters

f (GHz)	S_{11}	S_{21}	S_{12}	S_{22}	S_{21} (dB)	K
3	$0.74\angle-59°$	$2.92\angle 123°$	$0.028\angle 97°$	$0.76\angle-12°$	9.3	1.01
4	$0.63\angle-75°$	$2.51\angle 110°$	$0.036\angle 107°$	$0.77\angle-13°$	8.0	1.15
5	$0.55\angle-91°$	$2.20\angle 97°$	$0.050\angle 120°$	$0.76\angle-17°$	6.9	1.06

figure $\hat{F}_m$ of the interconnected network will remain approximately the same if

$$C \ll \frac{1}{\omega}|Y_{21}| = \frac{2Y_0}{\omega} \frac{|S_{21}|}{|(1+S_{11})(1+S_{22})-S_{12}S_{21}|} \tag{5.186}$$

and will decrease, that is, $\hat{F}_m < F_m$ as C approaches infinity, which means a short circuit, but this also decreases the gain to zero. In practice it is easier to employ the series inductance feedback, since it can be readily approximated by a short length of short-circuited stub. All the design formulas for these types of feedback amplifiers are derived in Appendix A8 and can be easily programmed on a digital computer.

5.9 NEGATIVE RESISTANCE REFLECTION AMPLIFIERS

When the operating frequency bandwidth is moderate (10–20% of the center frequency), most negative resistance devices such as IMPATT, Gunn, and tunnel diodes can be approximated by the dissipationless models shown in Figs. 5.34*a*–*b*. In the following discussion we will present an analytical technique to design negative resistance reflection amplifiers using these models. The general configuration of this type of amplifier is shown in Fig. 5.35. It has been shown in (4.39) that the transducer power gain from port 1 to port 3 of the amplifier is given by

$$G(\omega^2) = \left| \frac{Z_2(j\omega) - Z_d^*(j\omega)}{Z_2(j\omega) + Z_d(j\omega)} \right|^2 = |\Gamma_2(j\omega)|^2 \tag{5.187}$$

With the diode models shown in Figs. 5.34*a*–*b*, we see that

$$-Z_d^*(j\omega) = -Z_d(-j\omega) = Z(j\omega) \tag{5.188}$$

where Z is the passive impedance of these models in which $-R$ is replaced by

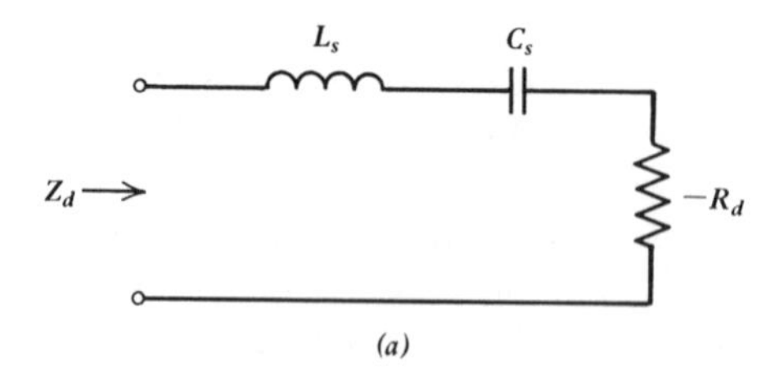

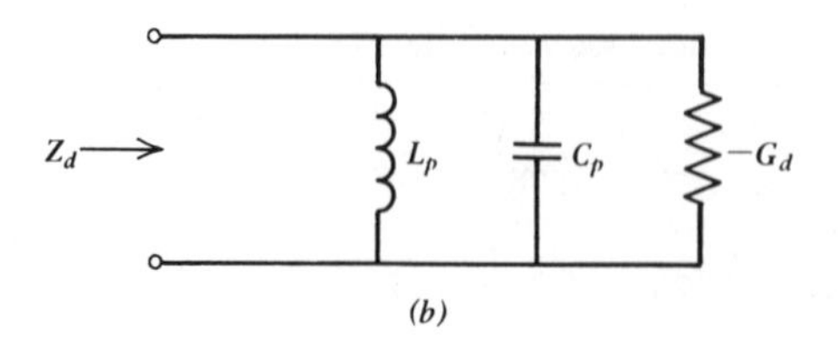

Fig. 5.34. Dissipationless models for negative resistance devices. (*a*) Series-resonant. (*b*) Parallel-resonant.

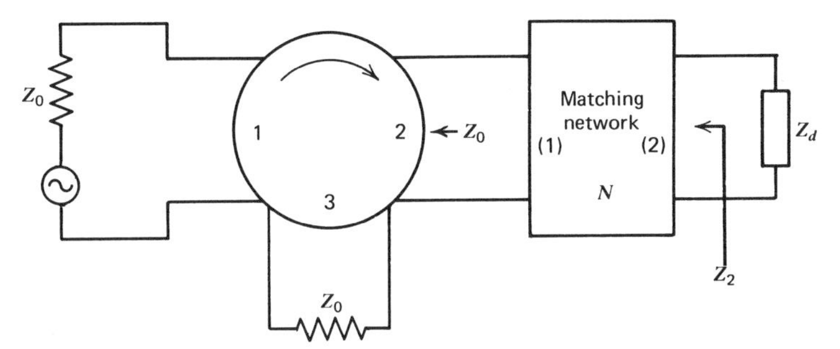

Fig. 5.35. Negative resistance reflection amplifier configuration.

R. Thus

$$G(\omega^2)=\left|\frac{Z_2(j\omega)+Z(j\omega)}{Z_2(j\omega)-Z(-j\omega)}\right|^2=\frac{1}{|S_2(j\omega)|^2} \tag{5.189}$$

where

$$S_2(j\omega)=\frac{Z_2(j\omega)-Z(j\omega)}{Z_2(j\omega)+Z(j\omega)} \tag{5.190}$$

is the reflection coefficient at port 2 of N when the diode is replaced by the passive impedance Z. Thus the problem of designing a lossless matching network N to maximize the reflection coefficient Γ_2 at the diode is equivalent to designing N to minimize the reflection coefficient S_2 when Z_d is replaced by Z, that is, N is a matching network for Z and Z_0 (Z_0 is the circulator port impedance which is 50 Ω in practice). We note that Z is the impedance of a series $R_dL_sC_s$ circuit for the diode model of Fig. 5.34a and is the impedance of a parallel $G_dL_pC_p$ circuit for the diode model of Fig. 5.34b. In practice $R_d \ll Z_0 = 50$ Ω. Therefore all the results derived for the input matching network of microwave transistors are applicable to this model. For example, for a Chebyshev matching network N, the gain-bandwidth constraint is given by (5.43) with R_i, L_i, and C_i replaced by R_d, L_s, and C_s respectively. Also the explicit formulas in section 5.4 for matching networks without an ideal transformer are applicable for this type of model. The model in Fig. 5.34b is dual to that in Fig. 5.34a, therefore the results can be obtained in a dual manner. Furthermore, as in the case of the design of transistor amplifiers, real frequency techniques in Section 5.8 can be applied to (5.187) to obtain the unknown resistance slope lines of $Z_2(j\omega)$.

5.10 CONCLUDING REMARKS

Explicit formulas for gain-bandwidth constraints for matching networks of nonsloped and sloped responses have been presented together with the synthesis procedure for microwave transistor and diode amplifiers. The nonsloped matching networks are easily designed and gain-bandwidth limitations can be determined before obtaining the minimum-phase reflection coefficient $\rho(s)$ through the spectral factorization of $1-G(-s)^2$. In the case of the sloped response, the gain-bandwidth constraints can only be determined by the coefficients of $\rho(s)$ with the preassigned parameters of $G(\omega^2)$ such as n, ϵ, and K_n, which can be selected iteratively. The conversion of the lumped amplifier to distributed form can cause the gain to deteriorate at the upper and lower frequency band if the fractional bandwidth is more than 50% of the center frequency. This usually requires optimization to meet the design goal. The techniques developed here can provide good initial circuits for octave bandwidth amplifiers and hence can guarantee fast convergence. The design of a microwave transistor amplifier from the device's unilateral model is simple although it has its own drawbacks since the input and output impedances of the transistor have to be modeled from its S-parameters. In many cases this can be achieved without much effort by using the Smith chart or an optimization scheme. Since there are only three elements in these impedances, convergence can be very fast. Another technique that can be very efficient in the design of very broadband amplifiers is to use semi-infinite constant slope lines to represent the resistive and reactive parts of the back-end equalizer impedance. The selection of the slope coefficients depends upon experience; they are best obtained through the use of least-square optimization. Once the semi-infinite slope characteristics are determined, they can be approximated by rational functions that are realizable by either lumped or distributed networks. The method is general in nature but relies heavily on least-square approximations and the designer's knowledge of lumped or distributed network synthesis.

PROBLEMS

5.1. Design a 4–7 GHz amplifier using the HFET-1102 whose scattering parameters are given in Table 5.1 with the input matching network having a sloped response and the output matching network having a nonsloped response.

5.2. Repeat Problem 5.1 with the input matching network design for a nonsloped response and the output network designed for a sloped response.

5.3. Design the output matching network for the low noise 6–8 GHz HFET-1101 amplifier of Example 5.6 to achieve a gain of at least 7 dB over the bandwidth with 0.5 dB ripple maximum. The HFET-1101 scattering parameters are given by

f(GHz)	S_{11}	S_{21}	S_{12}	S_{22}
6	$0.674\angle-152°$	$1.74\angle 36.4°$	$0.075\angle 6.2°$	$0.60\angle-93°$
7	$0.631\angle-176°$	$1.60\angle 15°$	$0.074\angle 1.3°$	$0.58\angle-111°$
8	$0.607\angle 163°$	$1.47\angle-4.4°$	$0.077\angle 0.5°$	$0.56\angle-130°$

5.4. Redesign the two-stage cascade amplifier in Example 5.5 by designing the input matching network for nonsloped response. Write a computer program to optimize the interstage matching network for an overall flat gain.

5.5. Convert the lumped 4–8 GHz amplifier in Example 5.3 to a distributed amplifier. Write a computer program to compute its gain and optimize the input and output matching networks to obtain a gain of 9 dB with 1 dB ripple maximum.

5.6. Convert the lumped 6–12 GHz amplifier in Example 5.4 to a distributed amplifier. Write a computer program to optimize its gain to 7.5±0.5 dB.

5.7. Design a matching network to match the parallel-series *RC-L* load, $R=299\ \Omega$, $C=0.387$ pF, $L=0.490$ nH, to 50 Ω from 5–8 GHz using the real frequency broadband technique described in Section 5.7.

5.8. Design a matching network to match the series *RLC* load, $R=5.93\ \Omega$, $L=0.56$ nH, $C=0.72$ pF, from 5–8 GHz using the technique described in Section 5.7.

5.9. Design a distributed low noise amplifier over the frequency band 3.7–4.2 GHz to achieve a noise figure of 1.4 dB maximum and a flat gain of 12±0.25 dB using a GaAs FET with the following scattering and noise parameters:

f (GHz)	S_{11}	S_{21}	S_{12}	S_{22}
3.7	$0.83\angle-81°$	$2.60\angle 91°$	$0.038\angle 29°$	$0.73\angle-54°$
3.8	$0.82\angle-84°$	$2.57\angle 87°$	$0.038\angle 26°$	$0.73\angle-55°$
3.9	$0.81\angle-86°$	$2.55\angle 84°$	$0.038\angle 24°$	$0.72\angle-57°$
4.0	$0.80\angle-89°$	$2.53\angle 81°$	$0.038\angle 22°$	$0.72\angle-59°$
4.1	$0.79\angle-92°$	$2.50\angle 78°$	$0.037\angle 21°$	$0.71\angle-61°$
4.2	$0.79\angle-94°$	$2.49\angle 75°$	$0.036\angle 20°$	$0.71\angle-62°$

F (dB)	$R_n/50$	Γ_m
0.76	1.0	$0.53\angle 61°$
0.69	1.1	$0.62\angle 63°$
0.67	1.2	$0.65\angle 60°$
0.82	1.0	$0.64\angle 62°$
1.04	1.1	$0.69\angle 62°$
1.06	1.1	$0.74\angle 63°$

5.10. Design a distributed low noise amplifier over the frequency band 3.6–4.4 GHz to achieve a noise figure of 2.2 dB maximum and a gain of 10±0.25 dB using a GaAs FET (potentially unstable) with the following scattering and noise parameters:

f (GHz)	S_{11}	S_{21}	S_{12}	S_{22}
3.6	$0.90\angle-80°$	$2.00\angle 99°$	$0.041\angle 32°$	$0.79\angle-53°$
3.8	$0.90\angle-86°$	$1.97\angle 94°$	$0.042\angle 29°$	$0.78\angle-57°$
4.0	$0.89\angle-91°$	$1.95\angle 89°$	$0.043\angle 26°$	$0.77\angle-61°$
4.2	$0.89\angle-96°$	$1.94\angle 84°$	$0.043\angle 23°$	$0.76\angle-65°$
4.4	$0.88\angle-102°$	$1.92\angle 79°$	$0.043\angle 19°$	$0.75\angle-69°$

F (dB)	$R_n/50$	Γ_m
1.4	2.0	$0.77\angle 65°$
1.3	2.1	$0.75\angle 68°$
1.5	2.2	$0.74\angle 70°$
1.6	2.0	$0.73\angle 73°$
1.8	2.1	$0.72\angle 74°$

5.11. Design a matching network to match a parallel-series RC-L load with $R=1\ \Omega$, $C=1.2$ F, $L=2.3$ H from $\omega=0$ to $\omega=1$ rad/sec to obtain a minimum transducer power gain $K_n=0.837$ with less than 0.25 dB ripple.

5.12. Design the matching network to 50 Ω for an IMPATT diode having the following admittance $Y_d=-G_d+jB_d$:

f (GHz)	$-G_d$ (mhos)	B_d (mhos)
4.30	0.002	0.014
4.80	0.004	0.040
5.30	0.005	0.050
5.80	0.006	0.060
6.25	0.007	0.070

to obtain a stable gain of 10±0.5 dB.

5.13. For Fig. P5.13 show that

$$\int_0^\infty \ln\left|\frac{1}{S_{22}(j\omega)}\right|^2 d\omega \leqslant \frac{2\pi}{RC}$$

and

$$\int_0^\infty \ln\left|\frac{1}{S_{11}(j\omega)}\right|^2 d\omega \leqslant \frac{2\pi}{RC}$$

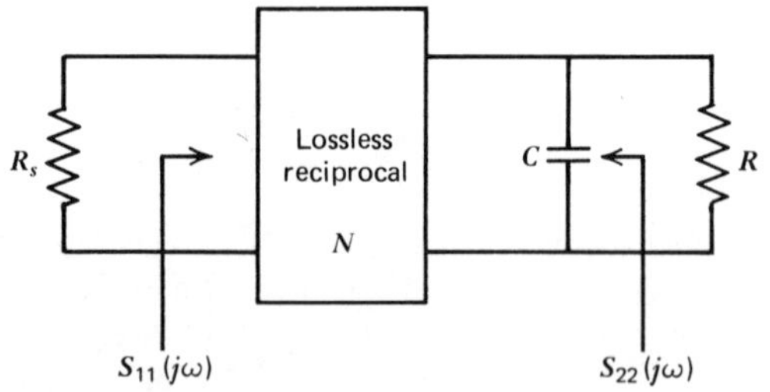

Fig. P5.13. Integral limitation of a parallel RC load.

5.14. For Fig. P5.14 show that

$$\int_0^\infty \ln\left|\frac{1}{S_{22}(j\omega)}\right|^2 d\omega = \frac{2\pi}{RC}$$

and

$$\int_0^\infty \omega^2 \ln\left|\frac{1}{S_{22}(j\omega)}\right|^2 d\omega \leqslant \frac{2\pi}{3} \cdot \frac{3R^2C-L}{LC^3R^3}$$

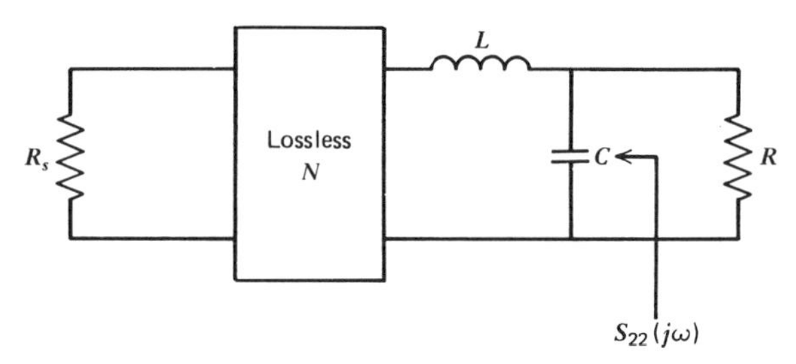

Fig. P5.14. Integral limitation of a parallel-series *RC-L* load.

5.15. For Fig. P5.15 show that

$$\int_0^{\infty} \frac{1}{\omega^2} \ln|S_{22}(j\omega)|^2 \, d\omega \geqslant \max\{-2\pi RC, -2\pi R/L\}$$

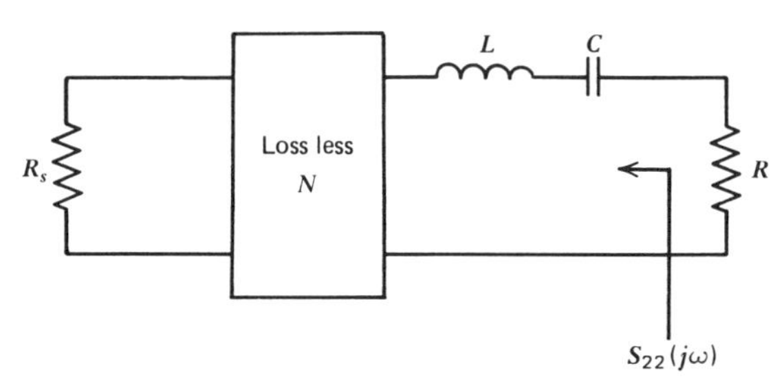

Fig. P5.15. Integral limitation of a series *RLC* load.

5.16. Obtain the condition for $K_n = 1$ in Equations (5.43) and (5.45).

5.17. Design a bandpass lossless Chebyshev matching network with nonsloped response for the parallel-series *RC-L* load where $R = 164\ \Omega$, $C = 0.428$ pF, $L = 0.474$ nH using $n = 3$ over the band 4–8 GHz.

5.18. Design a distributed amplifier from 3 to 5 GHz employing the potentially unstable HFET-1100 with source inductance series feedback as shown in Table 5.21.

5.19. Design a lumped amplifier from 3 to 5 GHz employing the potentially unstable HFET-1100 with drain-to-gate capacitance parallel feedback using the scattering parameters in Table 5.20.

5.20. Can the GaAs FET NE-38806 whose scattering parameters are given below be made stable using source inductance series feedback only? If not, use drain-to-gate capacitance parallel feedback or both types to make it stable.

f (GHz)	S_{11}	S_{21}	S_{12}	S_{22}	K
2.0	$0.93\angle -42°$	$2.90\angle 131°$	$0.025\angle 65°$	$0.67\angle -27°$	0.61
2.5	$0.90\angle -54°$	$3.04\angle 122°$	$0.035\angle 57°$	$0.67\angle -35°$	0.55
3.0	$0.87\angle -67°$	$3.13\angle 111°$	$0.045\angle 48°$	$0.65\angle -43°$	0.58
3.5	$0.83\angle -80°$	$3.12\angle 95°$	$0.053\angle 38°$	$0.60\angle -53°$	0.73
4.0	$0.78\angle -93°$	$3.03\angle 79°$	$0.058\angle 28°$	$0.55\angle -63°$	0.92

REFERENCES

1. H. Bode, *Network Analysis and Feedback Amplifier Design*. New York: Van Nostrand, 1947.
2. M. E. Van Valkenburg, *Introduction to Modern Network Synthesis*. New York: Wiley, 1960.
3. H. J. Carlin, "Distributed circuit design with transmission line elements," *Proc. IEEE*, **59**, no. 7, 1059–1081, 1971.
4. D. C. Youla, "A new theory of broadband matching," *IEEE Trans. Circuit Theory*, **CT-11**, no. 3, 30–50, 1964.
5. F. M. Fano, "Theoretical limitations on the broadband matching of arbitrary impedances," *J. Franklin Inst.*, **249**, 57–149, 1950.
6. H. J. Carlin and J. J. Komiak, "A new method of broadband equalization applied to microwave amplifiers," *IEEE Trans. Microwave Theory Tech.*, **MTT-27**, no. 2, 93–98, 1979.
7. D. J. Mellor and J. G. Linvill, "Synthesis of interstage networks of prescribed gain versus frequency slopes," *IEEE Trans. Microwave Theory Tech.*, **MTT-23**, no. 12, 1013–1020, 1975.
8. W. H. Ku and W. C. Peterson, "Optimum gain-bandwidth limitations for transistor amplifiers as reactively constrained active two-port networks," *IEEE Trans. Circuit Syst.*, **CAS-22**, no. 6, 523–533, 1975.
9. M. E. Mohari-Bohassan and W. H. Ku, "Tapered-magnitude bandpass distributed network transfer functions," *Internat. J. Circuit Theory Applic.*, **5**, 367–378, 1977.
10. T. T. Ha, "Broadband matching network design with sloped response for microwave GaAs FET amplifiers," *IEE J. Electron. Circuits Syst.*, **3**, no. 3, 97–102, 1979.
11. T. T. Ha and T. H. Dao, "Applications of Takahasi's results to broadband matching for microwave amplifiers," *IEEE Trans. Circuit Syst.*, **CAS-26**, no. 11, 970–973, 1979.
12. H. Takahasi, "On the ladder-type filter network with the Chebyshev response," *J. Inst. Elect. Commun. Engrs. Japan*, **74**, no. 2, 65–74, 1951.
13. L. Weinberg and P. Slepian, "Takahasi's results on Chebyshev and Butterworth ladder networks," *IRE Trans. Circuit Theory*, **CT-7**, no. 2, 88–101, 1960.
14. E. Green, "Synthesis of ladder networks to give Butterworth or Chebyshev response in the passband," *Proc. IEE (London)*, **101**, pt. 4, 192–203, 1954.
15. G. L. Matthaei, L. Young, and E. M. T. Jones, *Microwave Filters, Impedance Matching Networks and Coupling Structures*. New York: McGraw-Hill, 1964.
16. W. K. Chen, *Theory and Design of Broadband Matching Networks*. New York: Pergamon Press, 1976.
17. G. C. Temes and J. W. LaPatra, *Introduction to Circuit Synthesis and Design*. New York: McGraw-Hill, 1977.
18. G. C. Temes and D. A. Calahan, "Computer-aided network optimization. The state-of-the-art," *Proc. IEEE*, **55**, no. 11, 1832–1863, 1967.
19. R. Fletcher and M. J. D. Powell, "A rapid convergent descent method for minimization," *Computer J.*, **6**, 163–168, 1963.
20. W. C. Davidson, "Variable metric method for minimization." AEC Research Development Report ANL-5990 (Rev.), 1968.

21. W. C. Davidson, "Variance algorithm for minimization," *Computer J.*, **10**, 406–410, 1968.

22. G. A. Vincent, "Impedance transformation without a transformer," *Frequency Technology*, **7**, 15–21, 1969.

23. C. A. Liechti and R. L. Tillman, "Design and performance of microwave amplifiers with GaAs Schottky-gate field-effect transistors," *IEEE Trans. Microwave Theory Tech.*, **MTT-22**, no. 5, 510–517, 1974.

24. E. S. Kuh and R. A. Rohrer, *Theory of Linear Active Networks*. San Francisco: Holden-Day, 1967.

25. COMPACT, the program package available from COMPACT Engineering, Palo Alto, CA.

6

Signal Distortion Characterizations and Microwave Power Combining Techniques

Besides the presence of noise and the finite bandwidth of the system, signal distortion also occurs in communication systems when the amplitude $|H(j\omega)|$ and phase $\phi(\omega)$ of a linear network transfer function $H(j\omega)=|H(j\omega)|\exp[j\phi(\omega)]$ are frequency dependent functions. This type of distortion is often called "linear distortion" or "transmission deviation." For example, distortionless transmission occurs if the output signal $y(t)$ is a scaled version of a delay input signal $x(t)$, that is,

$$y(t)=Kx(t-t_0)$$

which implies that the system transfer function $H(j\omega)$ must be

$$H(j\omega)=K\exp(-j\omega t_0)$$

In other words, any deviation from the constant amplitude K and the negative linear phase shift $-j\omega t_0$ will result in amplitude and phase distortion as a function of frequency. Besides this type of distortion, other deviations can occur if the system possesses nonlinear elements. In this case, the system cannot be described by a single transfer function as in the linear case; instead the output is often expressed as a nonlinear function of the input, that is, $y(t)=T[x(t)]$ as is shown in Fig. 6.1. Signal distortion resulting from this type of deviation is often called "nonlinear distortion." This characterization of signal distortion is very important in a communication system since its usefulness is determined by its dynamic range (the range of signal levels that can be processed with high quality), which is dictated by its noise figure or its sensitivity as the lower limit and by its acceptable level of signal distortion as its upper limit. In Chapter 2 we studied the sensitivity of communication receivers by its noise characterization. In this chapter, part of the discussion will focus on the characterization of signal distortion and nonlinearities in microwave devices and systems. We begin with the discussion of "nonlinear distortion" in memoryless systems.

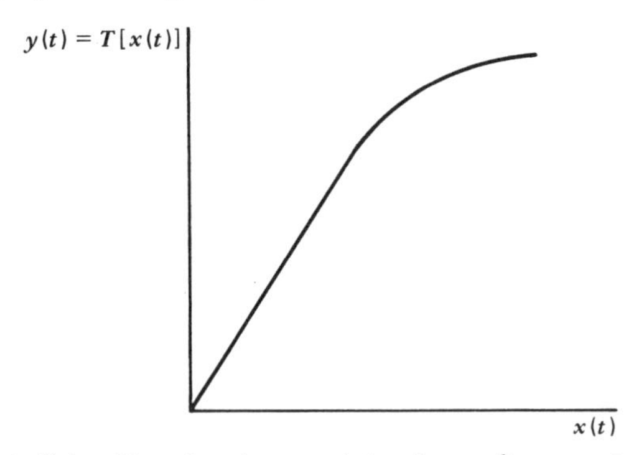

Fig. 6.1. Transfer characteristic of a nonlinear system.

6.1 AMPLITUDE NONLINEARITY IN MEMORYLESS SYSTEMS

One of the most common nonlinearity characterization of a two-port system is its amplitude distortion, which results in its nonlinear transfer characteristic. If we assume that the two-port is memoryless, that is, its output voltage is an instantaneous function of its input voltage, and its nonlinearity is weak (which is the case of most communication systems) then the output voltage $e_o(t)$ can be represented by a power series of the input voltage $e_i(t)$ as

$$e_o = k_1 e_i + k_2 e_i^2 + k_3 e_i^3 + \cdots \tag{6.1}$$

Note that for a linear two-port all k_i's are zero for $i=2,3,\ldots,$ and the representation of e_o in (6.1) neglects the phase characteristic of the two-port network, which results in phase variation in the output with respect to any input. Discussion of such characterization will be postponed to a later section.

Single-Frequency Input Test

Consider a two-port with a mild nonlinearity such that e_o can be represented by the first three terms of (6.1), that is,

$$e_o = k_1 e_i + k_2 e_i^2 + k_3 e_i^3 \tag{6.2}$$

Let $e_i = A\cos\omega_1 t$. Then e_o can be written as

$$\begin{aligned} e_o &= k_1 A\cos\omega_1 t + k_2 A^2\cos^2\omega_1 t + k_3 A^3\cos^3\omega_1 t \\ &= k_1 A\cos\omega_1 t + k_2 A^2\left(\tfrac{1}{2} + \tfrac{1}{2}\cos 2\omega_1 t\right) + k_3 A^3\left(\tfrac{3}{4}\cos\omega_1 t + \tfrac{1}{4}\cos 3\omega_1 t\right) \\ &= \tfrac{1}{2}k_2 A^2 + \left(k_1 A + \tfrac{3}{4}k_3 A^3\right)\cos\omega_1 t + \tfrac{1}{2}k_2 A^2\cos 2\omega_1 t + \tfrac{1}{4}k_3 A^3\cos 3\omega_1 t \end{aligned} \tag{6.3}$$

From (6.3) we see that the output signal consists of a component at the applied fundamental frequency ω_1 and spurious signals at dc, the second harmonic frequency $2\omega_1$, and the third harmonic frequency $3\omega_1$. From (6.3) we remark that the fundamental component of e_o has an amplitude of $k_1A[1+\frac{3}{4}(k_3/k_1)A^2]$, which is greater than k_1A (the gain if the two-port is linear) if $k_3>0$ and smaller than k_1A if $k_3<0$. This property is called gain expansion or gain compression. Most practical devices are compressive, that is, $k_3<0$ and their output power is usually characterized at the 1 dB gain compression point. From (6.3) we see that the gain at the fundamental frequency ω_1 is given by

$$G = 20\log\frac{k_1A+\frac{3}{4}k_3A^3}{A} = 20\log\left(k_1+\frac{3}{4}k_3A^2\right) \tag{6.4}$$

as compared to the linear gain G_0 defined as

$$G_0 = 20\log\frac{k_1A}{A} = 20\log k_1 \tag{6.5}$$

The 1 dB gain compression point is defined as the signal level where

$$G_{1\text{dB}} = G_0 - 1 \quad \text{dB} \tag{6.6}$$

or equivalently,

$$k_1+\frac{3}{4}k_3A^2 = 0.891k_1 \tag{6.7a}$$

Hence at the 1 dB gain compression point the amplitude of e_i is limited as

$$A^2 = 0.145\frac{k_1}{|k_3|}, \qquad k_3<0 \tag{6.7b}$$

Assume that the input and output impedance of the two-port are $Z_{\text{in}} = Z_{\text{out}} = R$ Ω. Then the input and output powers P_i and P_o at the fundamental frequency ω_1 are given in dBm as

$$P_i = 10\log\left\{\left(\frac{A}{\sqrt{2}}\right)^2\frac{10^3}{R}\right\} \quad \text{dBm} \tag{6.8}$$

$$\begin{aligned} P_o &= 10\log\left\{\left(\frac{k_1A+\frac{3}{4}k_3A^3}{\sqrt{2}}\right)^2\frac{10^3}{R}\right\} \quad \text{dBm} \\ &= G + P_i \quad \text{dBm} \end{aligned} \tag{6.9}$$

The output power at 1 dB gain compression point $P_{1\text{dB}}$ is

$$P_{1\text{dB}} = G_{1\text{dB}} + P_i = G_0 - 1 + P_i \quad \text{dBm} \tag{6.10}$$

Substituting (6.7b) and (6.8) into (6.10) yields

$$P_{1\text{dB}} = G_0 - 1 + 10\log\left[\frac{0.145k_1}{2|k_3|}\frac{10^3}{R}\right] \quad \text{dBm}$$

$$= 10\log\frac{57.70k_1^3}{|k_3|R} = 10\log\left\{\left(\frac{1}{17.33}\frac{k_1^3}{|k_3|}\right)\frac{10^3}{R}\right\} \quad \text{dBm} \qquad (6.11)$$

For $R = 50\ \Omega$ we have

$$P_{1\text{dB}} = 10\log\frac{k_1^3}{|k_3|} + 0.62 \quad \text{dBm} \qquad (6.12)$$

Example 6.1

Consider a two-port with the following transfer characteristic (assuming $R = 50\ \Omega$)

$$e_o = 15e_i - 2e_i^3$$

According to (6.5) and (6.6), the linear gain $G_o = 23.5$ dB and the 1 dB gain compression point is $G_{1\text{dB}} = 22.5$ dB. At this point the amplitude is limited to $A = (0.145k_1/|k_3|)^{1/2} = 1.044$ V, and the output power is $P_{1\text{dB}} = 32.89$ dBm. A plot of $P_{1\text{dB}}$ is shown in Fig. 6.2.

Two-Frequency Input Test

Now consider an input signal $e_i = A(\cos\omega_1 t + \cos\omega_2 t)$ that consists of two equal amplitude sinusoids at two different frequencies ω_1 and ω_2.

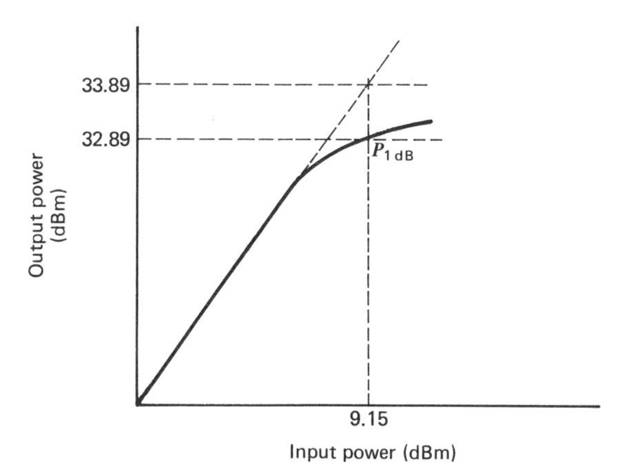

Fig. 6.2. A plot of output power versus input power for $e_o = 15e_i - 2e_i^2$.

Applying e_i to (6.2) yields

$$e_o = k_1 A(\cos\omega_1 t + \cos\omega_2 t) + k_2 A^2(\cos\omega_1 t + \cos\omega_2 t)^2 + k_3 A^3(\cos\omega_1 t + \cos\omega_2 t)^3$$

$$= k_2 A^2 + k_2 A^2 \cos(\omega_1 - \omega_2)t + \left(k_1 A + \tfrac{9}{4}k_3 A^3\right)\cos\omega_1 t$$

$$+ \left(k_1 A + \tfrac{9}{4}k_3 A^3\right)\cos\omega_2 t + \tfrac{3}{4}k_3 A^3 \cos(2\omega_1 - \omega_2)t$$

$$+ \tfrac{3}{4}k_3 A^3 \cos(2\omega_2 - \omega_1)t + k_2 A^2 \cos(\omega_1 + \omega_2)t + \tfrac{1}{2}k_2 A^2 \cos 2\omega_1 t$$

$$+ \tfrac{1}{2}k_2 A^2 \cos 2\omega_2 t + \tfrac{3}{4}k_3 A^3 \cos(2\omega_1 + \omega_2)t + \tfrac{3}{4}k_3 A^3 \cos(2\omega_2 + \omega_1)t$$

$$+ \tfrac{1}{4}k_3 A^3 \cos 3\omega_1 t + \tfrac{1}{4}k_3 A^3 \cos 3\omega_2 t \tag{6.13}$$

From (6.13) it is seen that the output signal consists of components at dc, the fundamental frequencies ω_1 and ω_2, the second and third harmonics $2\omega_1, 2\omega_2$, and $3\omega_1, 3\omega_2$ and the second-order intermodulation products at $\omega_1 \pm \omega_2$ (the sum of the coefficients of ω_1 and ω_2 is 2), and the third-order intermodulation products at $2\omega_1 \pm \omega_2$ and $2\omega_2 \pm \omega_1$ (the sum of the coefficients of ω_1 and ω_2 is 3). In systems where the operating frequency band is less than an octave, all the spurious signals at $\omega_1 \pm \omega_2$,, $2\omega_1$, $2\omega_2$, $2\omega_1 + \omega_2$, $2\omega_2 + \omega_1$, $3\omega_1$ and $3\omega_2$ fall outside the passband and can be filtered out by appropriate filters. But all the spurious signals at the frequencies $2\omega_1 - \omega_2$ and $2\omega_2 - \omega_1$ will fall within the passband and can distort the desired signal at the fundamental frequency ω_1 or ω_2. The input and output spectrum of e_i and e_o are shown in Fig. 6.3*a–b*. A useful measure of the third-order intermodulation distortion is the "intercept point," defined as the output power level P_I at which the output power $P_{(2\omega_1 - \omega_2)}$ at the frequency $2\omega_1 - \omega_2$ would intercept the output power P_o at ω_1 (when the two-port is linear) if low-level results were extrapolated into the higher-power region as shown in Fig. 6.4. We remark that at low level, the output power P_o is directly proportional to the amplitude of the input signal while the output power $P_{(2\omega_1 - \omega_2)}$ is directly proportional to the cube of the input amplitude. Thus the plot of each on a log-log scale (or dBm/dBm scale) will be a straight line with a

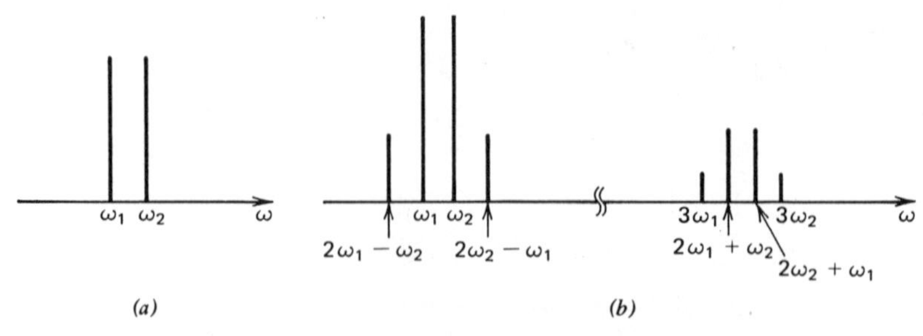

Fig. 6.3. (*a*) Input spectrum. (*b*) Output spectrum of a two-port with third-order distortion.

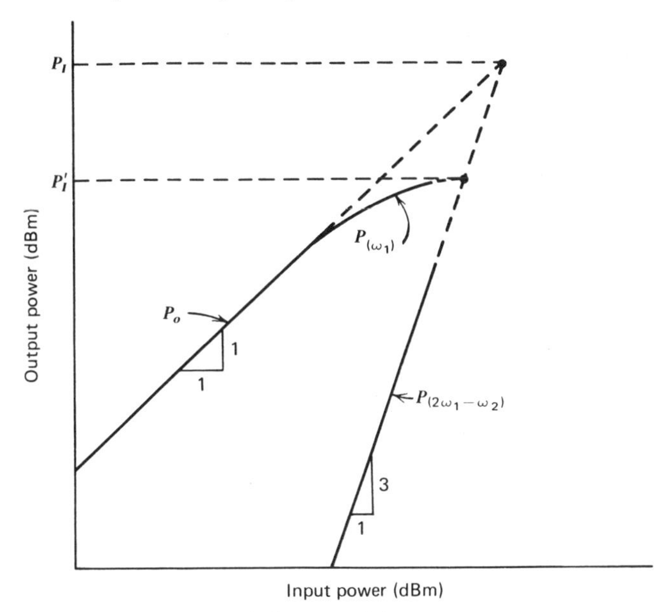

Fig. 6.4. Definition of the intercept point.

slope corresponding to the order of the response, that is, the response at ω_1 will have a slope of 1 and the response at $2\omega_1 - \omega_2$ will have a slope of 3. Their intersection is the intercept point. We note that the actual amplitude of the output signal at ω_1 is $k_1 A + \frac{9}{4} k_3 A^3$ where $k_3 < 0$ for compressive two-ports. Hence at lower power levels ($k_1 A \gg \frac{9}{4} k_3 A^3$), the response of the output power $P_{(\omega_1)}$ at ω_1 almost coincides with the response of the output power P_o at ω_1 when the two-port is assumed to be linear. At higher power levels the response of $P_{(\omega_1)}$ will be compressed and will deviate from the response of P_o as shown in Fig. 6.4. From (6.13) we have

$$P_o = 10\log\left\{\left(\frac{k_1 A}{\sqrt{2}}\right)^2 \frac{10^3}{R}\right\} \quad \text{dBm} \tag{6.14}$$

$$P_{(\omega_1)} = 10\log\left\{\left(\frac{k_1 A + \frac{9}{4}k_3 A^3}{\sqrt{2}}\right)^2 \frac{10^3}{R}\right\} \quad \text{dBm} \tag{6.15}$$

$$P_{(2\omega_1-\omega_2)} = 10\log\left\{\left(\frac{\frac{3}{4}k_3 A^3}{\sqrt{2}}\right)^2 \frac{10^3}{R}\right\} \quad \text{dBm} \tag{6.16}$$

Since at P_I by definition, $P_o = P_{(2\omega_1-\omega_2)}$, by comparing (6.14) and (6.16) we obtain the theoretical amplitude A at P_I as

$$A^2 \text{ (at } P_I) = \frac{4}{3} \frac{k_1}{|k_3|} \tag{6.17}$$

and therefore

$$P_I = 10 \log \left(\left(\frac{2}{3} \frac{k_1^3}{|k_3|} \right) \frac{10^3}{R} \right) \quad \text{dBm} \tag{6.18}$$

If $R = 50\ \Omega$

$$P_I = 10 \log \frac{k_1^3}{|k_3|} + 11.25 \quad \text{dBm} \tag{6.19}$$

And from Fig. 6.4 we note that the response of $P_{(\omega_1)}$ intersects the response of $P_{(2\omega_1 - \omega_2)}$ at the point P_I'. It is useful to relate P_I' to P_I. At P_I' we have $P_{(\omega_1)} = P_{(2\omega_1 - \omega_2)}$. By comparing (6.15) and (6.16) the amplitude A at P_I' is given as

$$A^2 \text{ (at } P_I') = \frac{2}{3} \frac{k_1}{|k_3|} \tag{6.20}$$

and hence

$$\begin{aligned} P_I' &= 10 \log \left(\left(\frac{1}{12} \frac{k_1^3}{|k_3|} \right) \frac{10^3}{R} \right) \quad \text{dBm} \\ &= P_I - 9 \quad \text{dBm} \end{aligned} \tag{6.21}$$

By comparing (6.11) and (6.18), a relationship between P_I and $P_{1\text{dB}}$ is given as

$$P_I = P_{1\text{dB}} + 10.63 \quad \text{dBm} \tag{6.22}$$

From (6.18) we note that the intercept power P_I is independent of the input power and is therefore a useful measure of the system nonlinearity. Also by comparing (6.14), (6.16), and (6.18), the outpower at $2\omega_1 - \omega_2$ can be given as

$$P_{(2\omega_1 - \omega_2)} = 3P_o - 2P_I \quad \text{dBm} \tag{6.23}$$

with lower level signal, $P_o \approx P_{(\omega_1)}$. Hence

$$P_{(2\omega_1 - \omega_2)} \approx 3P_{(\omega_1)} - 2P_I \quad \text{dBm} \tag{6.24}$$

At higher level input power, the relation in (6.24) no longer holds. Using (6.15),

(6.16), and (6.18) we obtain

$$P_{(2\omega_1-\omega_2)}-60\log\left[1-0.343R^{1/3}\frac{|k_3|^{1/3}}{k_1}\left(\log^{-1}\frac{P_{(2\omega_1-\omega_2)}}{10}\right)^{1/3}\right]$$

$$=3P_{(\omega_1)}-2P_I \quad \text{dBm} \tag{6.25}$$

Example 6.2

Consider the transfer characteristic of a two-port given in Example 6.1, that is, $e_o=15e_i-2e_i^3$; then the intercept power P_I for a 50 Ω source and load is given by (6.19) as $P_I=43.52$ dBm. Now if $P_{(\omega_1)}=-10$ dBm, which is well below the 1 dB gain compression point $P_{1\text{dB}}=32.89$ dBm, the third-order intermodulation power $P_{(2\omega_1-\omega_2)}$ can be evaluated by (6.24) as $P_{(2\omega_1-\omega_2)}=-117$ dBm, that is, the third-order intermodulation product is 107 dB down from the fundamental output power at ω_1.

In summary, when the nonlinearity of a system can be expressed by a power series of three terms, the intercept point is a very convenient way to predict the distortion level of the third-order intermodulation products, which usually fall into the operating frequency band even for narrowband systems. The intercept point can be easily measured indirectly using equation (6.24) by measuring the third-order intermodulation product power and the fundamental frequency power at a level well below the intercept point, that is, in the small-signal region.

6.2 DYNAMIC RANGE

The signal processing ability of a communication receiver is determined largely by the noise level as its lower limit and by the signal distortion as its upper limit. This range is usually called the dynamic range of a communication receiver. Achieving the optimum dynamic range involves tradeoffs between the input signal level with respect to the noise level and the output signal distortion. From (2.57) we see that the noise output power of a two-port with noise figure F is given as

$$N_o=kT_0BGF \quad \text{W} \tag{6.26}$$

where $k=1.374\times10^{-23}$ J/°K, $T_0=290$°K, B is the two-port bandwidth in Hz defined as

$$B=\frac{1}{G}\int_0^{\infty}G_f(f)\,df \tag{6.27}$$

$G_f(f)$ is the measured gain of the two-port and G is the maximum value of

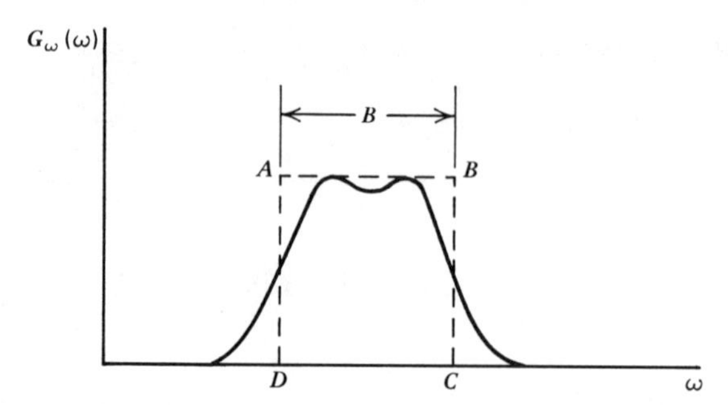

Fig. 6.5. Definition of the noise bandwidth.

$G_f(f)$. This is illustrated in Fig. 6.5, showing the measured gain $G_f(f)$ as a function of the frequency f and the rectangle $ABCD$ having the same area as the area under the curve $G_f(f)$. The width of the rectangle is the noise bandwidth B. In most applications, the overall bandwidth of the system is determined by its bandpass filters. Expressing N_o in dBm we have

$$N_o = -174 \text{ dBm} + 10 \log B + G(\text{dB}) + F(\text{dB}) \tag{6.28}$$

In order for an input signal to be detectable, its output power level must be above the level of the thermal noise power N_o. The higher the noise level is, the higher the input signal level must be used in order to obtain a predetermined signal-to-noise ratio. This in turns increases the third-order intermodulation distortion. From (6.25) we obtain

$$P_{(\omega_1)} - P_{(2\omega_1 - \omega_2)} = \tfrac{2}{3}\left(P_I - P_{(2\omega_1 - \omega_2)}\right) \tag{6.29}$$

If the third-order intermodulation product level is set equal to the noise level, then (6.29) determines the dynamic range of the two-port system defined as

$$\text{DR (dB)} = \tfrac{2}{3}\left[P_I + 174 \text{ dBm} - 10 \log B - G(\text{dB}) - F(\text{dB})\right] \tag{6.30}$$

Example 6.3

Consider a communication receiver with the following parameters: an input noise figure $F = 1.5$ dB, an output maximum gain of 50 dB, an equivalent noise bandwidth $B = 70$ MHz, and an intercept point $P_I = +15$ dBm. The receiver dynamic range is given by (6.30) as

$$\text{DR (dB)} = \tfrac{2}{3}\left[15 + 174 - 10 \log(70)(10^6) - 50 - 1.5\right] = 59 \text{ dB}$$

6.3 CROSS MODULATION IN MEMORYLESS SYSTEMS

Another type of distortion occurs in nonlinear systems when the modulation of one signal is transferred to another signal; in other words, new modulation unwantedly appears. This is called "cross modulation." Consider the case of an input signal that consists of two signals of equal amplitude A at frequencies ω_1 and ω_2, one modulated and the other initially unmodulated as follows:

$$e_i(t)=A(1+M\cos\omega_m t)\cos\omega_1 t+A\cos\omega_2 t \tag{6.31}$$

where M is the modulation index, ω_m is the modulation frequency and ω_1 and ω_2 are the carrier frequencies. Substituting (6.31) into (6.2) yields the following output terms:

$$\begin{aligned}
e_o &= k_1 A(1+M\cos\omega_m t)\cos\omega_1 t+k_1 A\cos\omega_2 t\\
&\quad+k_2A^2\Big[\big(1+2M\cos\omega_m t+\tfrac{1}{2}M^2+\tfrac{1}{2}M^2\cos 2\omega_m t\big)\big(\tfrac{1}{2}+\tfrac{1}{2}\cos 2\omega_1 t\big)\\
&\qquad+\tfrac{1}{2}+\tfrac{1}{2}\cos 2\omega_2 t+(1+M\cos\omega_m t)\{\cos(\omega_1-\omega_2)t+\cos(\omega_1+\omega_2)t\}\Big]\\
&\quad+k_3A^3\Big[\big\{1+3M\cos\omega_m t+3M^2\big(\tfrac{1}{2}+\tfrac{1}{2}\cos 2\omega_m t\big)+M^3\big(\tfrac{3}{4}\cos\omega_m t+\tfrac{1}{4}\cos 3\omega_m t\big)\big\}\\
&\qquad\times\big(\tfrac{3}{4}\cos\omega_1 t+\tfrac{1}{4}\cos 3\omega_1 t\big)+3\big(1+2M\cos\omega_m t+\tfrac{1}{2}M^2+\tfrac{1}{2}M^2\cos 2\omega_m t\big)\\
&\qquad\times\big(\tfrac{1}{2}+\tfrac{1}{2}\cos 2\omega_1 t\big)\cos\omega_2 t+3(1+M\cos\omega_m t)\big(\tfrac{1}{2}+\tfrac{1}{2}\cos 2\omega_2 t\big)\cos\omega_1 t\\
&\qquad+\tfrac{3}{4}\cos\omega_2 t+\tfrac{1}{4}\cos 3\omega_2 t\Big]
\end{aligned} \tag{6.32}$$

From (6.32) the cross modulation term e_{CM} is given as

$$e_{CM}=\big(k_1A+\tfrac{3}{4}k_3A^3\big)\cos\omega_2 t+\big(3k_3A^3M\cos\omega_m t\big)\cos\omega_2 t$$

$$=E_{CM}(1+M'\cos\omega_m t)\cos\omega_2 t \tag{6.33a}$$

where

$$E_{CM}=k_1A+\tfrac{3}{4}k_3A^3 \tag{6.33b}$$

$$M'=\frac{3k_3A^2M}{k_1+\tfrac{3}{4}k_3A^2} \tag{6.33c}$$

Thus the modulation has been transferred to the carrier frequency ω_2. The modulation index M' can be measured at the frequencies $\omega_2 \pm \omega_m$ since

$$e_{CM} = E_{CM}\left[\cos \omega_2 t + \tfrac{1}{2}M'\cos(\omega_2 - \omega_m)t + \tfrac{1}{2}M'\cos(\omega_2 + \omega_m)t\right] \quad (6.34)$$

and the cross modulation factor is defined as the ratio of M' to M, that is,

$$CM = \frac{M'}{M} = \frac{3k_3A^2}{k_1 + \frac{3}{4}k_3A^2} \quad (6.35)$$

At a lower level signal, $k_1A \gg \frac{3}{4}k_3A^3$ and (6.35) can be approximated by

$$CM \approx \frac{3k_3A^2}{k_1} \quad (6.36)$$

and in terms of dB

$$CM\ (\text{dB}) \approx 20 \log CM = 10 \log\left[\frac{3k_3A^2}{k_1}\right]^2 \quad (6.37)$$

Comparing this with (6.14) and (6.16) we obtain the relation between CM (dB) and the intermodulation distortion of the two unmodulated carriers at ω_1 and ω_2 with the same amplitude A

$$CM\ (\text{dB}) = P_{(2\omega_1 - \omega_2)} - P_o - 12 \quad \text{dB}$$

$$= 2(P_o - P_I) - 12 \approx 2\left(P_{(\omega_1)} - P_I\right) - 12 \quad \text{dB} \quad (6.38)$$

Example 6.4

Consider the transfer characteristic $e_o = 15e_i - 2e_i^3$ of a two-port system with 50 Ω source and load. We found in Example 6.2 that the intercept point $P_I = 43.52$ dBm. For a fundamental output power $P_{(\omega_1)} = 20$ dBm, the cross modulation factor is given by (6.38) as CM (dB) $= -79$ dB or $CM = 0.0011$.

6.4 LINEAR DISTORTION. GROUP DELAY

As mentioned in Section 6.1, signal distortion can result in a linear network when the amplitude and phase of the transfer function deviate from the ideal characteristics. If the transfer function $H(j\omega)$ is known, the output voltage $y(t)$ can be found by the inverse Fourier transform as

$$y(t) = \mathcal{F}^{-1}\{H(j\omega)X(j\omega)\} \quad (6.39)$$

where $X(j\omega)$ is the Fourier transform of the input $x(t)$. For example, consider a transfer function of the form

$$H(j\omega)=K(1+a\cos\tau\omega)\exp(-j\omega t_0)$$

whose amplitude is a constant with a superimposed cosinusoidal ripple of frequency τ and relative amplitude a, together with a linear phase shift. $H(j\omega)$ can be written as

$$H(j\omega)=K\left[\exp(-j\omega t_0)+\frac{a}{2}\exp\{j\omega(\tau-t_0)\}+\frac{a}{2}\exp\{j\omega(\tau+t_0)\}\right]$$

For an input $x(t)$, the resulting output $y(t)$ can be computed using (6.39) as

$$y(t)=Kx(t-t_0)+\frac{Ka}{2}x(t-t_0+\tau)+\frac{Ka}{2}x(t-t_0-\tau) \qquad (6.40)$$

Thus the amplitude ripple results in the generation of two additional scaled and delayed versions of $x(t)$, which are usually called echoes. As mentioned before, a linear phase shift implies a constant time delay for all the frequency components of the signal. If the phase shift is a nonlinear function of frequency, the various frequency components of the signal will experience different time delays and this results in phase distortion. A convenient indication of phase distortion is the "group delay," defined as the negative of the derivative of the phase shift versus frequency as follows:

$$t_d=-\frac{d\phi}{d\omega} \qquad (6.41)$$

When the phase shift of a system is a linear function of the frequency, the group delay is a constant, which indicates that there is no phase distortion. An example of linear and nonlinear phase shift versus frequency is shown in Fig. 6.6. The group delay can be measured by amplitude-modulating a carrier, transmitting it through the system and then measuring the phase shift of the envelope as shown

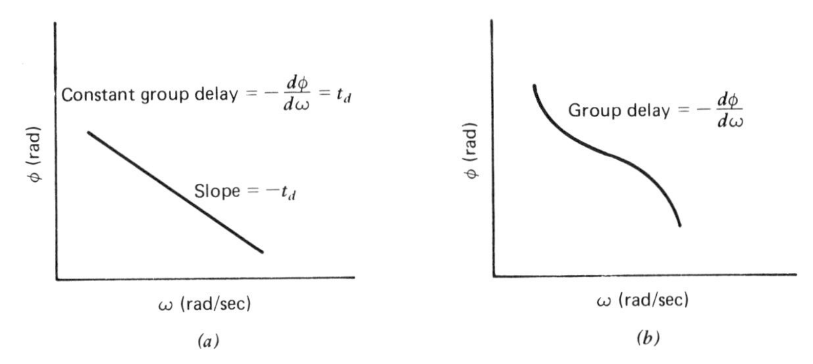

Fig. 6.6. Phase shift versus frequency. (*a*) Linear. (*b*) Nonlinear.

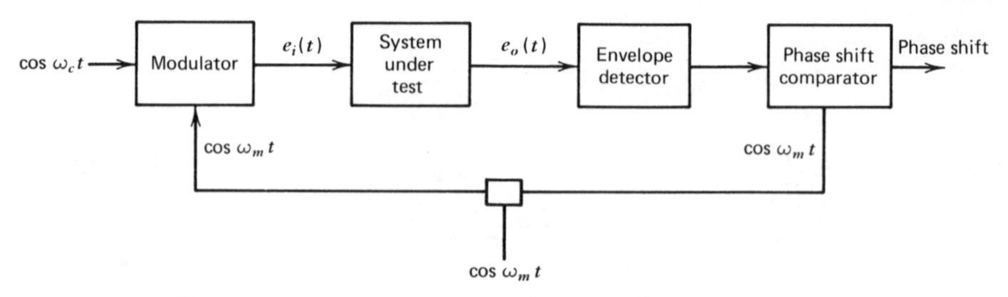

Fig. 6.7. Schematic diagram of group delay measurement.

in Fig. 6.7. The amplitude-modulated input signal can be expressed as

$$e_i = E_i(1 + M\cos\omega_m t)\cos\omega_c t$$

$$= E_i\cos\omega_c t + \frac{ME_i}{2}\cos(\omega_c + \omega_m)t + \frac{ME_i}{2}\cos(\omega_c - \omega_m)t \qquad (6.42)$$

where M is the modulation index, ω_m is the modulating frequency, ω_c is the carrier frequency and E_i is the amplitude selected such that e_i is a small signal for the system. The output signal of the system under test with small-signal gain K can be expressed as

$$e_o = KE_i\cos\omega_c(t - t_d) + \frac{KME_i}{2}\cos(\omega_c + \omega_m)(t - t_d)$$

$$+ \frac{KME_i}{2}\cos(\omega_c - \omega_m)(t - t_d)$$

$$= KE_i[1 + M\cos\omega_m(t - t_d)]\cos\omega_c(t - t_d) \qquad (6.43)$$

where t_d is the group delay of the system under test, assumed to be constant from $\omega_c - \omega_m$ to $\omega_c + \omega_m$. The modulation frequency ω_m determines the bandwidth of the system within which fluctuations in group delay can be measured. The smaller ω_m, the more accurate t_d can be measured. However, a smaller ω_m results in a smaller phase shift and decreases the resolution. In practice ω_m can be chosen between 1 and 10 MHz for small group delay measurement and 20 KHz for larger group delay. Since the phase shift is measured only at ω_m, the carrier frequency ω_c can be swept over the entire bandwidth of the system, and thus the group delay over the system bandwidth can be measured. A relationship between group delay t_d and the phase shift of the modulation signal that is the envelope of the signal e_i can be expressed as

$$t_d = \frac{\phi_e}{\omega_m} \qquad (6.44)$$

where ϕ_e is the modulation envelope phase shift in radians.

6.5 AM-TO-PM CONVERSION

Besides the distortion caused by a nonlinear phase characteristic as a function of frequency, a two-port system can have another type of phase distortion where the phase shift is a function of the instantaneous amplitude of the signal such as that of amplitude-modulated signals. For example, the output phase can have a ripple around a mean phase value as is shown in Fig. 6.8. This effect is called AM-to-PM conversion. Let θ_p be the peak phase deviation from the mean phase cause by an amplitude-modulated carrier with modulation index M. The peak phase error K_p can be expressed in degrees/dB of AM as

$$K_p = \frac{\theta_p(180°/\pi)}{20\log(1+M)} \quad \text{degree/dB} \tag{6.45}$$

Example 6.5

Consider a system whose output phase shift θ is a function of the instantaneous input amplitude $a(t)=A(1+M\cos\omega_m t)$ of an amplitude-modulated signal $a(t)\cos\omega_c t$ as follows:

$$\theta(a)=ca^2(t)$$

Thus

$$\begin{aligned}\theta(a)&=cA^2\left(1+2M\cos\omega_m t+M^2\cos^2\omega_m t\right) \quad \text{radians}\\ &\approx cA^2(1+2M\cos\omega_m t), \quad M\ll 1\\ &\approx \theta_0+\theta_p\cos\omega_m t\end{aligned}$$

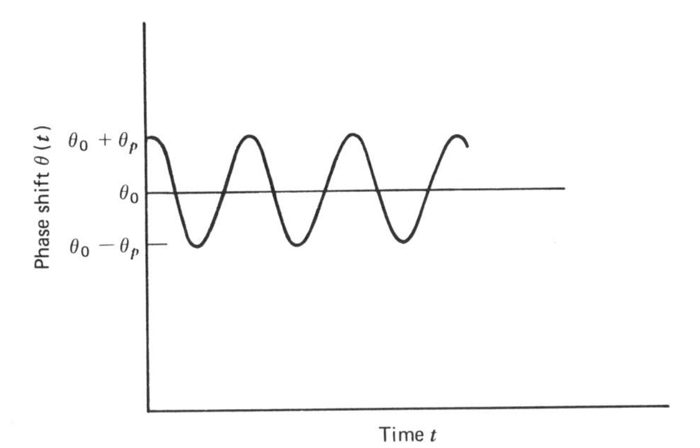

Fig. 6.8. Phase modulation induced by amplitude modulation.

where $\theta_0 = cA^2$ and $\theta_p = 2cMA^2$. The peak phase error K_p is given by (6.45) as

$$K_p = \frac{2cMA^2(180°/\pi)}{20\log(1+M)} \approx \frac{2cMA^2(180°/\pi)}{8.69M} = 13.2cA^2 \quad \text{degree/dB}$$

6.6 FREQUENCY-DOMAIN DISTORTION ANALYSIS: VOLTERRA FUNCTIONAL SERIES APPROACH

In Section 6.2 we studied the "nonlinear distortion" of a two-port system using the power series analysis method by assuming the system is memoryless. In this case, only amplitude nonlinearities are considered, since any phase information such as an AM-to-PM conversion cannot be represented by a power series alone. In many applications, the power series representation of signal distortion is adequate for analysis if the AM-to-PM conversion of the system is negligible, that is, the amplitude nonlinearity dominates the AM-to-PM conversion. In communication systems where the phase distortion is a main concern, such representation is inadequate. In practice, both amplitude and phase nonlinearities exist in systems with memory, that is, systems whose distortion is frequency dependent. The nonlinearity analysis can be best represented by a Volterra functional series, which has been described as a "power series with memory." Volterra series are very useful in calculating small but troublesome distortions in communication receivers where the first few terms are sufficient to characterize them.

6.6.1 Input-Output Representation by Volterra Functional Series

It has been shown that every functional $y(t) = T[x(t)]$ that is continuous in the field of continuous functions can be represented by a series of the form

$$y(t) = \sum_{n=1}^{\infty} y_n(t) = \int_{-\infty}^{\infty} h_1(u_1)x(t-u_1)\,du_1$$

$$+ \iint_{-\infty}^{\infty} h_2(u_1, u_2)x(t-u_1)x(t-u_2)\,du_1\,du_2$$

$$+ \iiint_{-\infty}^{\infty} h_3(u_1, u_2, u_3)x(t-u_1)x(t-u_2)x(t-u_3)\,du_1\,du_2\,du_3 + \cdots + \tag{6.46a}$$

where

$$y_n(t)=\int_{-\infty}^{\infty}\cdots\int h_n(u_1,u_2,\ldots,u_n)x(t-u_1)x(t-u_2)\cdots x(t-u_n)\,du_1\,du_2\cdots du_n \tag{6.46b}$$

The nth-order kernel $h_n(u_1, u_2,\ldots, u_n)$ in (6.46b) is also called the "nth-order impulse response." Note that the first-order kernel $h_1(u_1)$ is simply the familiar impulse response of a linear network and is a generalization of the first term in the power series expansion (6.1). The second term in (6.46a) is of a quadratic nature and $y_2(t)$ is a two-dimensional convolution of $x(t)$ and the second order impulse $h_2(u_1, u_2)$; this term is a generalization of the second term in (6.1). Similarly, the third term in (6.46a) is of a cubic nature and its response $y_3(t)$ is a three-dimensional convolution of $x(t)$ and the third order impulse response $h_3(u_1, u_2, u_3)$; this term is a generalization of the third term in (6.1). Thus the use of Volterra functional series in the analysis of nonlinear system is a generalization of convolution integral used in linear system analysis. Furthermore, in each term of the Volterra series, the output depends on the past values of the input; therefore the Volterra series do indeed represent frequency dependent systems. In practice, the first three terms of (6.46) are adequate for weak nonlinearity, which is the case for most communication receivers. An equivalent block diagram for a nonlinear system based upon the functional expansion is shown in Fig. 6.9*a–b* for the first n terms.

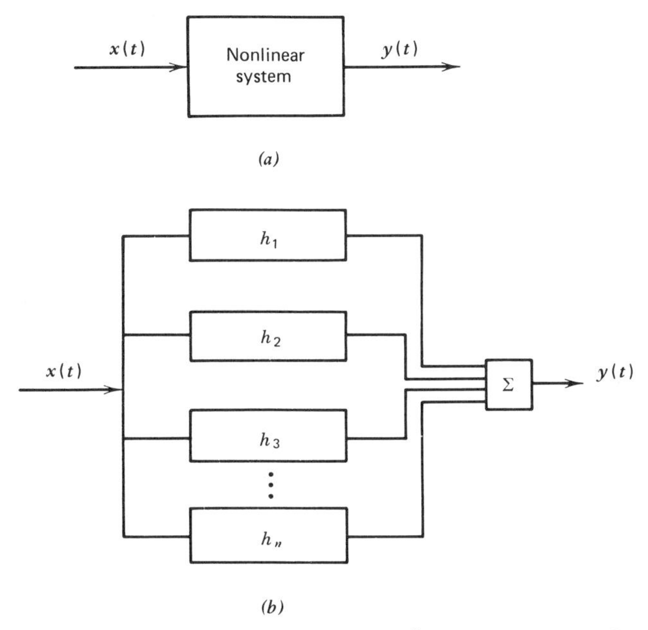

Fig. 6.9. (*a*) A nonlinear system. (*b*) Functional expansion of a nonlinear system.

The n-dimensional Fourier transform of the nth order impulse response $h_n(u_1, u_2, \ldots, u_n)$ is called the "nth-order nonlinear transfer function" and is given by

$$H_n(\omega_1, \omega_2, \ldots \omega_n) = \int_{-\infty}^{\infty} \cdots \int h_n(u_1, u_2, \ldots, u_n) \exp[-j(\omega_1 u_1 + \omega_2 u_2 + \cdots + \omega_n u_n)]\, du_1 du_2 \ldots du_n \qquad (6.47)$$

Conversely,

$$h_n(u_1, u_2, \ldots, u_n) = \int_{-\infty}^{\infty} \cdots \int H_n(\omega_1, \omega_2, \ldots, \omega_n) \exp[j(\omega_1 u_1 + \omega_2 u_2 + \cdots + \omega_n u_n)]\, d\omega_1 d\omega_2 \ldots d\omega_n \qquad (6.48)$$

From (6.46b) we observe that the nth-order kernel h_n is not unique in the sense that several distinct nth-order kernel may give the same nth order output y_n for the same input x, since an interchange of the argument of $h_n(u_1, u_2, \ldots, u_n)$ does not affect the input-output relationship. Hence there may be up to $n!$ distinct nth-order kernels corresponding to the $n!$ permutation of n variables. However, the symmetrized kernel h'_n defined by

$$h'_n(u_1, u_2, \ldots, u_n) = \frac{1}{n!} \sum_{\begin{bmatrix}\text{all permutations}\\ \text{of } u_1, u_2, \ldots, u_n\end{bmatrix}} h_n(u_1, u_2, \ldots, u_n) \qquad (6.49)$$

is unique. The corresponding nth-order transfer function H'_n of h'_n can also be shown to be symmetrized as seen from (6.47) as

$$H'_n(\omega_1, \omega_2, \ldots \omega_n) = \frac{1}{n!} \sum_{\begin{bmatrix}\text{all permutations}\\ \text{of } \omega_1, \omega_2, \ldots, \omega_n\end{bmatrix}} H_n(\omega_1, \omega_2, \ldots, \omega_n) \qquad (6.50)$$

For the rest of the discussion, we will assume that all the nth-order impulse response and transfer functions of a nonlinear system are symmetrized.

Substituting (6.47) into (6.46b) yields

$$y_n(t) = \int_{-\infty}^{\infty} \cdots \int H_n(\omega_1, \omega_2, \ldots, \omega_n) \prod_{i=1}^{n} X(\omega_i) \exp(j\omega_i t)\, d\omega_i \qquad (6.51)$$

where $X(\omega_i)$ is the Fourier transform of $x(t)$ with frequency ω_i

$$X(\omega_i) = \int_{-\infty}^{\infty} x(t) \exp(-j\omega_i t)\, dt \qquad (6.52)$$

Taking the Fourier transform of both sides of (6.51) yields the "nth-order output spectrum"

$$Y_n(\omega)=\int_{-\infty}^{\infty}\cdots\int H_n(\omega_1,\omega_2,\ldots,\omega_n)\,\delta(\omega-\omega_1-\omega_2-\cdots-\omega_n)\prod_{i=1}^{n}X(\omega_i)\,d\omega_i \tag{6.53}$$

where $\delta(\cdot)$ is the familiar Dirac function. The input-output spectrum relationship is obtained by taking the Fourier transform of both sides of (6.46a) as

$$Y(\omega)=\sum_{n=1}^{\infty}Y_n(\omega) \tag{6.54}$$

6.6.2 Computation of Nonlinear Transfer Functions

In this section we are primarily interested in the calculation of the nonlinear transfer functions when the system equations are known and the system's input-output description can be represented by a Volterra series. One way to achieve this is to use a harmonic input method. This relies on the fact that a harmonic input must result in a harmonic output for (6.46a). Let the input to the system $x(t)$ be the sum of exponentials

$$x(t)=\sum_{k=1}^{K}A_k\exp(j\omega_k t) \tag{6.55}$$

where A_k may be complex and ω_k may be any positive or negative real number. Substituting (6.55) into (6.46b) yields

$$y_n(t)=\int_{-\infty}^{\infty}\cdots\int h_n(u_1,u_2,\ldots,u_n)\prod_{i=1}^{n}\sum_{k=1}^{K}A_k\exp[j\omega_k(t-u_i)]\,du_i$$

$$=\sum_{k_1=1}^{K}\sum_{k_2=1}^{K}\cdots\sum_{k_n=1}^{K}\int_{-\infty}^{\infty}\cdots\int h_n(u_1,u_2,\ldots,u_n)\sum_{i=1}^{n}A_{k_i}\exp[j\omega_{k_i}(t-u_i)]\,du_i$$

$$=\sum_{k_1=1}^{K}\sum_{k_2=1}^{K}\cdots\sum_{k_n=1}^{K}\prod_{i=1}^{n}A_{k_i}\exp(j\omega_{k_i}t)$$

$$\times\int_{-\infty}^{\infty}\cdots\int h_n(u_1,u_2,\ldots,u_n)\prod_{i=1}^{n}\exp(-j\omega_{k_i}u_i)\,du_i \tag{6.56}$$

By combining (6.47) and (6.56) we obtain

$$y_n(t) = \sum_{k_1=1}^{K} \sum_{k_2=1}^{K} \cdots \sum_{k_n=1}^{K} \left[\prod_{i=1}^{n} A_{k_i} \exp(j\omega_{k_i} t) \right] H_n(j\omega_{k_1}, j\omega_{k_2}, \ldots, j\omega_{k_n})$$

$$= \sum_{k_1=1}^{K} \sum_{k_2=1}^{K} \cdots \sum_{k_n=1}^{K} \left[\left(\prod_{i=1}^{n} A_{k_i} \right) H_n(j\omega_{k_1}, j\omega_{k_2}, \ldots, j\omega_{k_n}) \times \exp\left[j(\omega_{k_1} + \omega_{k_2} + \cdots + \omega_{k_n}) t \right] \right] \tag{6.57}$$

Example 6.6

Let $x(t) = A_1 \exp(j\omega_1 t) + A_2 \exp(j\omega_2 t)$, that is, $K=2$ in (6.55). We wish to compute $y_2(t)$, that is, $n=2$ in (6.46b). From (6.57) we have

$$y_2(t) = \sum_{k_1=1}^{2} \sum_{k_2=1}^{2} A_{k_1} A_{k_2} H_2(j\omega_{k_1}, j\omega_{k_2}) \exp\left[j(\omega_{k_1} + \omega_{k_2}) t \right]$$

$$= \sum_{k_1=1}^{2} A_{k_1} A_1 H_2(j\omega_{k_1}, j\omega_1) \exp\left[j(\omega_{k_1} + \omega_1) t \right] + A_{k_1} A_2 H_2(j\omega_{k_1}, j\omega_2) \exp\left[j(\omega_{k_1} + \omega_2) t \right]$$

$$= A_1 A_1 H_2(j\omega_1, j\omega_1) \exp(j2\omega_1 t) + A_1 A_2 H_2(j\omega_1, j\omega_2) \exp\left[j(\omega_1 + \omega_2) t \right] + A_2 A_1 H_2(j\omega_2, j\omega_1) \exp\left[j(\omega_2 + \omega_1) t \right] + A_2 A_2 H_2(j\omega_2, j\omega_2) \exp(j2\omega_2 t)$$

From (6.57) we note that in the specific case where $K=n$ and $A_k=1$ for $k=1,2,\ldots,n$, that is, $x(t) = \prod_{k=1}^{n} \exp(j\omega_k t)$, then

$$y_n(t) = \sum_{k_1=1}^{n} \sum_{k_2=1}^{n} \cdots \sum_{k_n=1}^{n} H_n(j\omega_{k_1}, j\omega_{k_2}, \ldots, j\omega_{k_n}) \times \exp\left[j(\omega_{k_1} + \omega_{k_2} + \cdots + \omega_{k_n}) t \right] \tag{6.58}$$

There are exactly $n!$ terms in $y_n(t)$ in (6.58) with $\exp[j(\omega_1 + \omega_2 + \cdots + \omega_n)t]$. Each corresponds to a permutation of $\omega_1, \omega_2, \ldots, \omega_n$ in the argument of $H_n(j\omega_1, j\omega_2, \ldots, j\omega_n)$. If $y(t)$ contains no terms associated with $\exp[j(\omega_1 + \omega_2 + \cdots + \omega_n)t]$ than these $n!$ terms in $y_n(t)$, then it follows from (6.50) that

$$H_n'(j\omega_1, j\omega_2, j\omega_n) = \frac{1}{n!} \sum \text{coefficient of } \exp\left[j(\omega_1 + \omega_2 + \cdots + \omega_n) t \right] \text{ in } y(t) \tag{6.59}$$

Thus the symmetrized nonlinear transfer function can be found using harmonic inputs if the frequencies $\omega_1, \omega_2, \ldots, \omega_n$ are linearly independent, that is, there are no rational numbers $m_1, m_2, \ldots, m_n$ (not all zero) such that

$$\sum_{i=1}^{n} m_i \omega_i = 0 \tag{6.60}$$

For example, let $\omega_1 = \sqrt{3}$, $\omega_2 = 2\sqrt{3} - 2$ and $\omega_3 = 2$, then $\omega_1 + \omega_2 + \omega_3 = 3\omega_1 = 3\sqrt{3}$. Since $\omega_1 + \omega_1 + \omega_1 = 3\omega_1$ is not one of the 3! frequencies obtained by the permutation of $(\omega_1, \omega_2, \omega_3)$, the use of (6.59) will not be valid. Note that ω_1, ω_2, and ω_3 are linearly dependent, since $2\omega_1 - \omega_2 - \omega_3 = 0$. In conclusion, if the input frequencies $\omega_1, \omega_2, \ldots, \omega_n$ are linearly independent, then $y(t)$ does not contain output terms with frequency $\omega_1 + \omega_2 + \cdots + \omega_n$ other than $n!$ terms in $y_n(t)$.

Since any output frequency $\omega_{k_1} + \omega_{k_2} + \cdots + \omega_{k_n}$ in (6.57) can be expressed as $m_1\omega_1 + m_2\omega_2 + \cdots + m_k\omega_k$ where m_i, $i = 1, 2, \ldots, K$ are non-negative integers, we see that if $m_1 + m_2 + \cdots + m_k = n$, then $\omega_0 = m_1\omega_1 + m_2\omega_2 + \cdots + m_k\omega_k$ is the frequency generated by the nth-order nonlinear transfer function. Furthermore, if the nth-order transfer function is symmetrized and all the input frequencies are linearly independent, then the sum of all terms with frequency $\omega_0 = m_1\omega_1 + m_2\omega_2 + \cdots + m_K\omega_K$ in (6.58) is given by

$$y_{\omega_0}(t) = n! \left[\prod_{k=1}^{K} \frac{A_k^{m_k}}{m_k!} \right] H_n(m_1[j\omega_1], m_2[j\omega_2], \ldots, m_K[j\omega_K]) \exp(j\omega_0 t) \tag{6.61}$$

where

$$\sum_{k=1}^{K} m_k = n \qquad \text{and} \qquad m_i[j\omega_i] = (\overbrace{j\omega_i, j\omega_i, \ldots, j\omega_i}^{i \text{ times}})$$

that is, m_i consecutive arguments in $H_n(\cdot)$ having the same frequency ω_i.

Example 6.7

Let $K = 2$, $m_1 = 1$, $m_2 = 2$. Then $n = m_1 + m_2 = 3$ and

$$y_{\omega_0}(t) = 3! \left(\prod_{k=1}^{2} \frac{A_k^{m_k}}{m_k!} \right) H_3(j\omega_1, j\omega_2, j\omega_2) \exp[j(\omega_1 + 2\omega_2)t]$$

$$= 3! \frac{A_1}{1!} \frac{A_2^2}{2!} H_3(j\omega_1, j\omega_2, j\omega_2) \exp[j(\omega_1 + 2\omega_2)t]$$

Note that the harmonic input method for finding the nonlinear transfer functions is an analytical method and cannot serve as a basis for measurements. The method can be used recursively once the system equation is known. The determination of $H_1(j\omega_1)$ is made by letting the input $x(t)=\exp(j\omega_1 t)$. Then the sum of two exponential inputs is applied to determine $H_2(j\omega_1, j\omega_2)$. This procedure is continued with an exponential added to the input at each step.

Example 6.8

Consider a nonlinear network contained a linear capacitor, a linear resistor, a nonlinear resistor, and a current source, all in parallel as shown in Fig. 6.10. The current through the nonlinear resistor is given as $G_2 v^2$ where v is the output voltage. The system equation is given by

$$i(t)=C\frac{dv(t)}{dt}+G_1 v(t)+G_2 v^2(t) \tag{6.62}$$

Here the input is $x(t)=i(t)$ and the output is $y(t)=v(t)$. In order to determine the nonlinear transfer function, first let $i(t)=\exp(j\omega_1 t)$. Then using (6.57) with $n=1$, we get, with $y_1(t)=v(t)=H_1(j\omega_1)\exp(j\omega_1 t)$ by equating the coefficients of $\exp(j\omega_1 t)$,

$$1=(j\omega_1 C+G_1)H_1(j\omega_1)$$

and hence

$$H_1(j\omega_1)=\frac{1}{j\omega_1 C+G_1} \tag{6.63}$$

Thus the first-order nonlinear transfer function is simply the solution of (6.62) as if the nonlinear resistor were an open circuit. Now let $i(t)=\exp(j\omega_1 t)+\exp(j\omega_2 t)$ and substitute $y_1(t)+y_2(t)=v(t)$ into (6.62) where $y_1(t)$ is obtained from (6.57) by letting $K=2$, $n=1$ and $A_1=A_2=1$, that is,

$$\begin{aligned} y_1(t) &= \sum_{k_1=1}^{2} H_1(j\omega_{k_1})\exp(j\omega_{k_1} t) \\ &= H_1(j\omega_1)\exp(j\omega_1 t)+H_1(j\omega_2)\exp(j\omega_2 t) \end{aligned} \tag{6.64}$$

and $y_2(t)$ is obtained from (6.57) by letting $K_2=2$, $n=2$ and $A_1=A_2=1$

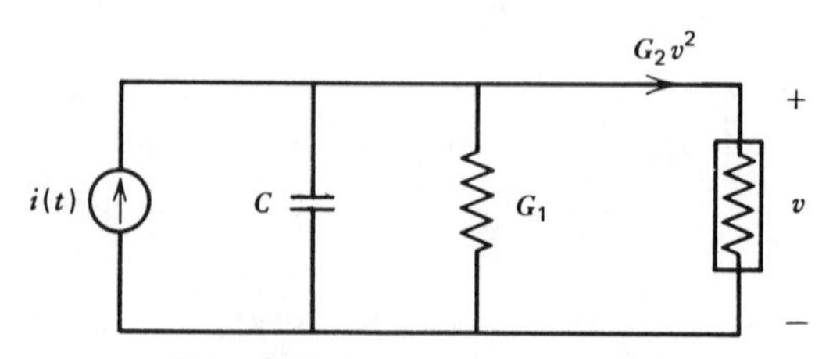

Fig. 6.10. A nonlinear network.

(Example 6.6). After equating the coefficients of $\exp[j(\omega_1+\omega_2)t]$ we get

$$0=2[j(\omega_1+\omega_2)C+G_1]H_2(j\omega_1,j\omega_2)+2G_2H_1(j\omega_1)H_2(j\omega_2) \quad (6.65)$$

Solving for $H_2(j\omega_1,j\omega_2)$ using (6.63) we get

$$H_2(j\omega_1,j\omega_2)=-G_2H_1(j\omega_1)H_2(j\omega_2)H_1[j(\omega_1+\omega_2)]$$
$$=\frac{-G_2}{(j\omega_1C+G_1)(j\omega_2C+G_1)[j(\omega_1+\omega_2)C+G_1]} \quad (6.66)$$

The procedure can be continued indefinitely to find higher-order $H(\cdot)$ in terms of lower-order nonlinear transfer functions.

6.6.3 Multitone Measurements

In this section we relate the analytical results of the previous section to the common method of multitone measurement of nonlinear systems as discussed in the case of memoryless systems. Consider a K-tone real input $x(t)$ given by

$$x(t)=\sum_{k=1}^{K}|A_k|\cos(\omega_kt+\angle A_k)$$
$$=\sum_{k=1}^{K}\left[\frac{A_k}{2}\exp(j\omega_kt)+\frac{A_k^*}{2}\exp(-j\omega_kt)\right] \quad (6.67)$$

where A_k can be a complex constant. Denote $A_{-k}=A_k^*$ and $\omega_{-k}=-\omega_k$. Then (6.67) can be written as

$$x(t)=\sum_{\substack{k=-K\\k\neq 0}}^{K}\frac{A_k}{2}\exp(j\omega_kt) \quad (6.68)$$

Now consider the output terms that contain the frequency

$$\omega_0=\sum_{\substack{k=-K\\k\neq 0}}^{K}m_k\omega_k=\sum_{k=1}^{K}(m_k\omega_k+m_{-k}\omega_{-k})=\sum_{k=1}^{K}(m_k-m_{-k})\omega_k$$

where

$$\sum_{\substack{k=-K\\k\neq 0}}^{K}m_k=n$$

and m_k $(k=\pm1,\pm2,\ldots,\pm K)$ are nonnegative integers (note that the input frequencies $\omega_{-k},\ldots,\omega_{-1},\omega_1,\ldots,\omega_k$ are not linearly independent since $\omega_{-k}+\omega_k=0$ for $k=\pm1,\pm2,\ldots,\pm K$). Using this information in (6.58) and (6.61) we obtain

$$y_{\omega_0}(t)=n!\left[\prod_{\substack{k=-K\\k\neq0}}^{K}\frac{(A_k/2)^{m_k}}{m_k!}\right]H_n(m_{-k}[j\omega_{-k}],\ldots,m_{-1}[j\omega_{-1}],$$

$$m_1[j\omega_1],\ldots,m_k[j\omega_k])\exp(j\omega_0 t) \quad (6.69)$$

Now assume that $\omega_1,\omega_2,\ldots,\omega_k$ are linearly independent. We see that for any output frequency

$$\omega_0=\sum_{k=1}^{K}\alpha_k\omega_k$$

the lowest nth-order nonlinear transfer function that gives this frequency is given by

$$n=\sum_{k=1}^{K}|\alpha_k|$$

and in this case the m_k are unique and given as, for $k=1,\ldots,K$

$$\begin{cases} m_k=\alpha_k & \text{and } m_{-k}=0 \quad \text{if } \alpha_k\geqslant0\\ m_k=0 & \text{and } m_{-k}=-\alpha_k \quad \text{if } \alpha_k<0\end{cases} \quad (6.70)$$

The proof is as follows. Suppose there exist m'_k, $k'=\pm1,\pm2,\ldots,\pm K$ such that

$$\sum_{\substack{k=-K\\k\neq0}}^{K} m'_k\leqslant n,\ m_k\geqslant0$$

and that

$$\omega_0=\sum_{\substack{k=-K\\k\neq0}}^{K} m'_k.$$

Then it is seen that $m'_k-m'_{-k}=m_k-m_{-k}=\alpha_k$ for $k=1,2,\ldots,K$. This implies that $m'_k-m_k=m'_{-k}-m_k$. Since m'_k and m'_{-k} are nonnegative integers and either m_k or m_{-k} or both are equal to zero by (6.70), $m'_k-m_k=m'_{-k}-m_{-k}\geqslant0$. Thus $m'_k+m'_{-k}\geqslant m_k+m_{-k}$. Since by assumption $m'_k\neq m_k$, there exists an i such that $m'_i-m_i=m'_{-i}-m_{-i}>0$, that is, $m'_i+m'_{-i}>m_i+m_{-i}$ and

$$\sum_{k=1}^{K} m'_k+m'_{-k}>\sum_{k=1}^{K} m_k+m_{-k}>n$$

This contradicts the assumption that

$$\sum_{k=1}^{K} m'_k + m'_{-k} \leqslant n.$$

It can also be proved that all nth-order output frequencies are present in the $(n+2)$th-order output.

Example 6.9

Consider the case $K=4$ and

$$\omega_0 = \sum_{k=1}^{4} \alpha_k \omega_k = 2\omega_1 + 0\omega_2 + (-1)\omega_3 + (-1)\omega_4$$

Thus the lowest order n is given by $n=2+|-1|+|-1|=4$ and the unique m_k that give rise to ω_0 is $m_{-1}=m_{-2}=m_2=m_3=m_4=0$, $m_1=2$, $m_{-3}=m_{-4}=1$.

Returning to (6.67) and considering the two-frequency input test with $K=2$, we have

$$x(t) = \sum_{\substack{k=-2 \\ k\neq 0}}^{2} |A_k| \cos(\omega_k t + \angle A_k)$$

Then the nonlinear transfer function $H_n(\cdot)$ can be found from (6.69) for $n=1,2,3$ as shown in Table 6.1 (H_n^* is replaced by H_n with a negative argument). The real responses at ω_1 and $2\omega_1-\omega_2$ are given as $y_{(\omega_1)}(t)=y_{\omega_1}(t)+y_{-\omega_1}(t)$ and $y_{(2\omega_1-\omega_2)}(t)=y_{2\omega_1-\omega_2}(t)+y_{-2\omega_1+\omega_2}(t)$

$$\begin{aligned} y_{(\omega_1)}(t) = {} & \left[\tfrac{1}{2}A_1 H_1(j\omega_1) + \tfrac{3}{8}|A_1|^2 A_1 H_3(-j\omega_1, j\omega_1, j\omega_1)\right. \\ & \left. + \tfrac{3}{4}|A_2|^2 A_1 H_3(-j\omega_2, j\omega_1, j\omega_2)\right] \exp(j\omega_1 t) \\ & + \left[\tfrac{1}{2}A_1^* H_1(-j\omega_1) + \tfrac{3}{8}|A_1|^2 A_1^* H_3(j\omega_1, -j\omega_1, -j\omega_1)\right. \\ & \left. + \tfrac{3}{4}|A_2|^2 A_1^* H_3(j\omega_2, -j\omega_1, j\omega_2)\right] \exp(-j\omega_1 t) \end{aligned} \tag{6.71a}$$

$$\begin{aligned} y_{(2\omega_1-\omega_2)}(t) = {} & \left[\tfrac{3}{8}A_1^2 A_2^* H_3(j\omega_1, j\omega_1, -j\omega_2)\right] \\ & \times \exp[j(2\omega_1-\omega_2)t] + \left[\tfrac{3}{8}A_1^{*2} A_2 H_3(-j\omega_1, -j\omega_1, j\omega_2\right] \\ & \times \exp[-j(2\omega_1-\omega_2)t] \end{aligned} \tag{6.71b}$$

Table 6.1 Nonlinear transfer functions from two-frequency input test

n	Item	m_{-2}	m_{-1}	m_1	m_2	ω_0	Coefficient of $\exp(j\omega_0 t)$
1	1	0	0	0	1	ω_2	$\frac{1}{2}A_2H_1(j\omega_2)$
	2	0	0	1	0	ω_1	$\frac{1}{2}A_1H_1(j\omega_1)$
	3	0	1	0	0	$-\omega_1$	$\frac{1}{2}A_1^*H_1(-j\omega_1)$
	4	1	0	0	0	$-\omega_2$	$\frac{1}{2}A_2^*H_1(-j\omega_2)$
2	1	0	0	0	2	$2\omega_2$	$\frac{1}{4}A_2^2H_2(j\omega_2, j\omega_2)$
	2	0	0	1	1	$\omega_1+\omega_2$	$\frac{1}{2}A_1A_2H_2(j\omega_1, j\omega_2)$
	3	0	0	2	0	$2\omega_1$	$\frac{1}{4}A_1^2H_2(j\omega_1, j\omega_1)$
	4	0	1	0	1	$-\omega_1+\omega_2$	$\frac{1}{2}A_1^*A_2H_2(-j\omega_1, j\omega_2)$
	5	0	1	1	0	0	$\frac{1}{2}\lvert A_1\rvert^2H_2(-j\omega_1, j\omega_1)$
	6	0	2	0	0	$-2\omega_1$	$\frac{1}{4}A_1^*H_2(-j\omega_1, j\omega_1)$
	7	1	0	0	1	0	$\frac{1}{2}\lvert A_2\rvert^2H_2(-j\omega_2, -j\omega_2)$
	8	1	0	1	0	$-\omega_2+\omega_1$	$\frac{1}{2}A_1A_2^*H_2(j\omega_1, -j\omega_2)$
	9	1	1	0	0	$-\omega_1-\omega_2$	$\frac{1}{2}A_1^*A_2^*H_2(-j\omega_1, -j\omega_2)$
	10	2	0	0	0	$-2\omega_2$	$\frac{1}{4}A_2^{*2}H_2(-j\omega_2, -j\omega_2)$
3	1	0	0	0	3	$3\omega_2$	$\frac{1}{8}A_2^3H_3(j\omega_2, j\omega_2, j\omega_2)$
	2	0	0	1	2	$\omega_1+2\omega_2$	$\frac{3}{8}A_1A_2^2H_3(j\omega_1, j\omega_2, j\omega_2)$
	3	0	0	2	1	$2\omega_1+\omega_2$	$\frac{3}{8}A_1^2A_2H_3(j\omega_1, j\omega_1, j\omega_2)$
	4	0	0	3	0	$3\omega_1$	$\frac{1}{8}A_1^3H_3(j\omega_1, j\omega_1, j\omega_1)$
	5	0	1	0	2	$-\omega_1+2\omega_2$	$\frac{3}{8}A_1^*A_2^2H_3(-j\omega_1, j\omega_2, j\omega_2)$
	6	0	1	1	1	ω_2	$\frac{3}{4}\lvert A_1\rvert^2A_2H_3(-j\omega_1, j\omega_1, j\omega_2)$
	7	0	1	2	0	ω_1	$\frac{3}{8}\lvert A_1\rvert^2A_1H_3(-j\omega_1, j\omega_1, j\omega_1)$
	8	0	2	0	1	$-2\omega_1+\omega_2$	$\frac{3}{8}A_1^{*2}A_2H_3(-j\omega_1, -j\omega_1, j\omega_1)$
	9	0	2	1	0	$-\omega_1$	$\frac{3}{8}\lvert A_1\rvert^2A_1^*H_3(-j\omega_1, -j\omega_1, j\omega_1)$
	10	0	3	0	0	$-3\omega_1$	$\frac{1}{8}A_1^{*3}H_3(-j\omega_1, -j\omega_1, -j\omega_1)$
	11	1	0	0	2	ω_2	$\frac{3}{8}\lvert A_2\rvert^2A_2H_3(-j\omega_2, j\omega_2, j\omega_2)$
	12	1	0	1	1	ω_1	$\frac{3}{4}\lvert A_2\rvert^2A_1H_3(j\omega_1, -j\omega_2, j\omega_2)$
	13	1	0	2	0	$2\omega_1-\omega_2$	$\frac{3}{8}A_1^2A_2^*H_3(j\omega_1, j\omega_1, -j\omega_2)$
	14	1	1	0	1	$-\omega_1$	$\frac{3}{4}\lvert A_2\rvert^2A_1^*H_3(-j\omega_1, -j\omega_2, j\omega_2)$
	15	1	1	1	0	$-\omega_2$	$\frac{3}{4}\lvert A_1\rvert^2A_2^*H_3(-j\omega_1, j\omega_1, -j\omega_2)$
	16	1	2	0	0	$-\omega_2-2\omega_1$	$\frac{3}{8}A_1^{*2}A_2^*H_3(-j\omega_1, -j\omega_1, -j\omega_2)$
	17	2	0	0	1	$-\omega_2$	$\frac{3}{8}\lvert A_2\rvert^2A_2^*H_3(-j\omega_2, -j\omega_2, j\omega_2)$
	18	2	0	1	0	$-2\omega_2+\omega_1$	$\frac{3}{8}A_2^{*2}A_1H_3(j\omega_1, -j\omega_2, -j\omega_2)$
	19	2	1	0	0	$-2\omega_2-\omega_1$	$\frac{3}{8}A_2^{*2}A_1^*H_3(-j\omega_1, -j\omega_2, -j\omega_2)$
	20	3	0	0	0	$-3\omega_2$	$\frac{1}{8}A_2^{*3}H_3(-j\omega_2, -j\omega_2, -j\omega_2)$

or equivalently,

$$y_{(\omega_1)}(t) = |B_1|\cos(\omega_1 t + \angle B_1) \tag{6.72a}$$

where

$$B_1 = A_1 H_1(j\omega_1) + \tfrac{3}{4}A_1^2 A_1 H_3(-j\omega_1, j\omega_1, j\omega_1) + \tfrac{3}{2}|A_2|^2 A_1 H_3(-j\omega_2, j\omega_1, j\omega_2) \tag{6.72b}$$

$$y_{(2\omega_1-\omega_2)}(t) = |B_2|\cos[(2\omega_1 - \omega_2)t + \angle B_2] \tag{6.72c}$$

where

$$B_2 = \tfrac{3}{4}A_1^2 A_2^* H_3(j\omega_1, j\omega_1, -j\omega_2) \tag{6.72d}$$

For 50 Ω input and output impedances, the power at ω_1 and $2\omega_1 - \omega_2$ are given by

$$P_{(\omega_1)} = 10\log\left\{\left[\frac{|B_1|}{\sqrt{2}}\right]^2 \frac{10^3}{50}\right\} \quad \text{dBm} \tag{6.73}$$

$$P_{(2\omega_1-\omega_2)} = 10\log\left\{\left[\frac{|B_2|}{\sqrt{2}}\right]^2 \frac{10^3}{50}\right\} \quad \text{dBm} \tag{6.74}$$

When the signals are small, there is little nonlinearity distortion and the power $P_{(\omega_1)}$ is approximately equal to the power P_o when the two-port system is linear, where

$$P_o = 10\log\left\{\left[\frac{|A_1 H_1(j\omega_1)|}{\sqrt{2}}\right]^2 \frac{10^3}{50}\right\} \quad \text{dBm} \tag{6.75}$$

By definition, at the third-order intercept point P_I, we have $P_o = P_{(2\omega_1-\omega_2)}$, assuming two equal-amplitude signals ($A_1 = A_2 = A$), that is,

$$|A||H_1(j\omega_1)| = \tfrac{3}{4}|A|^3|H_3(j\omega_1, j\omega_1, -j\omega_2)|$$

which yields

$$|A|^2 \text{ (at } P_I) = \frac{4}{3}\frac{|H_1(j\omega_1)|}{|H_3(j\omega_1, j\omega_1, -j\omega_2)|} \tag{6.76}$$

Therefore

$$P_I = 10\log\left\{\frac{2}{3}\frac{|H_1(j\omega_1)|^3}{|H_3(j\omega_1, j\omega_1, -j\omega_2)|}\frac{10^3}{50}\right\} \quad \text{dBm}$$

$$= 10\log\frac{|H_1(j\omega_1)|^3}{|H_3(j\omega_1, j\omega_1, -j\omega_2)|} + 11.25 \quad \text{dBm} \tag{6.77}$$

We see that if the system is memoryless, (6.77) would reduce to (6.19). Note that the larger the third-order intercept point, the smaller the third-order response and the less the third-order intermodulation distortion.

6.6.4 AM-to-PM Conversion

Consider the single-frequency input $x(t)=|A|\cos(\omega t+\angle A)$. From (6.69) the output at frequency ω can be computed with $n=1$ and $n=3$, respectively, as

$$y_{(\omega)}(t)=\left[\tfrac{1}{2}AH_1(j\omega)+\tfrac{3}{8}|A|^2AH_3(-j\omega,j\omega,j\omega)\right]\exp(j\omega t)$$
$$+\left[\tfrac{1}{2}A^*H_1(-j\omega)+\tfrac{3}{8}|A|^2A^*H_3(j\omega,-j\omega,-j\omega)\right]\exp(-j\omega t)$$
$$=|C|\cos(\omega t+\angle C) \tag{6.78a}$$

where

$$C=AH_1(j\omega)+\tfrac{3}{4}|A|^2AH_3(-j\omega,j\omega,j\omega) \tag{6.78b}$$

Let $A=a(t)=A_m(1+M\cos\omega_m t)$ be the instantaneous amplitude of the input $x(t)$ where A_m, M, and ω_m are the peak amplitude, the modulation index, and the modulating frequency. Then (6.78b) can be written as $(M<1)$

$$C=a(t)H_1(j\omega)+\tfrac{3}{4}a^3(t)H_3(-j\omega,j\omega,j\omega) \tag{6.79}$$

From (6.79) we note that if the phase of $H_1(j\omega)$ is not equal to or 180° out of phase with respect to the phase of $H_3(-j\omega,j\omega,j\omega)$ then the phase of C will be a function of $a(t)$, that is, the phase of the output signal $y_{(\omega)}(t)$ will vary with the input signal level. This is the AM-to-PM conversion effect of the nonlinear system and is entirely described by the phase of the nonlinear transfer functions.

6.6.5 Cross Modulation

To determine the cross modulation, consider the input signal $x(t)$

$$x(t)=A(1+M\cos\omega_m t)\cos\omega_1 t+A\cos\omega_2 t \tag{6.80}$$

where M is the modulation index, ω_m is the modulation frequency and ω_1 and ω_2 are the carrier frequencies. Using complex exponentials, we can write $x(t)$ as

$$x(t)=\sum_{\substack{k=-4\\k\neq 0}}^{4} A_k\exp(j\omega_k t) \tag{6.81}$$

where $\omega_3=\omega_1+\omega_m$, $\omega_4=\omega_1-\omega_m$, $A_{-1}=A_1=A_{-2}=A_2=A/2$, $A_{-3}=A_3=A_{-4}=A_4=AM/4$, $(\omega_{-k}=-\omega_k)$.

From (6.69), the output terms that contain the frequencies ω_2, $\omega_2-\omega_m$, and $\omega_2+\omega_m$ are given by

(a) For $\omega_0=\omega_2$ the lowest order n is $n=1$, which yields $m_2=1$. The next order n that gives rise to $\omega_0=\omega_2$ is $n=3$. If the distortion is small, it can be ignored. Thus with this assumption we have the real response $y_{(\omega_2)}(t)=y_{\omega_2}(t)+y_{-\omega_2}(t)$ as

$$\begin{aligned} y_{(\omega_2)}(t) &= \tfrac{1}{2}AH_1(j\omega_2)\exp(j\omega_2 t)+\tfrac{1}{2}AH_1(-j\omega_2)\exp(-j\omega_2 t) \\ &= A|H_1(j\omega_2)|\cos\{\omega_2 t+\angle[H_1(j\omega_2)]\} \end{aligned} \tag{6.82}$$

(b) For $\omega_0=\omega_2+\omega_m=\omega_3+\omega_2-\omega_1=-\omega_4+\omega_2+\omega_1$, $n=3$ is the lowest order with two distinct sets of $(m_{-4},\ldots,m_{-1},m_1,\ldots,m_4)$ namely (0, 0, 0, 1, 0, 1, 1, 0) and (1, 0, 0, 0, 1, 1, 0, 0,). In all case we have

$$\begin{aligned} y_{\omega_2+\omega_m}(t) = 3!\left(\sum_{k=-4}^{4}\frac{A_k^{m_k}}{m_k!}\right)H_3(m_{-4}[j\omega_{-4}],\ldots,m_{-1}[j\omega_{-1}], \\ m_1[j\omega_1],\ldots,m_4[j\omega_4])\exp[j(\omega_2+\omega_m)t] \end{aligned} \tag{6.83}$$

For (0,0,0,1,0,1,1,0) we obtain (note that $0!=1$)

$$\begin{aligned} y_{\omega_2+\omega_m}(t) &= 3!\left(\frac{A_{-1}}{1!}\frac{A_2}{1!}\frac{A_3}{1!}\right)H_3(j\omega_{-1},j\omega_2,j\omega_3)\exp[j(\omega_2+\omega_m)t] \\ &= \frac{3A^3M}{8}H_3[-j\omega_1,j\omega_2,j(\omega_1+\omega_m)]\exp[j(\omega_2+\omega_m)t] \end{aligned} \tag{6.84a}$$

For (1,0,0,0,1,1,0,0) we obtain

$$\begin{aligned} y_{\omega_2+\omega_m}(t) &= 3!\left(\frac{A_{-4}}{1!}\frac{A_1}{1!}\frac{A_2}{1!}\right)H_3(j\omega_{-4},j\omega_1,j\omega_2)\exp[j(\omega_2+\omega_m)t] \\ &= \frac{3A^3M}{8}H_3[-j(\omega_1-\omega_m),j\omega_1,j\omega_2]\exp[j(\omega_2+\omega_m)t] \end{aligned} \tag{6.84b}$$

The real response $y_{(\omega_2+\omega_m)}(t)=y_{\omega_2+\omega_m}(t)+y_{-\omega_2-\omega_m}(t)$ is given by

$$\begin{aligned} y_{(\omega_2+\omega_m)}(t) = {} & \frac{3A^3M}{4}|H_3[-j\omega_1,j\omega_2,j(\omega_1+\omega_m)]| \\ & \times\cos((\omega_2+\omega_m)t+\angle\{H_3[-j\omega_1,j\omega_2,j(\omega_1+\omega_m)]\}) \\ & +\frac{3A^3M}{4}|H_3[-j(\omega_1-\omega_m),j\omega_1,j\omega_2]|\cos((\omega_2+\omega_m)t \\ & +\angle\{H_3[-j(\omega_1-\omega_m),j\omega_1,j\omega_2]\}) \end{aligned} \tag{6.85}$$

(c) For $\omega_0 = \omega_2 - \omega_m = \omega_4 - \omega_1 + \omega_2 = -\omega_3 + \omega_2 + \omega_1$, the lowest order is $n=3$ with two distinct sets of $(m_{-4}, \ldots, m_{-1}, m_1, \ldots, m_4)$ namely (0, 0, 0, 1, 0, 1, 0, 1) and (0, 1, 0, 0, 1, 1, 0, 0). In both cases we have

$$y_{\omega_2 - \omega_m}(t) = 3! \left(\prod_{k=-4}^{4} \frac{A_k^{m_k}}{m_k!} \right) H_3(m_{-4}[j\omega_{-4}], \ldots, m_{-1}[j\omega_{-1}], m_1[j\omega_1], \ldots, m_4[j\omega_4]) \exp[j(\omega_2 - \omega_m)t] \tag{6.86}$$

For (0,0,0,1,0,1,0,1) we get

$$y_{\omega_2 - \omega_m}(t) = \frac{3A^3M}{8} H_3[-j\omega_1, j\omega_2, j(\omega_1 - \omega_m)] \exp[j(\omega_2 - \omega_m)t] \tag{6.87a}$$

For (0,1,0,0,1,1,0,0) we obtain

$$y_{\omega_2 - \omega_m}(t) = \frac{3A^3M}{8} H_3[-j(\omega_1 + \omega_m), j\omega_1, j\omega_2] \exp[j(\omega_2 - \omega_m)t] \tag{6.87b}$$

Thus the real response $y_{(\omega_2 - \omega_m)}(t)$ is given as

$$\begin{aligned} y_{(\omega_2 - \omega_m)}(t) = {} & \frac{3A^3M}{4} |H_3[-j\omega_1, j\omega_2, j(\omega_1 - \omega_m)]| \\ & \times \cos\big((\omega_2 - \omega_m)t + \angle\{H_3[-j\omega_1, j\omega_2, j(\omega_1 - \omega_m)]\}\big) \\ & + \frac{3A^3M}{4} |H_3[-j(\omega_1 + \omega_m), j\omega_1, j\omega_2]| \\ & \times \cos\big((\omega_2 - \omega_m)t + \angle\{H_3[-j(\omega_1 + \omega_m), j\omega_1, j\omega_2]\}\big) \end{aligned} \tag{6.88}$$

The cross modulation component associated with the frequencies ω_2 and $\omega_2 \pm \omega_m$ are then given by (6.82), (6.85), and (6.88)

$$y_{CM}(t) = y_{(\omega_2)}(t) + y_{(\omega_2 + \omega_m)}(t) + y_{(\omega_2 - \omega_m)}(t) \tag{6.89}$$

If $\omega_m \ll (\omega_1, \omega_2)$, then by continuity of $H_3(\cdot)$ we have $H_3[-j\omega_1, j\omega_2, j(\omega_1 + \omega_m)] \approx H_3[-j(\omega_1 - \omega_m), j\omega_1, j\omega_2] \approx H_3[-j\omega_1, j\omega_2, j(\omega_1 - \omega_m)] \approx H_3[-j(\omega_1 + \omega_m), j\omega_1, j\omega_2] \approx H_3(-j\omega_1, j\omega_2, j\omega_1) \approx H_3(-j\omega_1, j\omega_1, j\omega_2)$ with the assumption that $H_3(\cdot)$ are symmetrized. In this case we get

$$\begin{aligned} y_{CM}(t) & \approx A|H_1(j\omega_2)| \cos\{\omega_2 t + \angle[H_1(j\omega_2)]\} \\ & \quad + 3A^3M|H_3(-j\omega_1, j\omega_1, j\omega_2)| \cos \omega_m t \cos\{\omega_2 t + \angle[H_3(-j\omega_1, j\omega_1, j\omega_2)]\} \\ & \approx A|H_1(j\omega_2)|[\cos(\omega_2 t + \beta_1) + b_M \cos \omega_m t \cos(\omega_2 t + \beta_2)] \end{aligned} \tag{6.90a}$$

where

$$\beta_1 = \angle\left[H_1(j\omega_2)\right] \tag{6.90b}$$

$$\beta_2 = \angle\left[H_3(-j\omega_1, j\omega_1, j\omega_2)\right] \tag{6.90c}$$

$$b_M = 3A^3M\,|H_3(-j\omega_1, j\omega_1, j\omega_2)|/|H_1(j\omega_2)| \tag{6.90d}$$

From (6.90a) we see that if $\beta_1 = \beta_2$, then amplitude cross modulation occurs in $y_{CM}(t)$, that is,

$$y_{CM}(t) \approx A\,|H_1(j\omega_2)|(1 + b_M \cos\omega_m t)\cos(\omega_2 t + \beta_1) \tag{6.91}$$

If $\beta_2 - \beta_1 = \phi$, then (6.90a) can be expressed as

$$\begin{aligned} y_{CM}(t) &\approx A\,|H_1(j\omega_2)|\left[\cos(\omega_2 t + \beta_1) + b_M \cos\omega_m t \cos(\omega_2 t + \beta_1 + \phi)\right] \\ &\approx A\,|H_1(j\omega_2)|\left[\cos(\omega_2 t + \beta_1) + b_M \cos\omega_m t \cos\phi \cos(\omega_2 t + \beta_1)\right. \\ &\quad \left. - b_M \cos\omega_m t \sin\phi \sin(\omega_2 t + \beta_1)\right] \\ &\approx A\,|H_1(j\omega_2)|\left[(1 + b_M \cos\phi \cos\omega_m t)\cos(\omega_2 t + \beta_1)\right. \\ &\quad \left. - b_M \sin\phi \cos\omega_m t \sin(\omega_2 t + \beta_1)\right] \end{aligned} \tag{6.92}$$

If $b_M \ll 1$, $y_{CM}(t)$ can be written as

$$y_{CM}(t) \approx A\,|H_1(j\omega_2)|(1 + b_M \cos\phi \cos\omega_m t)\cos(\omega_2 t + \beta_1) \tag{6.93}$$

6.6.6 Cascade Connection of Nonlinear Systems

In this section, the nonlinear transfer functions of two nonlinear systems in cascade as shown in Fig. 6.11 are derived in terms of the individual system's nonlinear transfer functions. Let $F_n(\cdot)$ be the nonlinear transfer functions of the overall system of order n, in order to evaluate $F_n(\cdot)$ we will repeatedly use (6.57). It is obvious that

$$F_1(j\omega) = H_1(j\omega)G(j\omega) \tag{6.94}$$

To compute $F_2(j\omega_1, j\omega_2)$ we must use the two-frequency harmonic input $x(t) = \exp(j\omega_1 t) + \exp(j\omega_2 t)$. The output $z_n(t)$ of the first system is given by (6.57) (Example 6.6) for $n = 1, 2$ as

$$\begin{aligned} z_1(t) &= H_1(j\omega_1)\exp(j\omega_1 t) + H_1(j\omega_2)\exp(j\omega_2 t) \\ z_2(t) &= H_2(j\omega_1, j\omega_1)\exp(j2\omega_1 t) + \left[H_2(j\omega_1, j\omega_2) + H_2(j\omega_2, j\omega_1)\right] \\ &\quad \times \exp\left[j(\omega_1 + \omega_2)t\right] + H_2(j\omega_2, j\omega_2)\exp(j2\omega_2 t) \end{aligned} \tag{6.95}$$

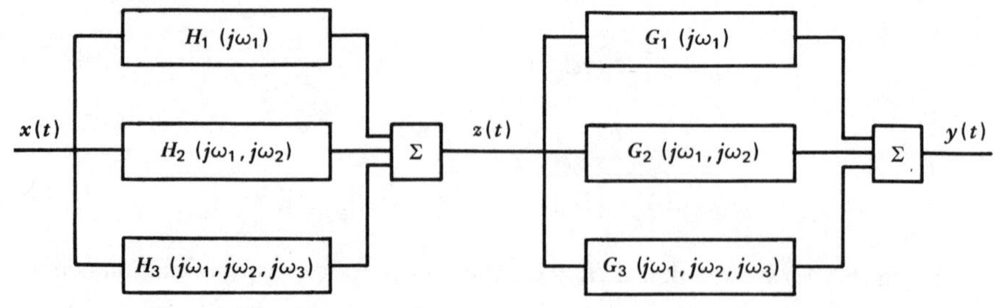

Fig. 6.11. The cascade connection of two nonlinear systems.

Applying

$$z(t)=z_1(t)+z_2(t)=\sum_{i=1}^{5} A_i \exp(j\omega_i' t)$$

where $A_1=H_1(j\omega_1)$, $A_2=H_1(j\omega_2)$, $A_3=H_2(j\omega_1, j\omega_1)$, $A_4=H_2(j\omega_1, j\omega_2)+H_2(j\omega_2, j\omega_1)$, $A_5=H_2(j\omega_2, j\omega_2)$ and $\omega_1'=\omega_1$, $\omega_2'=\omega_2$, $\omega_3'=2\omega_1$, $\omega_4'=\omega_1+\omega_2$, $\omega_5'=2\omega_2$, to the input of the second system yields, for $n=1,2$

$$\begin{aligned}
y_1(t)&=\sum_{k_1=1}^{5} A_{k_1} G_1(j\omega_{k_1}') \exp(j\omega_{k_1}' t)\\
&=H_1(j\omega_1)G_1(j\omega_1)\exp(j\omega_1 t)+H_1(j\omega_2)G(j\omega_2)\exp(j\omega_2 t)\\
&\quad+H_2(j\omega_1, j\omega_1)G_1(j2\omega_1)\exp(j2\omega_1 t)+\left[H_2(j\omega_1, j\omega_2)+H_2(j\omega_2, j\omega_1)\right]\\
&\quad\times G_1\left[j(\omega_1+\omega_2)\right]\exp\left[j(\omega_1+\omega_2)t\right]+H_2(j\omega_2, j\omega_2)G_1(j2\omega_2)\exp(j2\omega_2 t)
\end{aligned} \tag{6.96}$$

From (6.96) we see that the coefficient of the term that contains $\exp[j(\omega_1+\omega_2)t]$ is given as $[H_2(j\omega_1, j\omega_2)+H_2(j\omega_2, j\omega_1)]G_1[j(\omega_1+\omega_2)]$.

$$\begin{aligned}
y_2(t)&=\sum_{k_1=1}^{5}\sum_{k_2=1}^{5} A_{k_1}A_{k_2}G_2(j\omega_{k_1}', j\omega_{k_2}')\exp\left[j(\omega_{k_1}'+\omega_{k_2}')t\right]\\
&=\sum_{k_1=1}^{5} A_{k_1}A_1G_2(j\omega_{k_1}', j\omega_1')\exp\left[j(\omega_{k_1}'+\omega_1')t\right]\\
&\quad+A_{k_1}A_2G_2(j\omega_{k_1}', j\omega_2')\exp\left[j(\omega_{k_1}'+\omega_2')t\right]\\
&\quad+A_{k_1}A_3G_2(j\omega_{k_1}', j\omega_3')\exp\left[j(\omega_{k_1}'+\omega_3')t\right]\\
&\quad+A_{k_1}A_4G_2(j\omega_{k_1}', j\omega_4')\exp\left[j(\omega_{k_1}'+\omega_4')t\right]\\
&\quad+A_{k_1}A_5G_2(j\omega_{k_1}', j\omega_5')\exp\left[j(\omega_{k_1}'+\omega_5')t\right]
\end{aligned} \tag{6.97}$$

From (6.97) we see that the coefficients of the terms that contain $\exp[j(\omega_1+\omega_2)t]$ are $H_1(j\omega_2)G_2(j\omega_2, j\omega_1)$ and $H_1(j\omega_1)H_1(j\omega_2)G_2(j\omega_1, j\omega_2)$. Therefore by combining (6.96) and (6.97) we see that the second order nonlinear transfer function $F_2(j\omega_1, j\omega_2)$ of the cascade system is given by

$$F_2(j\omega_1, j\omega_2)=\left[H_2(j\omega_1, j\omega_2)+H_2(j\omega_2, j\omega_1)\right]G_1\left[j(\omega_1+\omega_2)\right] \\ +H_1(j\omega_1)H_1(j\omega_2)\left[G_2(j\omega_1, j\omega_2)+G_2(j\omega_2, j\omega_1)\right] \quad (6.98)$$

If $H_2(\cdot)$ and $G_2(\cdot)$ are all symmetrized, (6.98) becomes

$$F_2(j\omega_1, j\omega_2)=2H_2(j\omega_1, j\omega_2)G_1\left[j(\omega_1+\omega_2)\right] \\ +2H_1(j\omega_1)H_1(j\omega_2)G_2(j\omega_1, j\omega_2) \quad (6.99)$$

The third-order nonlinear transfer function $F_3(j\omega_1, j\omega_2, j\omega_3)$ can be found in a similar way using a three-frequency harmonic input $x(t)=\exp(j\omega_1 t)+\exp(j\omega_2 t)+\exp(j\omega_3 t)$. The sum of the coefficients of $\exp[j(\omega_1+\omega_2+\omega_3)t]$ in the output $y(t)=y_1(t)+y_2(t)+y_3(t)$ of the second system with its corresponding input $z(t)=z_1(t)+z_2(t)+z_3(t)$ gives $F_3(j\omega_1, j\omega_2, j\omega_3)$ for symmetrized $H_3(\cdot)$ and $G_3(\cdot)$

$$F_3(j\omega_1, j\omega_2, j\omega_3)=6\{H_1(j\omega_1)H_1(j\omega_2)H_1(j\omega_3)G_3(j\omega_1, j\omega_2, j\omega_3) \\ +H_1(j\omega_1)H_2(j\omega_2, j\omega_3)G_2\left[j\omega_1, j(\omega_2+\omega_3)\right] \\ +H_1(j\omega_2)H_2(j\omega_1, j\omega_3)G_2\left[j\omega_2, j(\omega_1+\omega_3)\right] \\ +H_1(j\omega_3)H_2(j\omega_1, j\omega_2)G_2\left[j\omega_3, j(\omega_1+\omega_2)\right] \\ +H_3(j\omega_1, j\omega_2, j\omega_3)G_3\left[j(\omega_1+\omega_2+\omega_3)\right]\} \quad (6.100)$$

6.7 INTERMODULATION DISTORTION IN GaAs FET AMPLIFIERS

With GaAs FETs increasingly used in microwave communication receivers, it is essential to understand their nonlinear distortion behavior, for example, their intermodulation characteristics. For small-signal high gain or low noise GaAs FET amplifiers, their distortion is basically mild and thus can be analyzed effectively by Volterra functional series. The purpose of this section is to present a nonlinear model for the GaAs FET obtained from the small-signal scattering parameters measured over a range of bias points as shown in Fig. 6.12 where the nonlinear intrinsic elements are functions of the bias state. Five nonlinearities are considered in this model. Most of the distortion arises from the nonlinear dependence of the drain current on gate bias; to a lesser extent it stems from the nonlinear bias dependence of the gate-source capacitance. The

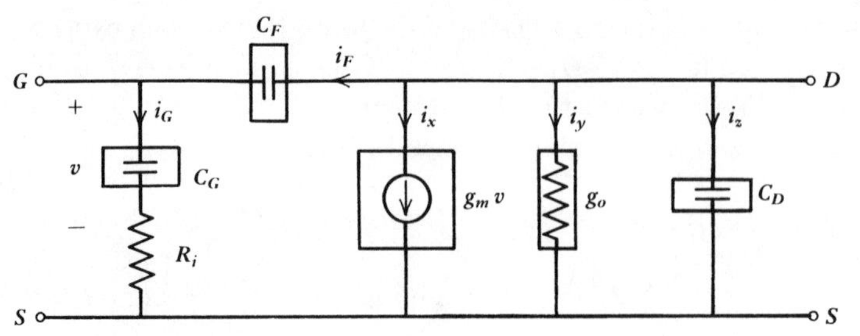

Fig. 6.12. A GaAs FET large-signal nonlinear model.

nonlinearities of the feedback capacitance C_F, the drain-source capacitance C_D and the output conductance g_o are considered to be small compared to the other two nonlinearities. From Fig. 6.12, the admittance parameters of the model are given by

$$Y_{11}=\frac{\omega^2 C_G^2 R_i}{1+(\omega C_G R_i)^2}+j\left(C_F+\frac{C_G}{1+(\omega C_G R_i)^2}\right) \tag{6.101a}$$

$$Y_{12}=-j\omega C_F \tag{6.101b}$$

$$Y_{21}=\frac{g_{m1}-g_{m2}\omega C_G R_i}{1+(\omega C_G R_i)^2}-j\left(\omega C_F+\frac{g_{m1}\omega C_G R_i+g_{m2}}{1+(\omega C_G R_i)^2}\right) \tag{6.101c}$$

$$Y_{22}=g_o+j\omega(C_F+C_D) \tag{6.101d}$$

The element values of the model can be obtained from the above admittance parameters, which in turn can be obtained directly from the measured scattering parameters as follows

$$C_F=\frac{-\operatorname{Im} Y_{12}}{\omega} \tag{6.102a}$$

$$C_D=\frac{\operatorname{Im} Y_{22}}{\omega}-C_F \tag{6.102b}$$

$$g_o=\operatorname{Re} Y_{22} \tag{6.102c}$$

$$g_{mo}=\operatorname{Re} Y_{21}|_{\omega\to 0} \tag{6.102d}$$

$$C_G=\frac{\operatorname{Im} Y_{11}}{\omega}-C_F \tag{6.102e}$$

$$R_i=\frac{\operatorname{Re} Y_{11}}{(\omega C_G)^2} \tag{6.102f}$$

$$\tau=\left(-\frac{\operatorname{Im} Y_{21}}{\omega}-g_{mo}R_i C_g-C_F\right)\Big/g_{mo} \tag{6.102g}$$

The above equations apply at frequencies where $(\omega R_i C_G)^2 \ll 1$. It has been shown that this model is very good for GaAs FETs up to 10 GHz. With the model of Fig. 6.12, the intrinsic nonlinear elements can be determined by the scattering parameters measured at different bias points. The transconductance g_m and the gate-source capacitance C_G exhibit a significant dependence on the gate bias voltage V_{GS} and can be expressed as a power series around its operating point

$$g_m(V_{GS}) = G_{m1} + G_{m2}V_{GS} + G_{m3}V_{GS}^2 + \cdots \tag{6.103}$$

$$C_G(V_{GS}) = C_{G1} + C_{G2}V_{GS} + C_{G3}V_{GS}^2 + \cdots \tag{6.104}$$

The output conductance g_o is in general a function of both V_{GS} and the drain-source voltage V_{DS}. However in high gain amplifiers with $V_{GS} \approx 0$, it can be expressed explicitly as a function of V_{DS} as

$$g_o(V_{DS}) = G_{o1} + G_{o2}V_{DS} + G_{o3}V_{DS}^2 + \cdots \tag{6.105}$$

The feedback capacitance C_F depends mostly on the drain-gate voltage V_{DG} and the drain capacitance C_D depends on V_{DS}

$$C_F = C_{F1} + C_{F2}V_{DG} + C_{F3}V_{DG}^2 + \cdots \tag{6.106}$$

$$C_D = C_{D1} + C_{D2}V_{DS} + C_{D3}V_{DS}^2 + \cdots \tag{6.107}$$

In practice the effect of C_D can usually be ignored. Some values of these nonlinear intrinsic elements of small-signal GaAs FETs are shown in Figs. 6.13

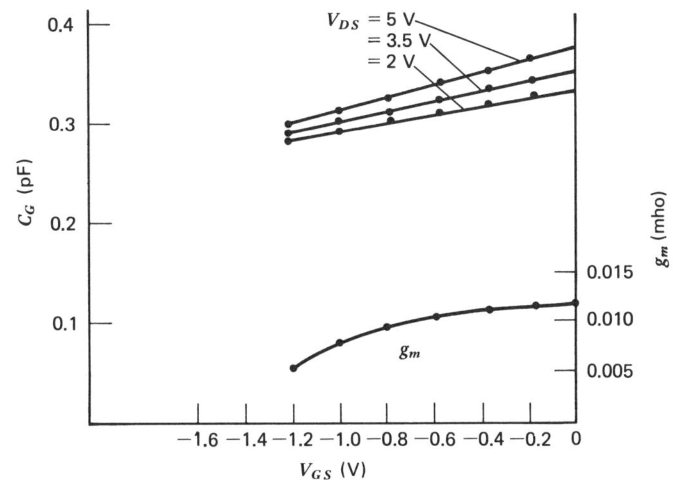

Fig. 6.13. Bias-dependent gate capacitance and transconductance, $G_{m1} = 0.012$, $G_{m2} = -0.0002$, $G_{m3} = -0.005$, $C_{G1} = 0.367$, $C_{G2} = 0.08$ [8].

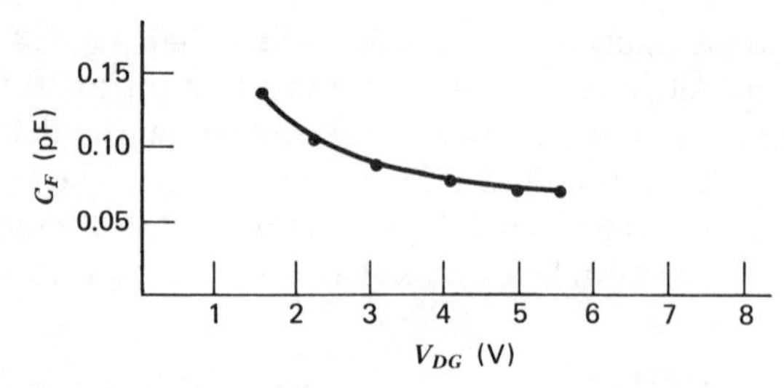

Fig. 6.14. Bias-dependent feedback capacitance, $C_{F1}=0.11$, $C_{F2}=-0.006$ [8].

and 6.14. These values are obtained by least-square fitting the measured values to (6.103), (6.104), and (6.106). Once the bias condition is determined, we can express the small-signal quantities in Fig. 6.12 about some operating point defined by the gate-source bias voltage V_{GS0}, the drain-source bias voltage V_{DS0} and the drain current I_{D0} as (all the small-signal quantities are denoted by lower case letters).

$$i_G = C_{G1}\frac{dv_{GS}}{dt} + C_{G2}\frac{d^2v_{GS}}{dt^2} + C_{G3}\frac{d^3v_{GS}}{dt^3} + \cdots \tag{6.108}$$

$$i_x = G_{m1}v + G_{m2}v^2 + G_{m3}v^3 + \cdots \tag{6.109}$$

$$i_F = C_{F1}\frac{dv_{DG}}{dt} + C_{F2}\frac{d^2v_{DG}}{dt^2} + C_{F3}\frac{d^3v_{DG}}{dt^3} + \cdots \tag{6.110}$$

$$i_y = G_{o1}v_{DS} + G_{o2}v_{DS}^2 + G_{o3}v_{DS}^3 + \cdots \tag{6.111}$$

$$i_z = C_{D1}\frac{dv_{DS}}{dt} + C_{D2}\frac{d^2v_{DS}}{dt^2} + C_{D3}\frac{d^3v_{DS}}{dt^3} + \cdots \tag{6.112}$$

In general, the first three terms of (6.108)–(6.112) are enough to represent mild distortion. To analyze the distortion, consider the complete single stage amplifier shown in Fig. 6.15 and its equivalent circuit shown in Fig. 6.16 where Z_L and Z_s represent the Thevenin impedance at the gate and at the drain of the GaAs

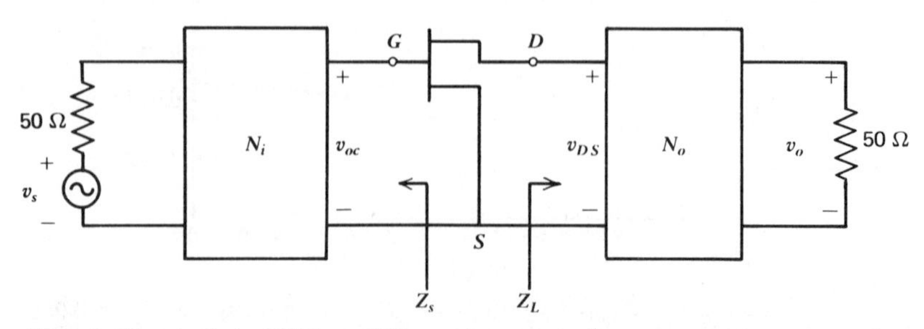

Fig. 6.15. A GaAs FET amplifier with input and output matching networks.

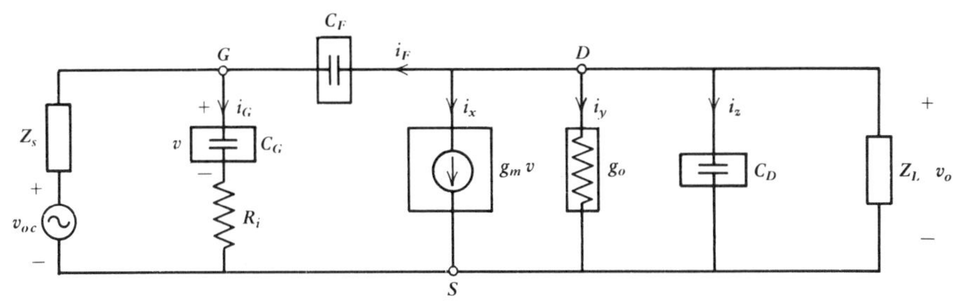

Fig. 6.16. Equivalent circuit of the GaAs FET amplifier for distortion analysis.

FET. The open-circuited voltage v_{oc} represents the Thevenin voltage at the output of N_i. The output voltage v_o can be evaluated once v_{DS} is known. For any input voltage v_s

$$v_s = \sum_{k=1}^{K} V \cos \omega_k t \tag{6.113}$$

the Thevenin voltage v_{oc} can be calculated in terms of v_s and the element values of N_i and can be represented in steady-state as

$$v_{oc} = \sum_{k=1}^{K} |A_k| \cos(\omega_k t + \angle A_k) \tag{6.114}$$

where A_k is a complex constant. To evaluate the intermodulation distortion, $K=2$ is chosen, and the results of the previous section such as the intercept point P_I in (6.77), the AM to PM conversion in (6.78), and the cross modulation factor in (6.90d) can all be evaluated in terms of the appropriate output voltage v_o. In the following discussion we will present a procedure for such computation that can be implemented on digital computers using nodal analysis. In order to do this we have to determine the current sources associated with each nonlinear elements.

First consider a nonlinear conductance described by the relationship

$$i = g_1 v + g_2 v^2 + g_3 v^3 + \cdots \tag{6.115}$$

where the first three terms are sufficient to represent the nonlinearity. Since harmonics for both current and voltage exist simultaneously, v can be expressed as

$$v = v_1 + v_2 + v_3 \tag{6.116}$$

where

$$v_i = V_i \cos(i\omega t + \phi_i), \qquad i = 1,2,3 \tag{6.117}$$

Substituting (6.116) into (6.115) and expressing i in the form

$$i = i_1 + i_2 + i_3 \tag{6.118}$$

we obtain

$$i_1 = g_1 V_1 \cos(\omega t + \phi_1) \tag{6.119a}$$

$$i_2 = g_1 V_2 \cos(2\omega t + \phi_2) + \tfrac{1}{2} g_2 V_1^2 \cos(2\omega t + 2\phi_1) \tag{6.119b}$$

$$i_3 \approx g_1 V_3 \cos(3\omega t + \phi_3) + g_2 V_1 V_2 \cos(3\omega t + \phi_1 + \phi_2) + \tfrac{1}{4} g_3 V_1^3 \cos(3\omega t + 3\phi_1) \tag{6.119c}$$

From (6.119b) and (6.119c) we see that the second and third harmonics of the current are represented by a linear conductance in shunt with a linear current source. The linear current sources for the second and third harmonic analysis are given respectively as

$$i_2' = \tfrac{1}{2} g_2 V_1^2 \cos(2\omega t + 2\phi_1) = \mathrm{Re}\left[V_{2\omega} \exp(j2\omega t)\right] \tag{6.120}$$

$$i_3' = g_1 V_1 V_2 \cos(3\omega t + \phi_1 + \phi_2) + \tfrac{1}{4} g_3 V_1^3 \cos(3\omega t + 3\phi_1)$$

$$= \sum_{k=1}^{2} \mathrm{Re}\left[V_{k,3\omega} \exp(j3\omega t)\right] \tag{6.121}$$

where $V_{2\omega} = (\frac{1}{2}) g_2 V_1^2 \exp(j2\phi_1)$, $V_{1,3\omega} = g_1 V_1 V_2 \exp[j(\phi_1 + \phi_2)]$, and $V_{2,3\omega} = (\frac{1}{4}) g_3 V_1^3 \exp(j3\phi_1)$ are the phasor representation of the associated components in steady-state.

The nonlinear capacitance element can be treated the same way using the nonlinear charge characteristic as

$$q = C_1 v + C_2 v^2 + C_3 v^3 \tag{6.122}$$

then evaluate the current i according to

$$i = \frac{dq}{dt}$$

The linear current sources associated with the second and the third harmonics can be shown to be

$$i_2' = -\omega C_2 V_1^2 \sin(2\omega t + 2\phi_1) = \mathrm{Re}\left[V_{2\omega}' \exp(j2\omega t)\right] \tag{6.123}$$

$$i_3' = -3\omega C_2 V_1 V_2 \sin(3\omega t + \phi_1 + \phi_2) - \tfrac{3}{4} \omega C_3 V_1^3 \sin(3\omega t + 3\phi_1)$$

$$= \sum_{k=1}^{2} \mathrm{Re}\left[V_{k,3\omega}' \exp(j3\omega t)\right] \tag{6.124}$$

where $V'_{2\omega} = -\omega C_2 V_1^2 \exp[j(2\phi_1 - \pi/2)]$, $V'_{1,3\omega} = -3\omega C_2 V_1 V_2 \exp[j(\phi_1 + \phi_2 - \pi/2)]$ and $V'_{2,3\omega} = \frac{1}{4} g_3 V_1^3 \exp[j(3\phi_1 - \pi/2)]$ are the corresponding phasor representations in steady-state.

The previous discussion showed how to calculate the current sources for harmonic distortion analysis at the frequencies 2ω and 3ω. The procedure to find the current sources for the intermodulation distortion analysis is similar, except the voltage v now is expressed as

$$v = (v_1 + \hat{v}_1) + (v_2 + \hat{v}_2) + (v_3 + \hat{v}_3) \tag{6.125}$$

where

$$v_i = V_i \cos(i\omega_1 t + \phi_i), \qquad i = 1,2,3 \tag{6.126}$$

$$\hat{v}_i = \hat{V}_i \cos(i\omega_2 t + \theta_i), \qquad i = 1,2,3 \tag{6.127}$$

Substituting v in (6.125)–(6.126) into (6.115) and (6.122) yields the second-order and third-order intermodulation current sources i'_2 and i'_2 respectively as

$$i'_2 = \mathrm{Re}\left\{ V_{\omega_1+\omega_2} \exp\left[j(\omega_1 + \omega_2)t\right] + V_{\omega_1-\omega_2} \exp\left[j(\omega_1 - \omega_2)t\right] \right. \tag{6.128}$$

$$\begin{aligned} i'_3 = \mathrm{Re}\Big\{ & V_{2\omega_1+\omega_2} \exp\left[j(2\omega_1 + \omega_2)t\right] + V_{2\omega_1-\omega_2} \exp\left[j(2\omega_1 - \omega_2)t\right] \\ & + V_{2\omega_2+\omega_1} \exp\left[j(2\omega_2 + \omega_1)t\right] + V_{2\omega_2-\omega_1} \exp\left[j(2\omega_2 - \omega_1)t\right] \\ & + V'_{2\omega_1+\omega_2} \exp\left[j(2\omega_1 + \omega_2)t\right] + V'_{2\omega_1-\omega_2} \exp\left[j(2\omega_1 - \omega_2)t\right] \\ & + V'_{2\omega_2+\omega_1} \exp\left[j(2\omega_2 + \omega_1)t\right] + V'_{2\omega_2-\omega_1} \exp\left[j(2\omega_2 - \omega_1)t\right] \end{aligned} \tag{6.129}$$

where

For nonlinear conductance:

$$V_{\omega_1+\omega_2} = g_2 V_1 \hat{V}_1 \exp\left[j(\phi_1 + \phi_2)\right]$$

$$V_{\omega_1-\omega_2} = g_2 V_1 \hat{V}_1 \exp\left[j(\phi_1 - \theta_1)\right]$$

$$V_{2\omega_1+\omega_2} = g_2 V_2 \hat{V}_1 \exp\left[j(\phi_2 + \theta_1)\right]$$

$$V_{2\omega_1-\omega_2} = g_2 V_2 \hat{V}_1 \exp\left[j(\phi_2 - \theta_1\right]$$

$$V_{2\omega_2+\omega_1} = g_2 V_1 \hat{V}_2 \exp\left[j(\phi_1 + \theta_2)\right]$$

$$V_{2\omega_2-\omega_1} = g_2 V_1 \hat{V}_2 \exp\left[j(\theta_2 - \phi_1)\right]$$

$$V'_{2\omega_1+\omega_2}=\tfrac{3}{4}g_3V_1^2\hat{V}_1\exp[j(2\phi_1+\theta_1)]$$

$$V'_{2\omega_1-\omega_2}=\tfrac{3}{4}g_3V_1^2\hat{V}_1\exp[j(2\phi_1-\theta_1)]$$

$$V'_{2\omega_2+\omega_1}=(\tfrac{3}{4})g_3\hat{V}_1^2V_1\exp[j(2\theta_1+\phi_1)]$$

$$V'_{2\omega_2-\omega_1}=\tfrac{3}{4}g_3\hat{V}_1^2V_1\exp[j(2\theta_1-\phi_1)]$$

For nonlinear capacitance:

$$V_{\omega_1+\omega_2}=-(\omega_1+\omega_2)C_2V_1\hat{V}_1\exp[j(\phi_1+\theta_1-\pi/2)]$$

$$V_{\omega_1-\omega_2}=-(\omega_1-\omega_2)C_2V_1\hat{V}_1\exp[j(\phi_1-\theta_1-\pi/2)]$$

$$V_{2\omega_1+\omega_2}=-(2\omega_1+\omega_2)C_2V_2\hat{V}_1\exp[j(\phi_2+\theta_1-\pi/2)]$$

$$V_{2\omega_1-\omega_2}=-(2\omega_1-\omega_2)C_2V_2\hat{V}_1\exp[(\phi_2-\theta_1-\pi/2)]$$

$$V_{2\omega_2+\omega_1}=-(2\omega_2+\omega_1)C_2V_1\hat{V}_2\exp[j(\phi_1+\theta_2-\pi/2)]$$

$$V_{2\omega_2-\omega_1}=-(2\omega_2-\omega_1)C_2V_1\hat{V}_2\exp[j(\theta_2-\phi_1-\pi/2)]$$

$$V'_{2\omega_1+\omega_2}=-\tfrac{3}{4}(2\omega_1+\omega_2)C_3V_1^2\hat{V}_1\exp[j(2\phi_1+\theta_1-\pi/2)]$$

$$V'_{2\omega_1-\omega_2}=-\tfrac{3}{4}(2\omega_1-\omega_2)C_3V_1^2\hat{V}_1\exp[j(2\phi_1-\theta_1-\pi/2)]$$

$$V'_{2\omega_2+\omega_1}=-\tfrac{3}{4}(2\omega_2+\omega_1)C_3\hat{V}_1^2V_1\exp[j(2\theta_1-\phi_1-\pi/2)]$$

$$V'_{2\omega_2-\omega_1}=-\tfrac{3}{4}(2\omega_2-\omega_1)C_3\hat{V}_1^2V_1\exp[j(2\theta_1-\phi_1-\pi/2)]$$

For cross modulation analysis, let the fundamental voltage be

$$v=A[1+M\cos(\omega_mt+\phi_m)]\cos(\omega_1t+\phi_1)+A\cos(\omega_2t+\phi_2) \quad (6.130)$$

Then only the odd terms in (6.115) and (6.122) cause cross modulation distortion. For the nonlinear conductance, the cross modulation current source can be shown to be ($\omega_m \ll \omega_1$)

$$i_{CM}\approx g_1A\left[1+\frac{3g_3MA^2}{g_1}\cos(\omega_mt+\phi_m)\right]\cos(\omega_2t+\phi_2)$$

$$\approx \mathrm{Re}\{V_{\omega_2}\exp(j\omega_2t)+V_{\omega_2+\omega_m}\exp[j(\omega_2+\omega_m)t]+V_{\omega_2-\omega_m}\exp[j(\omega_2-\omega_m)t]\}$$

(6.131)

where $V_{\omega_2} = g_1 A \exp(j\phi_2)$, $V_{\omega_2+\omega_m} = \frac{3}{2} g_3 M A^3 \exp[j(\phi_2+\phi_m)]$, and $V_{\omega_2-\omega_m} = \frac{3}{2} g_3 M A^3 \exp[j(\phi_2-\phi_m)]$, and for the nonlinear capacitance,

$$i_{CM} \approx -\omega_2 C_1 A \left[1 + \frac{3C_3 MA^2}{C_1} \cos(\omega_m t + \phi_m)\right] \sin(\omega_2 t + \phi_2)$$

$$\approx \mathrm{Re}\left\{ V'_{\omega_2} \exp(j\omega_2 t) + V'_{\omega_2+\omega_m} \exp\left[j(\omega_2+\omega_m)t\right] + V'_{\omega_2-\omega_m} \exp\left[j(\omega_2-\omega_m)t\right] \right. \quad (6.132)$$

where $V'_{\omega_2} = -\omega_2 C_1 A \exp[j(\phi_2 - \pi/2)]$, $V'_{\omega_2+\omega_m} = \frac{3}{2}\omega_2 C_3 M A^3 \exp[j(\phi_2+\phi_m - \pi/2)]$ and $V'_{\omega_2-\omega_m} = -\frac{3}{2}\omega_2 C_3 M A^3 \exp[j(\phi_2 - \phi_m - \pi/2)]$.

Using these results, each nonlinear element can be replaced by a linear element in shunt with a linear current source that represent the distortion as shown in Fig. 6.17 (the subscript $n=2,3$ for each current source stands for the second- and third-order intermodulation or harmonics). The nodal equations for this circuit can be represented by the following matrix equation at steady-state:

$$YE = J \quad (6.133)$$

where

$$Y = \begin{bmatrix} \frac{1}{Z_s} + j\omega C_{F1} & -j\omega C_{F1} & j\omega C_{G1} \\ \frac{1}{R_i} & 0 & \frac{1}{R_i} + j\omega C_{G1} \\ -j\omega C_{F1} & G_{o1} + \frac{1}{Z_L} + j\omega(C_{F1} + C_{D1}) & G_{m1} \end{bmatrix} \quad (6.134)$$

$$E^T = \left[V_{GS}(j\omega), V_{DS}(j\omega), V(j\omega)\right]^T \quad (6.135)$$

$$J^T = \left[\frac{1}{Z_s} V_{oc}(j\omega) + I'_{Fn}(j\omega) - I'_{Gn}(j\omega),\ I'_{Gn}(j\omega),\right.$$
$$\left. -I'_{xn}(j\omega) - I'_{yn}(j\omega) - I'_{zn}(j\omega) - I'_{Fn}(j\omega)\right]^T \quad (6.136)$$

where all the voltages and currents are in phasor representation for steady-state responses.

The following algorithm for the computation of the second- and third-order intermodulation distortions at frequency $\omega_1 \pm \omega_2$ and $2\omega_1 \pm \omega_2$ is described below (algorithms for harmonic distortions at frequency $2\omega, 3\omega$ and for cross modulation distortion can be generated in a similar manner).

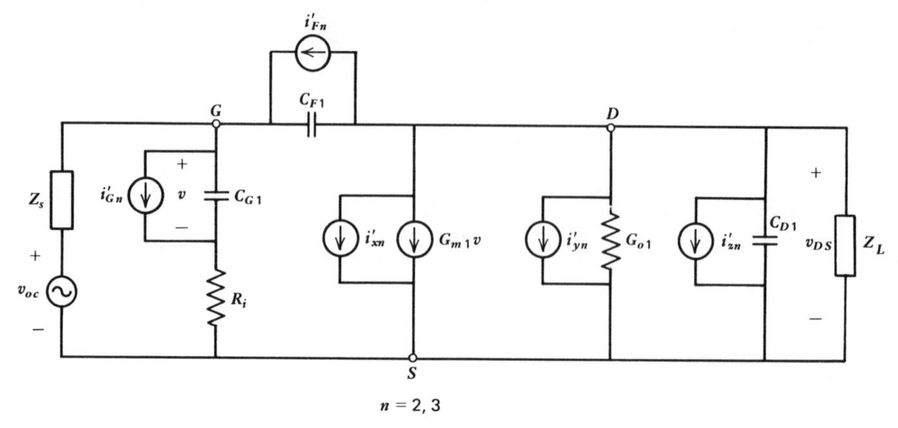

Fig. 6.17. A GaAs FET amplifier circuit for nodal analysis.

Step 1. Neglect the small nonlinearities by setting all the current sources associated with the nonlinear elements to zero and calculate the steady-state response in phasor representation of the output voltages $V_{DS,\omega}$, $V_{GS,\omega}$, and V_ω at frequencies $\omega=\omega_1, \omega_2$ for the input $V_{oc}=A_k$ given in (6.114) by using (6.133).

Step 2. Compute the second-order distortion current sources associated with each nonlinear element to be applied to the circuit in Fig. 6.17 to produce the intermodulation products at frequencies $\omega=\omega_1 \pm \omega_2, 2\omega_1, 2\omega_2$ by using (6.120), (6.123), and (6.128).

Step 3. Compute the second-order distortion voltages $V_{DS,\omega}$, $V_{GS,\omega}$, and V_ω at frequencies $\omega=\omega_1 \pm \omega_2, 2\omega_1, 2\omega_2$ using (6.133) at these frequencies by setting $V_{oc}=0$ and by using the current sources computed in Step 2.

Step 4. Compute the third-order distortion current sources associated with each nonlinear element that produce intermodulation products at frequencies $\omega=2\omega_1 \pm \omega_2, 2\omega_2 \pm \omega_1$ by using (6.129) and the voltages in Step 3.

Step 5. Compute the third-order intermodulation output voltage $V_{DS,\omega}$ at frequencies $\omega=2\omega_1 \pm \omega_2, 2\omega_2 \pm \omega_1$ using (6.133) at these frequencies by setting $V_{oc}=0$ and using the current sources computed in Step 4.

Note that the Volterra series analysis is valid only when the distortion is "weak" (only the first three terms in the Volterra series representation of the nodal voltages in terms of the input voltage are sufficient). Furthermore, we assume that the amplifier is stable so that the Volterra series exist.

6.8 INTERMODULATION DISTORTION IN IMPATT AMPLIFIERS

Amplifiers using IMPATT diodes are also used in microwave systems and with recent advances in technology, IMPATT amplifiers become practical in the millimeter wave region. In many applications where linear power amplification is required, it is important to be able to predict the intermodulation distortion arising from IMPATT diode nonlinearities. The purpose of this section is to investigate the nonlinear distortion produced in IMPATT amplifiers using the Volterra functional series approach. An IMPATT diode may be represented at a constant operating current by a parallel combination of a negative conductance $-G_d$ and a capacitance C_d; both depend heavily on the amplitude V_d of the sinusoidal voltage v_d across its terminal and can be expressed as

$$-G_d = -G_{d1} + G_{d2}V_d + G_{d3}V_d^2 + \cdots \tag{6.137}$$

$$C_d = C_{d1} + C_{d2}V_d + C_{d3}V_d^2 + \cdots \tag{6.138}$$

and $-G_{d1}, G_{d2}, \ldots$ and $C_{d1}, C_{d2}, \ldots$ can be obtained by least-square fitting (6.137) and (6.138) to the experimental measured diode admittance $Y_d(V_d) = -G_d(V_d) + j\omega C_d(V_d)$, which is a function of the amplitude of the rf voltage across the diode and frequency at a certain dc bias current I_{dc}. Once the operating point is fixed, the small-signal current i_d through the diode can be expressed as

$$i_d = -G_d v_d + \frac{d(C_d v_d)}{dt} \tag{6.139}$$

Using the same approach as in Section 6.7, it can be shown in the same way that the diode can be replaced by a parallel combination of the linear conductance G_{d1}, the linear capacitance C_{d1}, and a current source i'_{dn} ($n=2$ for second-order distortion and $n=3$ for third-order distortion). Since i_d in (6.139) is the sum of two terms, the current i'_{dn} must be the sum of two current source obtained by (6.120) and (6.123), and by (6.128) and (6.129) for $n=2$ and $n=3$, respectively, for harmonic currents. For intermodulation currents, i'_{dn} is obtained from (6.128) and (6.129).

To analyze the distortion, consider the reflection IMPATT amplifier shown in Fig. 6.18, where the circulator is assumed to be ideal with all ports having real characteristic impedance $Z_o = 50\ \Omega$. Then its scattering matrix $\mathbf{S}_c$ normalized to Z_o is given as

$$\begin{bmatrix} b_1 \\ b_2 \\ b_3 \end{bmatrix} = \mathbf{S}_c \begin{bmatrix} a_1 \\ a_2 \\ a_3 \end{bmatrix} = \begin{bmatrix} 0 & 0 & 1 \\ 1 & 0 & 0 \\ 0 & 1 & 0 \end{bmatrix} \begin{bmatrix} a_1 \\ a_2 \\ a_3 \end{bmatrix} \tag{6.140}$$

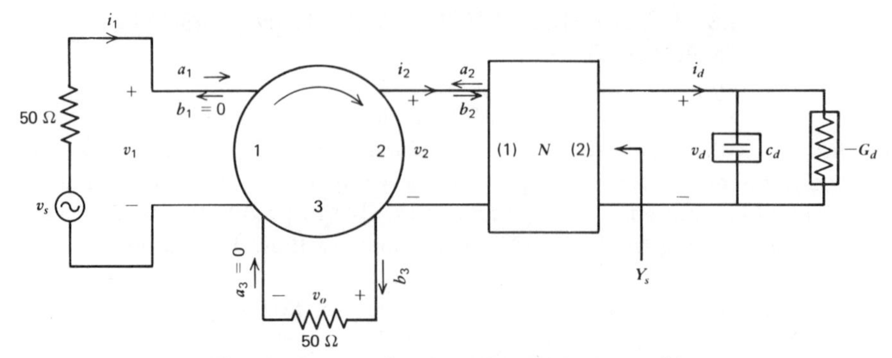

Fig. 6.18. A reflection IMPATT diode amplifier.

where a_i and b_i $(i=1,2,3)$ are the incident and reflected waves. In practice, the generator and load impedances are chosen to be Z_0, that is, $R_s=R_L=Z_0$, thus yielding $b_1=0$ and $a_3=0$. Note that from (6.140), $b_2=a_1=v_1+Z_0i_1=v_s/2+Z_0(v_s/2Z_0)=v_s$ and $b_3=a_2=v_o+Z_0(v_o/Z_0)=2v_o$ where v_o is the output voltage. Thus given an input voltage v_s or equivalently, an input wave a_1, our purpose is to compute the intermodulation output wave b_3 or equivalently, the output voltage v_o, at the intermodulation frequency. Once a_2 is calculated, v_o will be determined automatically as $v_o=a_2/2$. In the subsequent discussion, let all the voltages, currents, and waves be represented by phasors for steady-state analysis (denoted by upper case letters). It is seen that

$$A_2=V_2+Z_0I_2 \tag{6.141}$$

let $\mathcal{C}=[\mathcal{C}_{ij}]$ be the chain matrix of the two-port matching network N, then

$$\begin{bmatrix} V_2 \\ I_2 \end{bmatrix}=\begin{bmatrix} \mathcal{C}_{11} & \mathcal{C}_{12} \\ \mathcal{C}_{21} & \mathcal{C}_{22} \end{bmatrix}\begin{bmatrix} V_d \\ I_d \end{bmatrix} \tag{6.142}$$

Substituting V_2 and I_2 into (6.142) yields

$$\begin{aligned} A_2 &= \mathcal{C}_{11}V_d+\mathcal{C}_{12}I_d+Z_0(\mathcal{C}_{21}V_d+\mathcal{C}_{22}I_d) \\ &= (\mathcal{C}_{11}+Z_0\mathcal{C}_{21})V_d+(\mathcal{C}_{12}+Z_0\mathcal{C}_{22})I_d \end{aligned} \tag{6.143}$$

From (6.143) we see that once V_d and I_d are known, the output voltage V_o is automatically determined. In order to compute V_d and I_d, consider the equivalent circuit of Fig. 6.19 for nodal analysis where V_{oc} represents the Thevenin's voltage at port 2 of the circulator and I_{sc} represents the Norton's current at port 2 of the matching network N. It can be shown that V_{oc} is equal to $V_1=V_s/2$ and that I_{sc} can be computed from V_{oc} and the element values of N at steady-state.

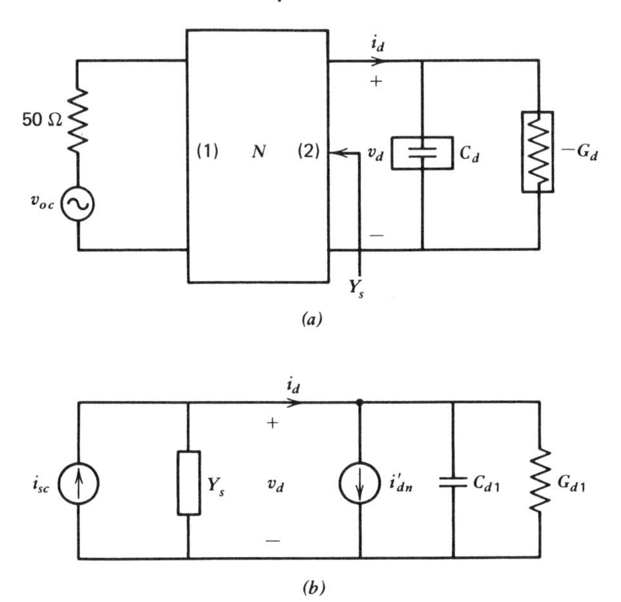

Fig. 6.19. (a)–(b) An equivalent circuit of the IMPATT diode amplifier for nodal analysis.

The admittance Y_s represents the admittance at port 2 of N when $V_{oc}=0$. Using the circuit in Fig. 6.19b, the voltage V_d can be expressed by

$$V_d = \frac{I_{sc} - I'_{dn}}{Y_s + G_{d1} + j\omega C_{d1}} \tag{6.144}$$

and the current I_d is

$$I_d = \frac{G_{d1} + j\omega C_{d1}}{Y_s + G_{d1} + j\omega C_{d1}} I_{sc} + \frac{Y_s}{Y_s + G_{d1} + j\omega C_{d1}} I'_{dn} \tag{6.145}$$

Substituting (6.144) and (6.145) into (6.143) yields

$$V_o = \frac{\mathcal{C}_{11} + Z_0\mathcal{C}_{21} + (\mathcal{C}_{12} + Z_0\mathcal{C}_{22})(G_{d1} + j\omega C_{d1})}{2(Y_s + G_{d1} + j\omega C_{d1})} I_{sc} + \frac{-(\mathcal{C}_{11} + Z_0\mathcal{C}_{21}) + Y_s(\mathcal{C}_{12} + Z_0\mathcal{C}_{22})}{2(Y_s + G_{d1} + j\omega C_{d1})} I'_{dn} \tag{6.146}$$

The following algorithm for the computation of the second- and third-order intermodulation distortions at frequencies $\omega_1 \pm \omega_2$ and $2\omega_1 \pm \omega_2$ is described below (other types of distortion such as cross modulation or AM-to-PM conversion can be achieved in a similar way by simply replaced the appropriate current sources I'_{dn}). We note that for a given input $v_s = \Sigma_{k=1}^{2} V \cos \omega_k t$, the short-circuited current i_{sc} can be represented by

$$i_{sc} = \sum_{k=1}^{2} |B_k| \cos(\omega_k t + \angle B_k) \tag{6.147}$$

Step 1. Set the current source I'_{dn} to zero and calculate the steady-state response in phasor representation of the output voltage $V_{o,\omega}$ at frequencies $\omega=\omega_1, \omega_2$ for the input phasor $I_{sc}=B_k$ given in (6.147).

Step 2. Compute the second-order distortion current source I'_{d2} that produces the intermodulation products at frequencies $\omega=\omega_1 \pm \omega_2, 2\omega_1, 2\omega_2$ by using (6.139).

Step 3. Compute the second-order distortion output voltages $V_{o,\omega}$ at frequencies $\omega=\omega_1 \pm \omega_2, 2\omega_1, 2\omega_2$ using (6.146) by setting $I_{sc}=0$ and by using I'_{d2} in step 2.

Step 4. Compute the third-order distortion current source I'_{d3} that produces the intermodulation products at frequencies $\omega=2\omega_1 \pm \omega_2, 2\omega_2 \pm \omega_1$ by using (6.139).

Step 5. Compute the third-order distortion output voltages $V_{o,\omega}$ at frequencies $\omega=2\omega_1 \pm \omega_2, 2\omega_2 \pm \omega_1$ using (6.146) by setting $I_{sc}=0$ and using I'_{d3} in step 4.

We note that the above procedure is valid for small-signal distortion such that the first three terms in the Volterra series representation of (6.146) are sufficient. For stronger distortion, five terms in the power series might be used, but the computation will be more involved. Finally, we also assume that the IMPATT amplifier is stable so that the Volterra representation is valid.

6.9 DISTORTION IMPROVEMENT BY BIAS COMPENSATION

In multiple-carrier telecommunications systems or in applications to amplitude modulation systems such as TV or SSB-AM transmission systems, the intermodulation distortion performance of high gain and high power GaAs FET and IMPATT diode amplifiers becomes an important factor. To improve the intermodulation distortion, feedforward amplifiers have been used. The technique results in a complex system and an increase in power consumption. In this section we present a much simpler approach that seems to be suited well for GaAs FET and IMPATT diode amplifiers. It is called the bias compensation technique. We will discuss the application of this scheme to the GaAs FET first as is shown in Fig. 6.20. For a constant drain voltage V_{DS}, it has been proved experimentally that the intermodulation voltage at frequencies $2\omega_1 - \omega_2$ and $2\omega_2 - \omega_1$ is reduced as the gate voltage is made more negative (which results in a decrease in the drain current). The negative gate-bias voltage that produces the minimum intermodulation product is denoted as V_{GSM}. We note that the more

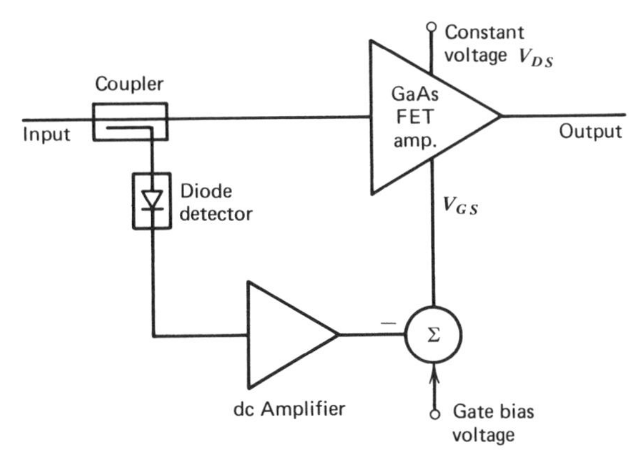

Fig. 6.20. The gate-bias compensation technique for the GaAs FET amplifier.

negative V_{GS}, the more the decrease in the gain of the amplifier. But there is a range of V_{GS} where the intermodulation product reduces more rapidly than the gain, resulting in a net improvement in the distortion. We note that if V_{GS} is made more negative than V_{GSM} as it approaches the pinch-off voltage, the intermodulation product will increase rapidly because of the nonlinear nature of the FET near pinch-off. In Fig. 6.20, a very small portion of the input signal is extracted through a directional coupler and detected by a diode detector. The dc voltage output of the diode detector is then amplified by a dc amplifier and added in a negative sense to the gate-bias voltage so that when the input level increase (which results in higher intermodulation products if no compensation exists) the net gate-bias voltage to the GaAs FET amplifier becomes more negative. This will result in a decrease in both gain and intermodulation product but with a net improvement. The technique applies only to a limited range of

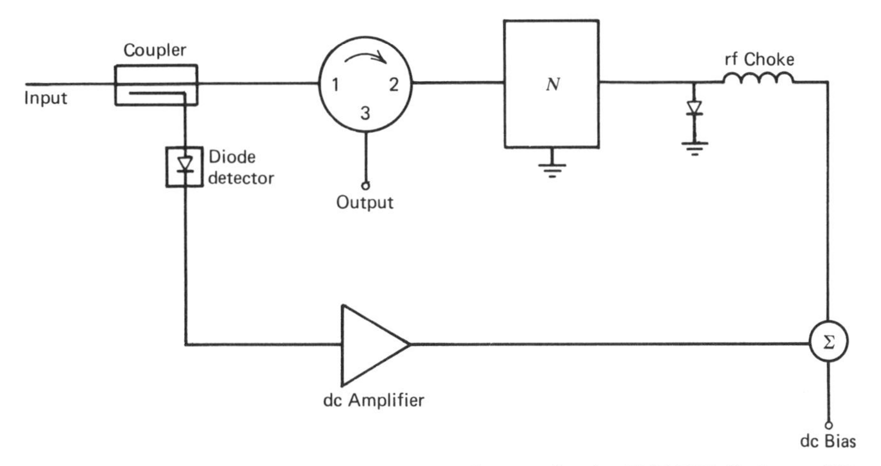

Fig. 6.21. The current-bias compensation technique for the IMPATT diode amplifier.

input signal levels, and the dc amplifier in the compensation loop can cause a degradation in the overall noise performance. If AM transmission is applied, the dc amplifier can be replaced by a video amplifier with constant gain in its frequency range, which is larger than the modulation frequency.

The bias compensation for the IMPATT diode amplifier is shown in Fig. 6.21, where the scheme works the same way as that of the GaAs FET amplifier. Here the gain of the dc amplifier is adjusted to supply an optimum compensation current to the diode to provide a net improvement in the intermodulation product. This type of amplifier can be used satisfactorily in multicarrier amplitude modulation transmission although pulse response may not be good because of the slow feedback response.

6.10 MICROWAVE POWER COMBINING TECHNIQUES

Microwave power combining is of particular importance in radar and satellite earth station transmitter applications, when many power sources are combined by various techniques to obtain high power, especially at microwave frequencies where a large amount of power might not be available from a single source. Some examples are the power combining of IMPATT diode oscillators at X- and Ku-bands, IMPATT power amplifiers in milimeter-wave systems, and power GaAs FETs. For economic reasons, it is often desirable to obtain a specific amount of microwave power by combining several small power sources rather than using a large power source. For example, low-cost helix-type traveling wave tubes are available at Ku-band whereas high power coupled cavity tubes in this band are expensive and have poorer characteristics than the helix-type tubes. Besides the generation of higher power, the power combining techniques can provide "graceful degradation" in the case of failure of one or more sources in the combined system. Graceful degradation refers to the fact that the output power is reduced but not completely lost. For example, if an individual amplifier in a balanced configuration fails, the output power drops to a quarter of, or 6 dB below, the original output power. In this section, existing and successful power combining techniques such as the chain (serial, nonbinary) structure, the corporate (tree, binary) structure are analyzed. Combining with different levels of power will be discussed in detail, taking into account the loss in each combining step.

6.10.1 Directional Couplers

The main circuit element in a power combining/dividing network is a three-port combiner/divider, which is a three-port reciprocal network that can be used to combine power from two isolated ports or to divide power between these two isolated ports. The three-port combiner/divider can be obtained from a directional coupler, which is a four-port reciprocal lossless network as shown in Fig. 6.22. The ideal directional coupler has the property

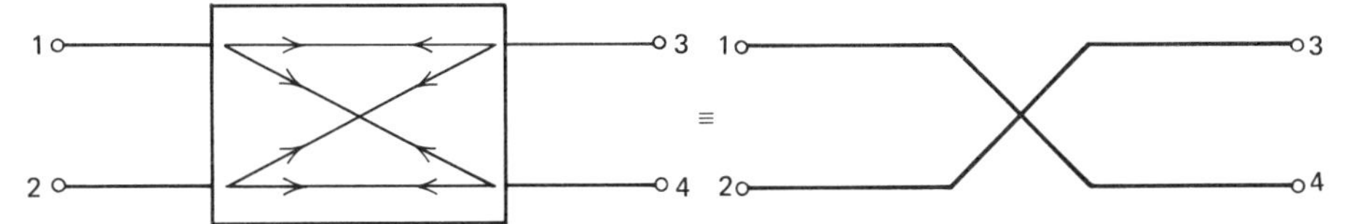

Fig. 6.22. A directional coupler.

that the wave incident on port 1 couples power into ports 3 and 4 but not port 2. Similarly, the power incident on port 2 couples power into ports 3 and 4 but not port 1. Thus ports 1 and 2 are isolated. Since the directional coupler is reciprocal, this property holds for ports 3 and 4. Furthermore, all ports are matched, that is, if three ports are terminated in matched loads, there is no reflection for an incident wave at the fourth port. Let $\mathbf{S}=[S_{ij}]$ be the symmetric and unitary scattering matrix of the four-port reciprocal lossless directional coupler.

$$\mathbf{S}=\begin{bmatrix} S_{11} & S_{12} & S_{13} & S_{14} \\ S_{12} & S_{22} & S_{23} & S_{24} \\ S_{13} & S_{23} & S_{33} & S_{34} \\ S_{14} & S_{24} & S_{34} & S_{44} \end{bmatrix} \tag{6.148}$$

From the above property, that is, port 1 is isolated from port 2 and port 3 is isolated from port 4 with all of them matched, we have

$$S_{11}=S_{22}=S_{33}=S_{44}=S_{12}=S_{34}=0 \tag{6.149}$$

Thus $\mathbf{S}$ in (6.148) can be written as

$$\mathbf{S}=\begin{bmatrix} 0 & 0 & S_{13} & S_{14} \\ 0 & 0 & S_{23} & S_{24} \\ S_{13} & S_{23} & 0 & 0 \\ S_{14} & S_{24} & 0 & 0 \end{bmatrix} \tag{6.150}$$

Since $\mathbf{S}$ is unitary for a lossless network, that is, $\mathbf{S}^{T*}\mathbf{S}=\mathbf{1}_4$ we obtain

$$S_{13}^{*}S_{23}+S_{14}^{*}S_{24}=0 \tag{6.151a}$$

$$S_{13}^{*}S_{14}+S_{23}^{*}S_{24}=0 \tag{6.151b}$$

Or equivalently,

$$|S_{13}||S_{23}|=|S_{14}||S_{24}| \tag{6.152a}$$

$$|S_{13}||S_{14}|=|S_{23}||S_{24}| \tag{6.152b}$$

Dividing (6.152a) by (6.152b) yields

$$\frac{|S_{23}|}{|S_{14}|} = \frac{|S_{14}|}{|S_{23}|} \tag{6.153}$$

or

$$|S_{23}| = |S_{14}| \tag{6.154}$$

Substituting (6.154) into (6.152a) yields

$$|S_{13}| = |S_{24}| \tag{6.155}$$

For (6.151) to be satisfied, one possible solution is

$$S_{13} = S_{24} = \alpha \tag{6.156a}$$

$$S_{23} = S_{14} = j\beta \tag{6.156b}$$

where α and β are real numbers. Therefore the scattering matrix of a direction coupler can be written as

$$\mathbf{S} = \left[\begin{array}{ccc|c} 0 & 0 & \alpha & j\beta \\ 0 & 0 & j\beta & \alpha \\ \alpha & j\beta & 0 & 0 \\ \hline j\beta & \alpha & 0 & 0 \end{array}\right] = \left[\begin{array}{c|c} \mathbf{S}_{11c} & \mathbf{S}_{21c}^T \\ \hline \mathbf{S}_{21c} & \mathbf{S}_{22c} \end{array}\right] \tag{6.157}$$

It can be easily seen from the unitary property that

$$\alpha^2 + \beta^2 = 1 \tag{6.158}$$

If port 4 is terminated by a load with reflection coefficient Γ_L, then the directional coupler becomes a three-port combiner with input ports 1 and 2 and output port 3, or a divider with input port 3 and output ports 1 and 2. To see how this works, let us use the cascade formula (1.21). The scattering matrix $\mathbf{\Sigma}$ of the combiner/divider is then given as

$$\begin{aligned} \mathbf{\Sigma} &= \mathbf{S}_{11c} + \mathbf{S}_{21c}^T (1 - \Gamma_L \mathbf{S}_{22c})^{-1} \Gamma_L \mathbf{S}_{21c} \\ &= \mathbf{S}_{11c} + \mathbf{S}_{21c}^T \Gamma_L \mathbf{S}_{21c} \end{aligned} \tag{6.159}$$

If the terminating load at port 4 is matched, that is, $\Gamma_L = 0$, then

$$\mathbf{\Sigma} = \mathbf{S}_{11c} = \begin{bmatrix} 0 & 0 & \alpha \\ 0 & 0 & j\beta \\ \alpha & j\beta & 0 \end{bmatrix} \tag{6.160}$$

Let a_i and b_i $(i=1,2,3)$ be the incident and reflected normalized scattering variables at ports 1, 2, and 3 of the three-port combiner/divider. For the combiner, let a_1 and a_2 be the only two incident waves at ports 1 and 2. Then

$$\begin{bmatrix} b_1 \\ b_2 \\ b_3 \end{bmatrix} = \begin{bmatrix} 0 & 0 & \alpha \\ 0 & 0 & j\beta \\ \alpha & j\beta & 0 \end{bmatrix} \begin{bmatrix} a_1 \\ a_2 \\ 0 \end{bmatrix} = \begin{bmatrix} 0 \\ 0 \\ \alpha a_1 + j\beta a_2 \end{bmatrix} \tag{6.161}$$

From (6.161) we see that there is no reflection at the input ports and the output power at port 3 is given by

$$|b_3|^2 = |\alpha a_1 + j\beta a_2|^2 \tag{6.162}$$

Note that the input power at port 1 is $|a_1|^2$ and at port 2 is $|a_2|^2$. Thus if we assume the combiner is ideal we expect the output power $|b_3|^2$ to be the sum of the input powers, that is, $|b_3|^2 = |a_1|^2 + |a_2|^2$. From (6.162) we see that this can be achieved with proper selection of β and phasing of a_1 and a_2. If we select a_1 and a_2 to be 90° out of phase, that is,

$$a_1 = |a_1| \exp(j\theta) \tag{6.163a}$$

$$a_2 = |a_2| \exp[j(\theta - \pi/2)] \tag{6.163b}$$

Then (6.162) can be rewritten as

$$|b_3|^2 = \big|\alpha|a_1| + \beta|a_2|\big|^2 \tag{6.164}$$

Now let

$$\beta = \alpha \frac{|a_2|}{|a_1|} \tag{6.165}$$

By noting that $\alpha^2 + \beta^2 = (\alpha^2 + \beta^2)^2 = 1$ we obtain

$$|b_3|^2 = \left|\alpha|a_1| + \frac{\beta^2}{\alpha}|a_1|\right|^2 = \frac{(\alpha^2+\beta^2)^2|a_1|^2}{\alpha^2}$$

$$= \frac{(\alpha^2+\beta^2)|a_1|^2}{\alpha^2} = |a_1|^2 + \frac{\beta^2}{\alpha^2}|a_1|^2 = |a_1|^2 + |a_2|^2 \tag{6.166}$$

Also from (6.165) we can derive (6.167) using $\alpha^2+\beta^2=1$, as follows:

$$\alpha=\pm\frac{1}{\left[1+\dfrac{|a_2|^2}{|a_1|^2}\right]^{1/2}}=\pm\frac{|a_1|}{\left[|a_1|^2+|a_2|^2\right]^{1/2}} \tag{6.167a}$$

$$\beta=\pm\frac{|a_2|}{\left[|a_1|^2+|a_2|^2\right]^{1/2}} \tag{6.167b}$$

For the divider, let a_3 be the only incident power at port 3. Then from (6.160) we get

$$\begin{bmatrix} b_1 \\ b_2 \\ b_3 \end{bmatrix}=\begin{bmatrix} 0 & 0 & \alpha \\ 0 & 0 & j\beta \\ \alpha & j\beta & 0 \end{bmatrix}\begin{bmatrix} 0 \\ 0 \\ a_3 \end{bmatrix}=\begin{bmatrix} \alpha a_3 \\ j\beta a_3 \\ 0 \end{bmatrix} \tag{6.168}$$

From (6.168) we see that there is no reflection at port 3 and the powers at port 1 and port 2 are

$$|b_1|^2=\alpha^2|a_3|^2 \tag{6.169a}$$

$$|b_2|^2=\beta^2|a_3|^2 \tag{6.169b}$$

Depending on the selection of α and β with $\alpha^2+\beta^2=1$, the input power $|a_3|^2$ can be split evenly or unevenly into the two output ports. If $\alpha^2=\beta^2=\frac{1}{2}$, then $|b_1|^2=|b_2|^2=|a_3|^2/2$, this is a 3 dB power divider since the output power at

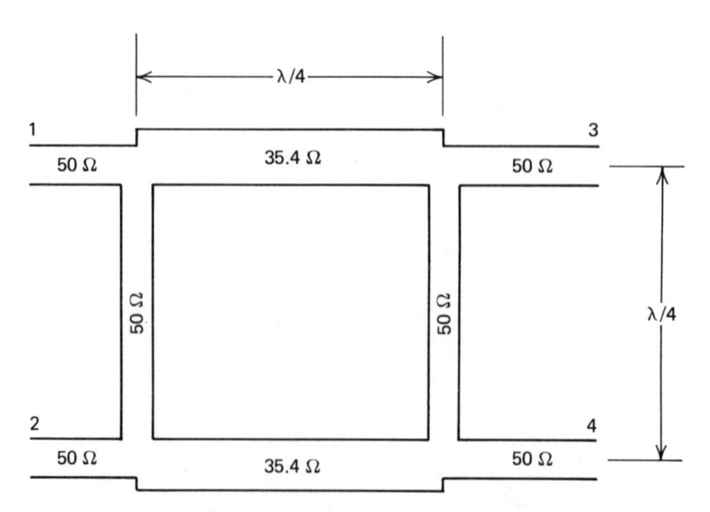

Fig. 6.23. A 90° hybrid coupler.

the two isolated output ports are half the input power (or 3 dB down from the input power). A 3 dB microstrip line coupler is shown in Fig. 6.23 (also called a 90° hybrid coupler or branch line coupler). Other types of combiner/dividers can be obtained from proximity directional couplers using parallel lines.

6.10.2 Split-T Power Combiner/Divider

We have discussed the construction of a three-port combiner/divider from a four-port reciprocal lossless directional coupler by terminating one port in a matched load. In this section we present a different type of power combiner/divider that provides two isolated, equal phase, equal or unequal amplitude input/outputs with each port matched. An equal phase, equal amplitude split-T power combiner/divider (also called a Wilkinson combiner/divider [1]) is shown in Fig. 6.24*a*. The equal phase, unequal amplitude split-T power combiner/divider is shown in Fig. 6.24*b* [2] where

$$K^2 = \frac{\text{power at port 1}}{\text{power at port 2}} = \frac{\alpha^2}{\beta^2} > 1 \tag{6.170}$$

$$Z_{01} = Z_0\left(\frac{K}{1+K^2}\right)^{1/4} \tag{6.171}$$

$$Z_{02} = Z_0 K^{3/4}(1+K^2)^{1/4} \tag{6.172}$$

$$Z_{03} = Z_0\left(\frac{1+K^2}{K^5}\right)^{1/4} \tag{6.173}$$

$$Z_{04} = Z_0\sqrt{K} \tag{6.174}$$

$$Z_{05} = \frac{Z_0}{\sqrt{K}} \tag{6.175}$$

$$R = Z_0\frac{1+K^2}{K} \tag{6.176}$$

The scattering matrix of the split-T combiner/divider is given by

$$\boldsymbol{\Sigma} = \begin{bmatrix} 0 & 0 & \alpha \\ 0 & 0 & \beta \\ \alpha & \beta & 0 \end{bmatrix} \tag{6.177}$$

where α and β are real numbers of the same sign. For example, if one wishes to design a split-T power combiner with port 1 power twice port 2 power, then from (6.177), the power output is [note that the input waves to a split-T

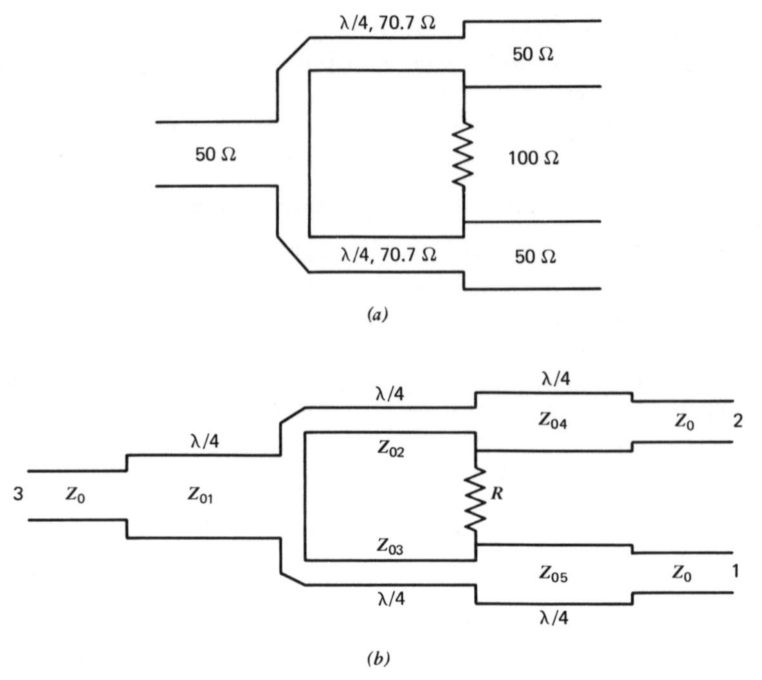

Fig. 6.24. (*a*) An equal phase, equal amplitude split-T combiner/divider (also called Wilkinson combiner/divider). (*b*) An equal phase, unequal amplitude split-T combiner/divider.

combiner must be in phase, that is, $a_1 = |a_1|\exp(j\theta)$, $a_2 = |a_2|\exp(j\theta)$],

$$|b_3|^2 = |\alpha|a_1| + \beta|a_2||^2 = \left|\alpha|a_2|\sqrt{2} + \beta|a_2|\right|^2$$
$$= |a_2|^2|\alpha\sqrt{2} + \beta|^2 \tag{6.178}$$

For $|b_3|^2 = |a_1|^2 + |a_2|^2 = 3|a_2|^2$, we must have

$$\left(\alpha\sqrt{2} + \beta\right)^2 = 3 \tag{6.179}$$

and

$$\alpha^2 + \beta^2 = 1 \tag{6.180}$$

This yields

$$\alpha = \frac{\sqrt{6}}{3}, \qquad \beta = \frac{1}{\sqrt{3}} \tag{6.181}$$

The construction of the split-T power combiner to achieve this can be obtained from (6.170)–(6.176). Since $K^2=2$, we get $Z_{01}=41.43\ \Omega$, $Z_{02}=85.43\ \Omega$, $Z_{03}=42.67\ \Omega$, $Z_{04}=59.46\ \Omega$, $Z_{05}=42.04\ \Omega$, and $R=106.07\ \Omega$.

6.10.3 *n*-Way Power Combining/Dividing Networks

Using the three-port combiner/divider as a building block, one can build an n-way power combining/dividing network with n inputs and one output, and one input and n outputs. Two popular n-way power combiner/dividers are the corporate and the chain structures shown in Figs. 6.25 and 6.26. In principle, with ideal three-port combiner/dividers, any amount of power could be combined/divided; however with practical combiners/dividers, insertion losses limit the efficiency of the overall structure. We note that $n=2^N-1$ and $n=N+1$ for corporate and chain structures, respectively. Using the result in Section 6.10.2 we can analyze these two types of combining/dividing networks. We will work explicitly with combiners with the understanding that dividers work just the opposite way. First we consider the chain combiner in Fig. 6.26, and let a_i $(i=1,2,\ldots,n)$ be the input incident power wave at port i (the individual three-port combiner will be denoted by C with appropriate subscript). From (6.166) we see that the output power of C_i with coupling coefficients α_i and β_i is given by

$$|b_i|^2=\frac{|a_1|^2}{\prod_{m=1}^{i}\alpha_m^2}=\sum_{m=1}^{i+1}|a_m|^2 \tag{6.182}$$

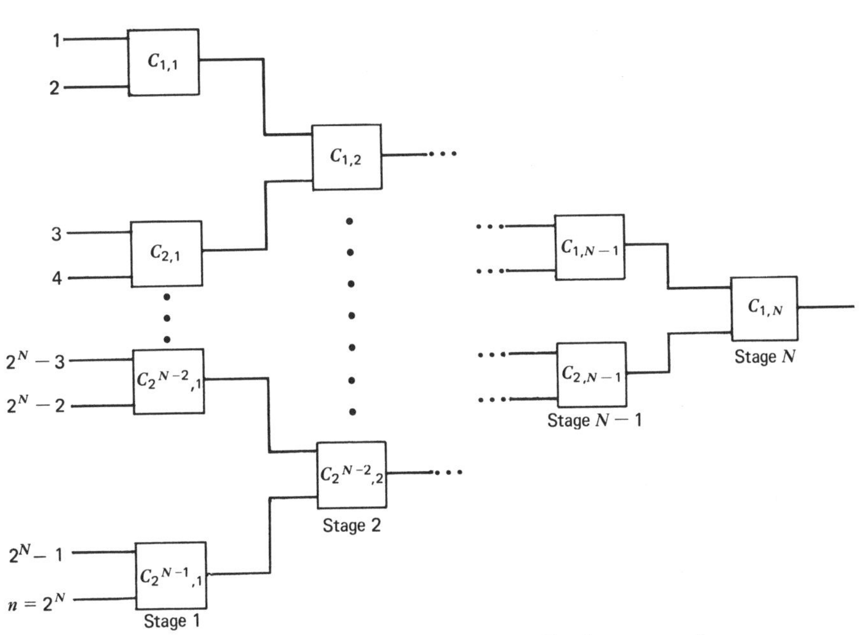

Fig. 6.25. A corporate combining/dividing network.

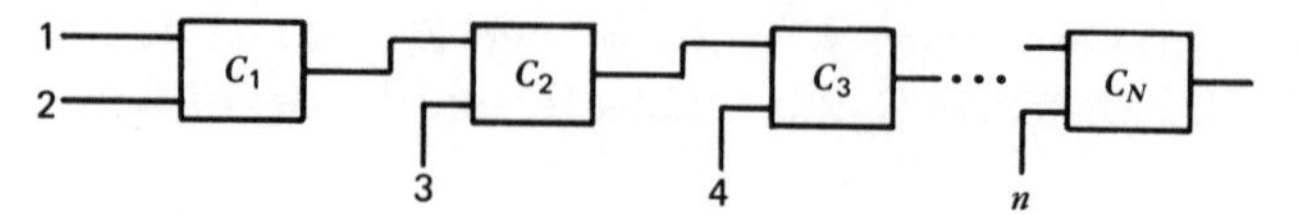

Fig. 6.26. A chain combining/dividing network.

where

$$\alpha_i = \pm \frac{1}{\left[1 + \dfrac{|a_{i+1}|^2}{\sum_{m=1}^{i} |a_m|^2}\right]^{1/2}} = \pm \left[\frac{\sum_{m=1}^{i} |a_m|^2}{\sum_{m=1}^{i+1} |a_m|^2}\right]^{1/2} \tag{6.183}$$

$$\beta_i = \pm \frac{|a_{i+1}|}{\left[\sum_{m=1}^{i+1} |a_m|^2\right]^{1/2}} \tag{6.184}$$

with the assumption that the phase of a_i $(i=1,2,\ldots,n)$ can be adjusted accordingly depending upon the type of the individual three-port combiners (in phase inputs for split-T combiners and 90° out-of-phase inputs if directional couplers are used).

Example 6.10

Let $n=4$ for a four-way chain power combiner with the assumption that a_1 is 90° out of phase with respect to a_2, a_3, and a_4 (C_i's are built from directional couplers). The input powers are $|a_1|^2=5$ W, $|a_2|^2=4$ W, $|a_3|^2=6$ W and $|a_4|^2=4.5$ W. From (6.183) and (6.184) the coupling coefficient α_i and β_i of the individual combiner C_i is given by

$$\alpha_1=0.745, \qquad \beta_1=0.667$$

$$\alpha_2=0.775, \qquad \beta_2=0.632$$

$$\alpha_3=0.877, \qquad \beta_3=0.480$$

If we define the coupling factor (in dB) of C_i as

$$K_i = -10\log \beta_i^2 \tag{6.185}$$

then $K_1=3.52$ dB, $K_2=3.98$ dB, and $K_3=6.38$ dB.

Now in order to take into account the loss in each nonideal three-port combiner C_i, let $L_i \geqslant 1$ be the coupling loss factor for each C_i (that is, the coupling loss of C_i in dB is $10\log L_i^2$, for example, if C_i has 0.1 dB coupling loss then $L_i = 1.0116$). Referring to Fig. 6.26, the output power of C_i is given recursively as

$$|b_i|^2 = \frac{|a_1|^2}{\prod_{m=1}^{i} L_m^2 \alpha_m^2} = \frac{1}{L_i^2}\left(|b_{i-1}|^2 + |a_{i+1}|^2\right) \tag{6.186a}$$

$$b_o = a_1 \tag{6.186b}$$

where

$$\alpha_i = \pm \frac{1}{\left[1 + \dfrac{|a_{i+1}|^2}{|b_{i-1}|^2}\right]^{1/2}} = \pm \frac{|b_{i-1}|}{\left[|b_{i-1}|^2 + |a_{i+1}|^2\right]^{1/2}} \tag{6.187}$$

$$\beta_i = \pm \frac{|a_{i+1}|}{\left[|b_{i-1}|^2 + |a_{i+1}|^2\right]^{1/2}} \tag{6.188}$$

Example 6.11

Let $|a_i|^2$, $i = 1,2,3,4$ take the values given in Example 6.10 and assume that C_1, C_2, and C_3 have a nominal loss of 0.3 dB each, that is, $L_i^2 = 1.0715$ $(i = 1,2,3)$. Then from (6.187) we obtain

$$\alpha_1 = 0.745, \qquad \beta_1 = 0.667$$

$$\alpha_2 = 0.764, \qquad \beta_2 = 0.646$$

$$\alpha_3 = 0.866, \qquad \beta_3 = 0.500$$

The output power is given by $|b_3|^2 = 16.74$ W compared to 19.5 W if all the combiners C_i are ideal.

For the corporate combiner in Fig. 6.25, it is easily seen that the output power of $C_{i,1}$ $(i = 1,2,\ldots,2^{N-1})$ is given by

$$|b_{i,1}|^2 = \frac{|a_{2i-1}|^2}{\alpha_{i,1}^2} \qquad i = 1,2,\ldots,2^{N-1} \tag{6.189}$$

where

$$\alpha_{i,1} = \pm \frac{1}{\left[1 + \dfrac{|a_{2i}|^2}{|a_{2i-1}|^2}\right]^{1/2}} \tag{6.190}$$

and hence the output of $C_{i,m}$ ($i = 1, 2, \ldots, 2^{N-m}$; $m = 1, 2, \ldots, N$) is given by the following recursive expression:

$$|b_{i,m}|^2 = \frac{|b_{2i-1,m-1}|^2}{\alpha_{i,m}^2}, \qquad i = 1, 2, \ldots, 2^{N-m}; \qquad m = 1, 2, \ldots, N \tag{6.191}$$

where

$$\alpha_{i,m} = \pm \frac{1}{\left[1 + \dfrac{|b_{2i,m-1}|^2}{|b_{2i-1,m-1}|^2}\right]^{1/2}} \tag{6.192}$$

If $L_{i,m}$ is the coupling loss factor of $C_{i,m}$. Then $\alpha_{i,m}$ will be replaced by $L_{i,m}\alpha_{i,m}$ in the equations (6.191)–(6.192). These expressions are derived with the assumption that the phase of $a_{i,m}$ can be adjusted accordingly depending upon the type of individual three-port combiners. If $|a_i| = |a|$, then

$$\alpha_i = \beta_i = \pm \frac{1}{\sqrt{2}}, \qquad i = 1, 2, \ldots, 2^N \tag{6.193}$$

for the corporate combining/dividing network and

$$\alpha_i = \pm \sqrt{\frac{i}{i+1}}, \qquad \beta_i = \pm \frac{1}{\sqrt{i+1}}, \qquad i = 1, 2, \ldots, N \tag{6.194}$$

for the chain combining/dividing network.

Example 6.12

An equal phase, equal amplitude four-way corporate combiner/divider can be built from a two-way Wilkinson combiner/divider as shown in Fig. 6.27. In the ideal case, the scattering matrix of the Wilkinson combiner/divider is given in (6.177) with $\alpha = \beta = 1/\sqrt{2}$ (that is, the coupling factor is $10 \log \beta^2 = 3$ dB). If $a_1 = a_2 = a_3 = a_4 = |a| \exp(j\theta)$, then the output b_o is given by (6.191) by

$$b_0 = 2|a| \exp(j\theta)$$

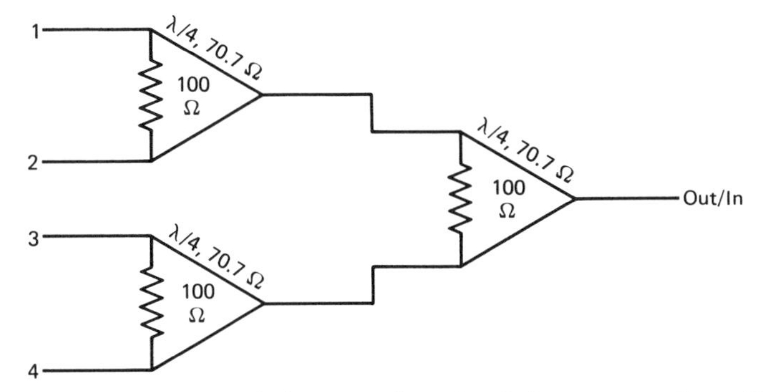

Fig. 6.27. A corporate equal phase, equal amplitude four-way combiner/divider.

and

$$|b_o|^2 = 4|a|^2$$

If the input and output roles are interchanged, it becomes a corporate four-way divider, that is, if $a_{in} = |a|\exp(j\theta)$ is the input wave, the output waves are given as

$$b_1 = b_2 = b_3 = b_4 = \tfrac{1}{2}|a|\exp(j\theta)$$

and

$$|b_1|^2 = |b_2|^2 = |b_3|^2 = |b_4|^2 = \tfrac{1}{4}|a|^2$$

Example 6.13

An equal phase, equal amplitude four-way chain combiner/divider can be built from the two-way split-T combiner/divider discussed in section 6.10.2 as shown in Fig. 6.28. The coupling factor of each C_i is given by (6.194) and (6.185) as

$$K_1 = -10\log\left(\frac{1}{\sqrt{2}}\right)^2 = 3\text{ dB} \qquad K_2 = -10\log\left(\frac{1}{\sqrt{3}}\right)^2 = 4.77\text{ dB}$$

$$K_3 = -10\log\left(\frac{1}{2}\right)^2 = 6\text{ dB}$$

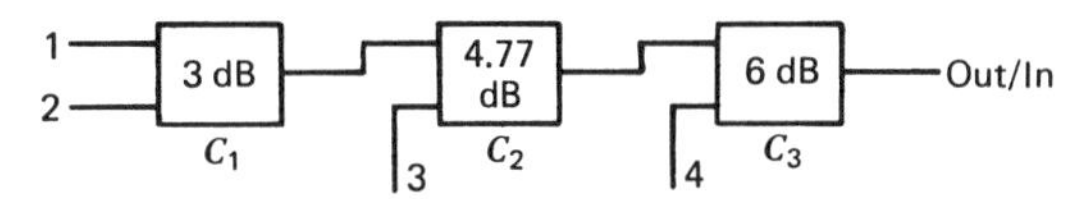

Fig. 6.28. A chain equal phase, equal amplitude four-way combiner/divider (C_1, C_2, C_3 are all split-T combiner/dividers).

If $|a_1|=|a_2|=|a_3|=|a_4|=|a|\exp(j\theta)$, then the output b_o is given by (6.182) as

$$b_o=\frac{|a|\exp(j\theta)}{\left[\prod_{m=1}^{3}(1-\beta_m^2)\right]^{1/2}}=\frac{|a|\exp(j\theta)}{\left[\left(\frac{1}{2}\right)\left(\frac{2}{3}\right)\left(\frac{3}{4}\right)\right]^{1/2}}=2|a|\exp(j\theta)$$

$$|b_o|^2=4|a|^2$$

If the input and output roles are interchanged we obtain a four-way chain divider.

6.10.4 *n*-Way Amplifiers

In many applications rf power levels are required that far exceed the capability of any single device (transistor, diode) or amplifier. Thus several devices or amplifiers must be coupled together to generate this power; that is, the input power is split and delivered to several devices or amplifiers by a dividing network, then the output power of these devices or amplifiers is combined by a combining network to generate the larger output power level. Such a system is called an n-way amplifier and is shown systematically in Fig. 6.29. The following example illustrates the use of n-way amplifiers.

Example 6.14

The gain and output power of microwave transistors at their 1 dB compression are usually specified by the manufacturer. Consider the following power GaAs FETs at 6 GHz with specifications given in Table 6.2.

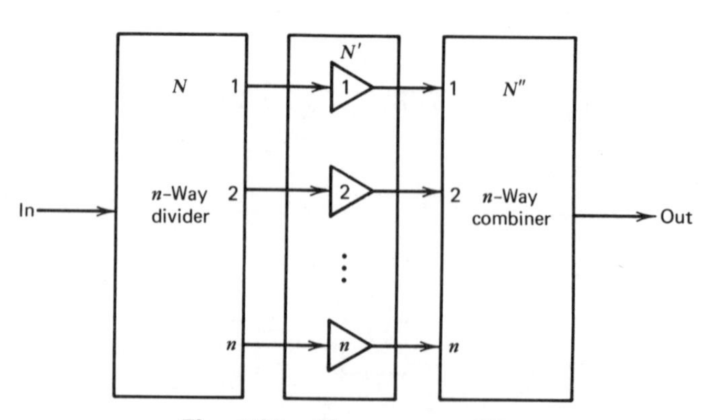

Fig. 6.29. The n-way amplifier.

Table 6.2 Specifications of power GaAs FETs

Device	P_{1dB} (dBm)	G_{1dB} (dB)	G (dB)
GaAs FET 1	28.5	6.74	7.74
GaAs FET 2	30.8	5.70	6.70
GaAs FET 3	35.1	5.00	6.00

We will demonstrate that a power amplifier that has $P_{1dB}=37.8$ dBm and $G_{1dB}=24.8$ dB can be built from the above GaAs FETs, although the output power at 1 dB gain compression exceeds their capability. With the above specification, the input power that produces $P_{1dB}=37.8$ dBm must be $P_{in}=37.8-24.8=13$ dBm. With this input power, FET 1 can be used in the first stage, which will produce an output power of 20.7 dBm. This input power can now be fed into FET 2, used in the second stage to produce an output power of 27.4 dBm, which is fed into FET 3 of the third stage to produce an output power of 33.4 dBm. In order to obtain the output power of 37.8 dBm, the output power of FET 3 is divided into two equal powers of 30.1 dBm by a two-way divider (assuming the coupling loss of the two-way divider is 0.3 dB). These powers are fed into two FET 3s to produce output powers of 35.1 dBm, which are then combined by a two-way combiner (assuming the coupling loss is again 0.3 dB) to obtain the final output power of 37.8 dBm. With a 13 dBm input (20 mW) this amplifier is capable of providing an output power of 37.8 dBm (6 W) with a gain of 24.8 dB at the 1 dB compression point, which exceeds the capability of FET 3. The amplifier is shown in Fig. 6.30. Note that in the above combining power amplifier we assume that the two FET 3 amplifiers are identical and that the two-way divider is identical to the two-way combiner except that the input and output roles are interchanged in such a way that they are the mirror images of each other. A deviation from this will decrease the combining efficiency, as will be discussed later.

In practice, the combining/dividing networks used in an n-way amplifier system are not ideal, that is, their ports are not perfectly matched (matched

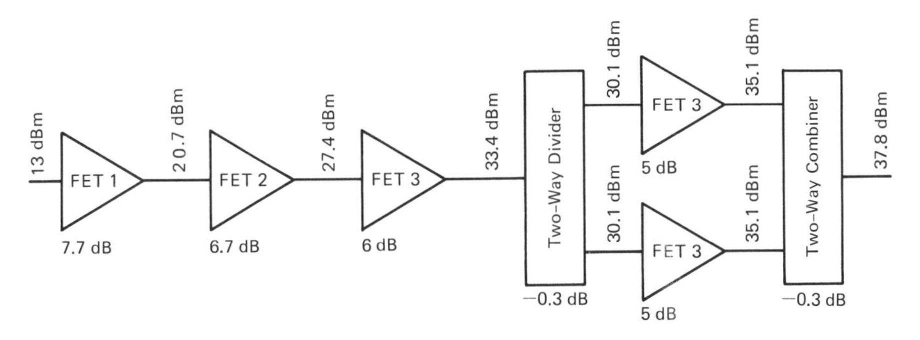

Fig. 6.30. Illustration of the power combining amplifier.

means no reflection or VSWR of one) and the isolation from port to port is not infinite. For example, the VSWR and isolation response of a two-way Wilkinson combiner/divider are shown in Fig. 6.31, where f_0 is the center frequency used to calculate $\lambda/4$ in Fig. 6.24. If these two-way combiner/dividers are used as building blocks for an n-way combining/dividing network such as the corporate or chain structures described previously, the performance over a broad frequency band can deviate significantly from the ideal one. In order to be able to analyze the performance of the practical n-way amplifier shown in Fig. 6.29 the following method is proposed. It is based entirely on the scattering matrices of the combining/dividing networks and the individual amplifiers in the system. Once their scattering parameters are known (the scattering parameters of individual two-way combiner/dividers can be obtained by odd-and-even mode analysis over a frequency band [3, 4]. From them, the scattering matrices of the n-way combining and dividing networks can be obtained by the equations to be discussed below. The scattering matrices of the n individual amplifiers can be obtained from their design or from measurement), the performance of the whole system such as its input VSWR, gain, power output, and degradation performance in the case of failure of a number of individual amplifiers, can all be predicted with computer-aided analysis. This is of particular importance in the design of very broadband GaAs FET amplifiers for electronic counter or support measure systems (for example amplifiers that operate from 2 to 10 or 6 to 16 GHz). With such wide bandwidth and high upper passband frequency, the GaAs

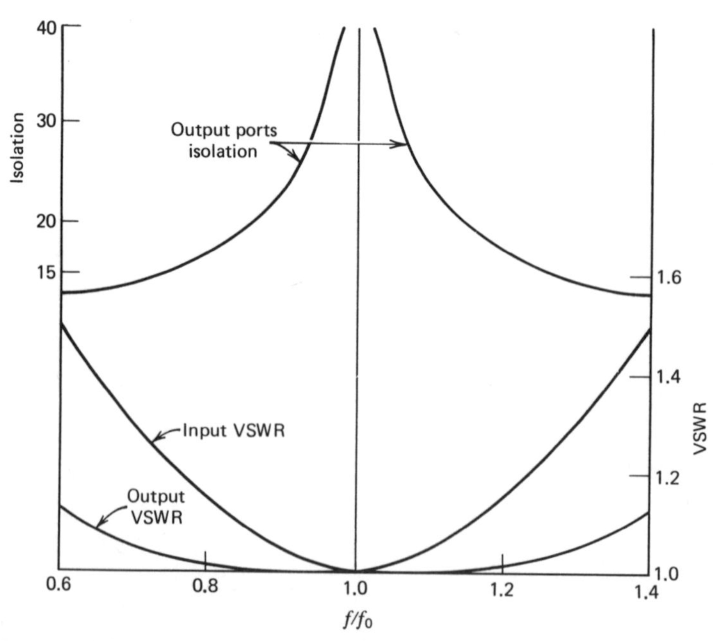

Fig. 6.31. VSWR and isolation as functions of frequency of a two-way Wilkinson combiner/divider.

FET power output is small enough (<26 dBm) to enable the network analyzer to be used to obtain its large-signal scattering parameters. In the case of high output power that exceeds the measurement capability of the network analyzer, the large-signal input and output reflection coefficients can be measured by conventional substitution methods together with the transmission coefficient, with the reverse transmission coefficient assumed to be zero (unilateral assumption).

Cascade-Connection with Scattering Matrices

Consider the $(n+l)$-port network N in direct cascade with the $(l+m)$-port network N' shown in Fig. 6.32. Let $\mathbf{S}=[S_{ij}]$ and $\mathbf{S}'=[S'_{ij}]$ be the scattering matrices of N and N', respectively, with their associated scattering variables $\mathbf{a}$, $\mathbf{b}$, $\mathbf{a}'$, and $\mathbf{b}'$. For the cascade connection to be valid we assume that the l output ports of N and the l input ports of N' are identically normalized (in practice, N and N' are both normalized to 50 Ω at each port). We can now write (note that $\mathbf{a}_1^T=[a_{1,1}, a_{1,2},\ldots,a_{1,k}]^T$, $\mathbf{a}_2^T=[a_{2,k+1}, a_{2,k+2},\ldots,a_{2,l}]^T$ etc.).

$$\begin{bmatrix}\mathbf{b}_1\\ \mathbf{b}_2\end{bmatrix} = \begin{matrix} k\{ \\ l\{ \end{matrix}\overset{\overbrace{\quad k \quad}\ \overbrace{\quad l \quad}}{\left[\begin{array}{c|c}\mathbf{S}_{11} & \mathbf{S}_{12}\\ \hline \mathbf{S}_{21} & \mathbf{S}_{22}\end{array}\right]}\begin{bmatrix}\mathbf{a}_1\\ \mathbf{a}_2\end{bmatrix} \tag{6.195}$$

$$\begin{bmatrix}\mathbf{b}'_1\\ \mathbf{b}'_2\end{bmatrix} = \begin{matrix} l\{ \\ m\{ \end{matrix}\overset{\overbrace{\quad l \quad}\ \overbrace{\quad m \quad}}{\left[\begin{array}{c|c}\mathbf{S}'_{11} & \mathbf{S}'_{12}\\ \hline \mathbf{S}'_{21} & \mathbf{S}'_{22}\end{array}\right]}\begin{bmatrix}\mathbf{a}'_1\\ \mathbf{a}'_2\end{bmatrix} \tag{6.196}$$

Our purpose is to find the scattering matrix $\mathbf{\Sigma}$ of the overall network N_Σ described by

$$\begin{bmatrix}\mathbf{b}_1\\ \mathbf{b}'_2\end{bmatrix} = \begin{bmatrix}\mathbf{\Sigma}_{11} & \mathbf{\Sigma}_{12}\\ \mathbf{\Sigma}_{21} & \mathbf{\Sigma}_{22}\end{bmatrix}\begin{bmatrix}\mathbf{a}_1\\ \mathbf{a}'_2\end{bmatrix} \tag{6.197}$$

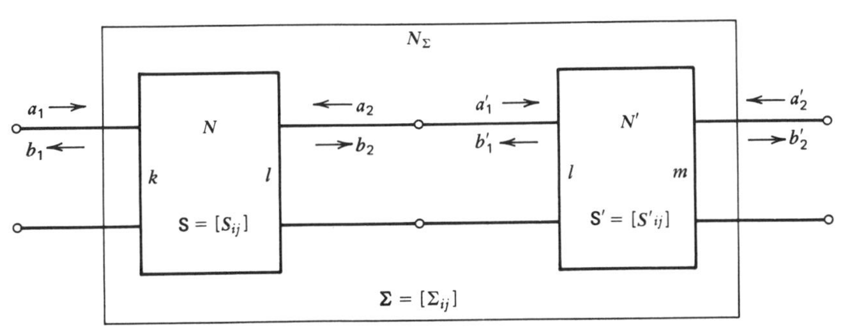

Fig. 6.32. The cascade connection representation of two networks for scattering matrix analysis.

From Fig. 6.32 we have

$$\mathbf{a}_1' = \mathbf{b}_2, \qquad \mathbf{a}_2 = \mathbf{b}_1' \tag{6.198}$$

owing to the same normalization at l cascaded ports. From (6.195), (6.196), and (6.198)

$$\begin{aligned}\mathbf{b}_1 &= \mathbf{S}_{11}\mathbf{a}_1 + \mathbf{S}_{12}\mathbf{b}_1' = \mathbf{S}_{11}\mathbf{a}_1 + \mathbf{S}_{12}(\mathbf{S}_{11}'\mathbf{a}_1' + \mathbf{S}_{12}'\mathbf{a}_2')\\ &= \mathbf{S}_{11}\mathbf{a}_1 + \mathbf{S}_{12}\mathbf{S}_{11}'\mathbf{a}_1' + \mathbf{S}_{12}\mathbf{S}_{12}'\mathbf{a}_2'\end{aligned} \tag{6.199}$$

$$\begin{aligned}\mathbf{a}_1' &= \mathbf{b}_2 = \mathbf{S}_{21}\mathbf{a}_1 + \mathbf{S}_{22}\mathbf{a}_2 = \mathbf{S}_{21}\mathbf{a}_1 + \mathbf{S}_{22}\mathbf{b}_1'\\ &= \mathbf{S}_{21}\mathbf{a}_1 + \mathbf{S}_{22}(\mathbf{S}_{11}'\mathbf{a}_1' + \mathbf{S}_{12}'\mathbf{a}_2')\\ &= \mathbf{S}_{21}\mathbf{a}_1 + \mathbf{S}_{22}\mathbf{S}_{11}'\mathbf{a}_1' + \mathbf{S}_{22}\mathbf{S}_{12}'\mathbf{a}_2'\end{aligned} \tag{6.200}$$

From (6.200) we obtain

$$\mathbf{a}_1' = (\mathbf{1}_l - \mathbf{S}_{22}\mathbf{S}_{11}')^{-1}(\mathbf{S}_{21}\mathbf{a}_1 + \mathbf{S}_{22}\mathbf{S}_{12}'\mathbf{a}_2') \tag{6.201}$$

Substituting (6.201) into (6.199) yields

$$\begin{aligned}\mathbf{b}_1 &= \left[\mathbf{S}_{11} + \mathbf{S}_{12}\mathbf{S}_{11}'(\mathbf{1}_l - \mathbf{S}_{22}\mathbf{S}_{11}')^{-1}\mathbf{S}_{21}\right]\mathbf{a}_1\\ &\quad + \left[\mathbf{S}_{12}\mathbf{S}_{12}' + \mathbf{S}_{12}\mathbf{S}_{12}'(\mathbf{1}_l - \mathbf{S}_{22}\mathbf{S}_{11}')^{-1}\mathbf{S}_{22}\mathbf{S}_{12}'\right]\mathbf{a}_2'\\ &= \left[\mathbf{S}_{11} + \mathbf{S}_{12}\mathbf{S}_{11}'(\mathbf{1}_l - \mathbf{S}_{22}\mathbf{S}_{11}')^{-1}\mathbf{S}_{21}\right]\mathbf{a}_1\\ &\quad + \left[\mathbf{S}_{12}\{\mathbf{1}_1 + \mathbf{S}_{11}'(\mathbf{1}_l - \mathbf{S}_{22}\mathbf{S}_{11}')^{-1}\mathbf{S}_{22}\}\mathbf{S}_{12}'\right]\mathbf{a}_2\\ &= \left[\mathbf{S}_{11} + \mathbf{S}_{12}\mathbf{S}_{11}'(\mathbf{1}_l - \mathbf{S}_{22}\mathbf{S}_{11}')^{-1}\mathbf{S}_{21}\right]\mathbf{a}_1\\ &\quad + \left[\mathbf{S}_{12}(\mathbf{1}_l - \mathbf{S}_{11}'\mathbf{S}_{22})^{-1}\mathbf{S}_{12}'\right]\mathbf{a}_2'\end{aligned} \tag{6.202}$$

We have used the expression $[\mathbf{1}_l + \mathbf{S}_{11}'(\mathbf{1}_l - \mathbf{S}_{22}\mathbf{S}_{11}')^{-1}\mathbf{S}_{22}]^{-1} = \mathbf{1}_l - \mathbf{S}_{11}'\mathbf{S}_{22}$ to simplify (6.202). To obtain $\mathbf{b}_2'$ we return to (6.196) and (6.198)

$$\begin{aligned}\mathbf{b}_2' &= \mathbf{S}_{21}'\mathbf{a}_1' + \mathbf{S}_{22}'\mathbf{a}_2' = \mathbf{S}_{21}'\mathbf{b}_2 + \mathbf{S}_{22}'\mathbf{a}_2' = \mathbf{S}_{21}'(\mathbf{S}_{21}\mathbf{a}_1 + \mathbf{S}_{22}\mathbf{a}_2) + \mathbf{S}_{22}'\mathbf{a}_2'\\ &= \mathbf{S}_{21}'\mathbf{S}_{21}\mathbf{a}_1 + \mathbf{S}_{21}'\mathbf{S}_{22}\mathbf{a}_2 + \mathbf{S}_{22}'\mathbf{a}_2'\end{aligned} \tag{6.203}$$

To eliminate $\mathbf{a}_2$ from (6.203) we note that

$$\begin{aligned}\mathbf{a}_2 &= \mathbf{b}_1' = \mathbf{S}_{11}'\mathbf{a}_1' + \mathbf{S}_{12}'\mathbf{a}_2' = \mathbf{S}_{11}'\mathbf{b}_2 + \mathbf{S}_{12}'\mathbf{a}_2'\\ &= \mathbf{S}_{11}'(\mathbf{S}_{21}\mathbf{a}_1 + \mathbf{S}_{22}\mathbf{a}_2) + \mathbf{S}_{12}'\mathbf{a}_2'\end{aligned} \tag{6.204}$$

which yields

$$\mathbf{a}_2 = \left[(\mathbf{1}_l - \mathbf{S}'_{11}\mathbf{S}_{22})^{-1}\mathbf{S}'_{11}\mathbf{S}_{21}\right]\mathbf{a}_1 + \left[(\mathbf{1}_l - \mathbf{S}'_{11}\mathbf{S}_{22})^{-1}\mathbf{S}'_{12}\right]\mathbf{a}_2 \tag{6.205}$$

Substituting (6.204) into (6.203) we obtain

$$\begin{aligned}
\mathbf{b}'_2 &= \left[\mathbf{S}'_{21}\mathbf{S}_{21} + \mathbf{S}'_{21}\mathbf{S}_{22}(\mathbf{1}_l - \mathbf{S}'_{11}\mathbf{S}_{22})^{-1}\mathbf{S}'_{11}\mathbf{S}_{21}\right]\mathbf{a}_1 \\
&\quad + \left[\mathbf{S}'_{22} + \mathbf{S}'_{21}\mathbf{S}_{22}(\mathbf{1}_l - \mathbf{S}'_{11}\mathbf{S}_{22})^{-1}\mathbf{S}'_{12}\right]\mathbf{a}_2 \\
&= \left[\mathbf{S}'_{21}\{\mathbf{1}_l + \mathbf{S}_{22}(\mathbf{1}_l - \mathbf{S}'_{11}\mathbf{S}_{22})^{-1}\mathbf{S}'_{11}\}\mathbf{S}_{21}\right]\mathbf{a}_1 \\
&\quad + \left[\mathbf{S}'_{22} + \mathbf{S}'_{21}\mathbf{S}_{22}(\mathbf{1}_l - \mathbf{S}'_{11}\mathbf{S}_{22})^{-1}\mathbf{S}'_{12}\right]\mathbf{a}_2 \\
&= \left[\mathbf{S}'_{21}(\mathbf{1}_l - \mathbf{S}_{22}\mathbf{S}'_{11})^{-1}\mathbf{S}_{21}\right]\mathbf{a}_1 \\
&\quad + \left[\mathbf{S}_{22} + \mathbf{S}'_{21}\mathbf{S}_{22}(\mathbf{1}_l - \mathbf{S}'_{11}\mathbf{S}_{22})^{-1}\mathbf{S}'_{12}\right]\mathbf{a}_2
\end{aligned} \tag{6.206}$$

Again we have employed the expression $[\mathbf{1}_l + \mathbf{S}_{22}(\mathbf{1}_l - \mathbf{S}'_{11}\mathbf{S}_{22})^{-1}\mathbf{S}'_{11}]^{-1} = (\mathbf{1}_l - \mathbf{S}'_{11}\mathbf{S}_{22})$ to simplify (6.206). From (6.202) and (6.206) the scattering matrix $\mathbf{\Sigma}$ of the cascade network N_{Σ} is given by

$$\mathbf{\Sigma} = \begin{bmatrix} \mathbf{S}_{11} + \mathbf{S}_{12}\mathbf{S}'_{11}(\mathbf{1}_l - \mathbf{S}_{22}\mathbf{S}'_{11})^{-1}\mathbf{S}_{21} & \mathbf{S}_{12}(\mathbf{1}_l - \mathbf{S}'_{11}\mathbf{S}_{22})^{-1}\mathbf{S}'_{12} \\ \mathbf{S}'_{21}(\mathbf{1}_l - \mathbf{S}_{22}\mathbf{S}'_{11})^{-1}\mathbf{S}_{21} & \mathbf{S}'_{22} + \mathbf{S}'_{21}\mathbf{S}_{22}(\mathbf{1}_l - \mathbf{S}'_{11}\mathbf{S}_{22})^{-1}\mathbf{S}'_{12} \end{bmatrix} \tag{6.207}$$

By applying (6.207) repeatedly, the scattering matrix of a chain of subnetworks can be computed. This equation forms the cornerstone for the computer-aided analysis of any n-way systems where the n-way amplifier is just a specific case.

Returning to the n-way amplifier in Fig. 6.29, we see that it can be represented by a cascade-connection of three networks: the $(1+n)$-port power divider N, the $(n+n)$-port uncoupled amplifier network N' and the $(n+1)$-port power combiner N''. The cascade connection forms a two-port n-way amplifier with one input and one output. In an n-way amplifier, the n-way power combiner and divider are selected to be identical, with the input and output roles interchanged. Their use in such a system must be in a mirror image fashion with proper phase adjustment, that is, if the jth port of the divider feeds the input of the jth amplifier, the output of this amplifier must feed the same jth port of the combiner obtained from the same divider with the input-output roles interchanged and with proper phase adjustment. Thus for example, assume the n-way power divider N in Fig. 6.29 is an equal amplitude, equal phase, ideal structure, then its scattering matrix **S** to be used in (6.207) is given by ($k=1$, $l=n$)

$$\mathbf{S}=\frac{1}{\sqrt{n}}\left[\begin{array}{c|cccc} 0 & 1 & 1 & \dots & 1 \\ \hline 1 & & & & \\ 1 & & & & \\ \vdots & & \mathbf{0}_n & & \\ 1 & & & & \end{array}\right]=\left[\begin{array}{c|c} \mathbf{S}_{11} & \mathbf{S}_{12} \\ \hline \mathbf{S}_{21} & \mathbf{S}_{22} \end{array}\right] \tag{6.208}$$

With the assumption that the n-way power combiner N'' in Fig. 6.29 is identical to N with input-output roles interchanged. Its scattering matrix $\mathbf{S}''$ to be employed in (6.207) is given by

$$\mathbf{S}''=\frac{1}{\sqrt{n}}\left[\begin{array}{cccc|c} & & & & 1 \\ & & & & 1 \\ & & \mathbf{0}_n & & \vdots \\ & & & & 1 \\ \hline 1 & 1 & \dots & 1 & 0 \end{array}\right]=\left[\begin{array}{c|c} \mathbf{S}''_{11} & \mathbf{S}''_{12} \\ \hline \mathbf{S}''_{21} & \mathbf{S}''_{22} \end{array}\right] \tag{6.209}$$

The scattering matrix $\mathbf{S}'$ of the $(n+n)$-port uncoupled amplifier network N' is obtained as

$$\mathbf{S}'=\left[\begin{array}{c|c} \mathbf{S}'_{11} & \mathbf{S}'_{12} \\ \hline \mathbf{S}'_{21} & \mathbf{S}'_{22} \end{array}\right] \tag{6.210a}$$

$$\mathbf{S}'_{ij}=\begin{bmatrix} s_{ij,1} & 0 & \dots & 0 \\ 0 & s_{ij,2} & \dots & 0 \\ \dots & \dots & \dots & \dots \\ 0 & 0 & \dots & s_{ij,n} \end{bmatrix}, \qquad i=1,2;\ j=1,2 \tag{6.210b}$$

where the scattering matrix of the pth individual amplifier is given by

$$\mathfrak{S}_p=[s_{ij,p}], \qquad i=1,2; \qquad j=1,2 \tag{6.211}$$

In order to obtain the two-port scattering matrix of the n-way amplifier, (6.207) is used to obtain $\boldsymbol{\Sigma}$ with **S**, and $\mathbf{S}'$ given by (6.208) and (6.210). Repeat (6.207) the second time with $\boldsymbol{\Sigma}$ replaced **S** and $\mathbf{S}''$ replaced $\mathbf{S}'$ where $\mathbf{S}''$ is given by (6.209). The two-port scattering matrix $\hat{\mathbf{S}}$ of the n-way amplifier employing an equal amplitude, equal phase power divider/combiner is therefore given by

$$\hat{\mathbf{S}}=\frac{1}{n}\begin{bmatrix} \sum_{p=1}^{n} s_{11,p} & \sum_{p=1}^{n} s_{12,p} \\ \sum_{p=1}^{n} s_{21,p} & \sum_{p=1}^{n} s_{22,p} \end{bmatrix} \tag{6.212}$$

If the n individual amplifiers are identical, that is, $\mathbb{S}_1=\mathbb{S}_2=\cdots=\mathbb{S}_n=\mathbb{S}=[s_{ij}]$, then the scattering matrix $\hat{\mathbb{S}}=\mathbb{S}$ as seen from (6.212). Furthermore, if the individual amplifiers are designed to have matched inputs and outputs, that is, $s_{11}=0$ and $s_{22}=0$ (the corresponding VSWR is 1), then the n-way amplifier also has matched input and output. A common situation with n-way amplifiers is the failure of one or more amplifiers in the system. Even if the individual amplifier is designed to have matched inputs and outputs, in the failure mode its input and output reflection coefficients are usually no longer zero (or no longer small for broadband matched amplifiers). It is not uncommon that a failed amplifier possesses input or output VSWR of 10 or more. To illustrate this, let the kth amplifier in a matched n-way amplifier possesses an input reflection coefficient $s_{11,k}$ in its failure mode. From (6.212) we see that the input reflection coefficient of the n-way amplifier in this failure mode is $s_{11,k}/n$ which yields the following input VSWR_i.

$$\text{VSWR}_i=\frac{1+|s_{11,k}/n|}{1-|s_{11,k}/n|}=\frac{n+|s_{11,k}|}{n-|s_{11,k}|} \tag{6.213}$$

Numerically let $|s_{11,k}|=0.8$ (the corresponding input VSWR of the failed kth amplifier is 9). Then for $n=2$, $\text{VSWR}_i=2.33$ compared to unity during normal operation.

From (6.212) the gain of the n-way amplifier is given by

$$G=\frac{1}{n^2}\left|\sum_{p=1}^{n} s_{21,p}\right|^2 \tag{6.214a}$$

Suppose $s_{21,1}=s_{21,2}=\cdots=s_{21,n}=s_{21}$, that is, all n individual amplifiers have identical transmission coefficients. Suppose in the failure mode, m amplifiers fail with $s_{21}=0$. Then the gain of the n-way amplifier becomes

$$G_F=\frac{1}{n^2}|(n-m)s_{21}|^2=\frac{(n-m)^2}{n^2}|s_{21}|^2 \tag{6.214b}$$

compared to $G=|s_{21}|^2$ in normal operation. This means that the output power of the n-way amplifier has dropped to $(n-m)^2/n^2$ its original level. Numerically let $m=1$ and $n=4$. Then in the failure mode, the output power drops to nine-sixteenths of its original level (or 2.5 dB below its original level). If $m=2$ and $n=4$, the loss of power is 6 dB. This type of failure in n-way amplifiers is called "graceful degradation" as opposed to "catastrophic failure" in the sense that the output power is decreased but not lost completely. In high power systems, the isolation resistors in the dividers must be large enough to handle their incident power when one amplifier fails, or a chain reaction can result. From this analysis we see that the failure of one or more amplifiers in an n-way amplifier system leads to two problems: an increase in input and output VSWR, and a graceful drop of output power. Communication systems are composed of many subsystems in direct cascade whose input and output VSWRs are usually kept small ($\leqslant 1.5$) so that their direct cascade interfaces do not cause multiple reflections that can seriously affect their transmission responses. Thus a failed amplifier with high input or output VSWR can cause serious degradation even if it has enough gain to transmit or receive signals. In order to avoid this problem, isolators can be employed at the input or output of each individual amplifier in an n-way system to provide small and constant input and output VSWR for both the normal and failed modes. If the isolation from port to port in the n-way combining/dividing networks is sufficiently high, then one might use one isolator at the input of the n-way divider and one at the output of the n-way combiner for the overall system. This can save $2n-2$ isolators. In practice, high isolation from port to port is difficult to obtain in low loss combiners/dividers (10–20 dB is typical) and if performance of an n-way amplifier in its partial failed mode is a main concern, isolators must be employed at inputs and outputs of individual amplifiers.

In conclusion, the n-way amplifier can provide an output power that is the sum of the individual amplifier output powers in the system minus the combining loss. It also provides a graceful degradation, that is, if m individual amplifiers fail, the output power drops to $(n-m)^2/n^2$ of the original power compared to the theoretical output power $(n-m)/n$ of the original power.

Balanced Amplifiers

One of the most commonly used microwave transistor amplifiers is a type of two-way amplifier that is traditionally called a "balanced amplifier." The balanced amplifier employs a 3 dB combiner/divider made from directional couplers such as the 3 dB 90° hybrids whose scattering matrix is given by (6.160), instead of the equal phase, equal amplitude split-T combiner/divider described previously. The balanced amplifier is shown schematically in Fig. 6.33. The 3 dB 90° hybrid produces a 90° phase shift at output ports. The arrangement is shown by the numbered ports in Fig. 6.33. The scattering matrix **S** of the 90° hybrid divider is given by

$$\mathbf{S}=\frac{1}{\sqrt{2}}\left[\begin{array}{c|cc}0 & j & 1\\ \hline j & 0 & 0\\ 1 & 0 & 0\end{array}\right]=\left[\begin{array}{c|c}\mathbf{S}_{11} & \mathbf{S}_{12}\\ \hline \mathbf{S}_{21} & \mathbf{S}_{22}\end{array}\right] \tag{6.215}$$

and the scattering matrix $\mathbf{S}''$ of the 90° hybrid combiner is given by

$$\mathbf{S}''=\frac{1}{\sqrt{2}}\left[\begin{array}{cc|c}0 & 0 & 1\\ 0 & 0 & j\\ \hline 1 & j & 0\end{array}\right]=\left[\begin{array}{c|c}\mathbf{S}''_{11} & \mathbf{S}''_{12}\\ \hline \mathbf{S}''_{21} & \mathbf{S}''_{22}\end{array}\right] \tag{6.216}$$

The network that comprises two uncoupled amplifiers whose scattering matrices

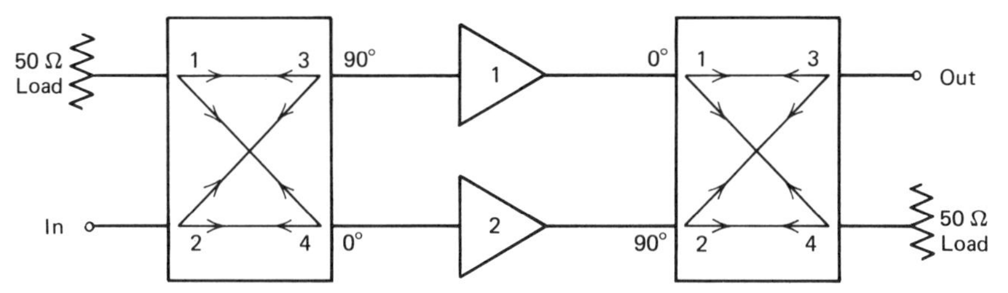

Fig. 6.33. The balanced amplifier.

are $\mathbb{S}_p = [s_{ij,p}]$, $p=1,2$ possesses the following scattering matrix $\mathbf{S}'$

$$\mathbf{S}' = \left[\begin{array}{cc|cc} s_{11,1} & 0 & s_{12,1} & 0 \\ 0 & s_{11,2} & 0 & s_{12,2} \\ \hline s_{21,1} & 0 & s_{22,1} & 0 \\ 0 & s_{21,2} & 0 & s_{22,2} \end{array}\right] = \left[\begin{array}{c|c} \mathbf{S}'_{11} & \mathbf{S}'_{12} \\ \hline \mathbf{S}'_{21} & \mathbf{S}'_{22} \end{array}\right] \tag{6.217}$$

Applying (6.207) twice yields the scattering matrix $\hat{\mathbf{S}}$ of the balanced amplifier

$$\hat{\mathbf{S}} = \frac{1}{2}\begin{bmatrix} -s_{11,1}+s_{11,2} & j(s_{12,1}+s_{12,2}) \\ j(s_{21,1}+s_{21,2}) & s_{22,1}-s_{22,2} \end{bmatrix} \tag{6.218}$$

From (6.218) we observe that if $s_{11,1}=s_{11,2}$ and $s_{22,1}=s_{22,2}$, that is, amplifiers 1 and 2 possess identical input and output reflection coefficients, then the balanced amplifier will have matched input and output ports (note that if the equal phase, equal amplitude combiner/dividers are used, the individual amplifiers must have matched inputs and outputs in order for the two-way amplifier to possess matched ports). This is an advantage of the balanced amplifier using 3 dB 90° hybrids. As long as $s_{11,1} \approx s_{11,2}$ and $s_{22,1} \approx s_{22,2}$, the balanced amplifier will have good input and output VSWR. The gain of the balanced amplifier is given as

$$G = \tfrac{1}{4}|s_{21,1}+s_{21,2}|^2 \tag{6.219}$$

If $s_{21,1}=s_{21,2}=s_{21}$, then $G=|s_{21}|^2$. Now let us assume that $s_{21,1}=|s_{21}|\exp(j\theta)$,

Table 6.3 Phase error combining loss in a balanced amplifier

$\Delta\theta$	$-10\log\frac{1}{4}\lvert 1+\exp(j\Delta\theta)\rvert^2$ (dB)
0°	0.00
5.63°	0.009
11.25°	0.040
22.5°	0.170
45°	0.690
90°	3.00

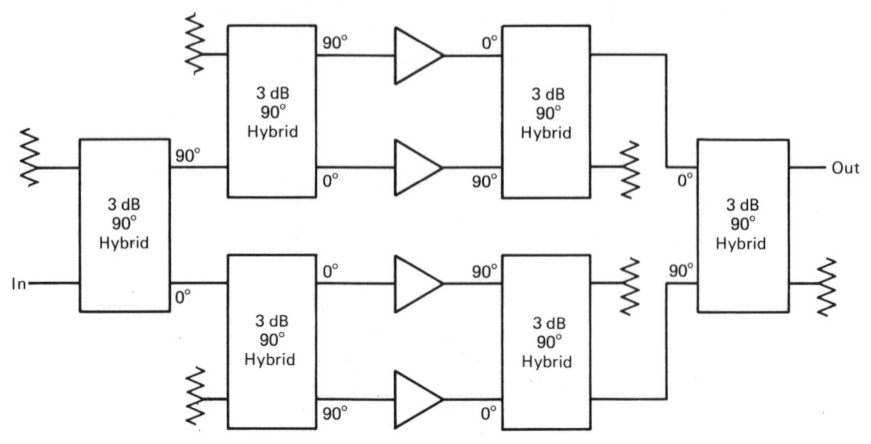

Fig. 6.34. A four-way amplifier employing 3 dB 90° hybrids.

and $s_{21,2}=|s_{21}|\exp[j(\theta+\Delta\theta)]$, that is, the two amplifiers in the configuration have the same gain but with a slight phase shift, then

$$G=\tfrac{1}{4}|s_{21}|^2|1+\exp(j\Delta\theta)|^2 \tag{6.220}$$

The factor $\frac{1}{4}|1+\exp(j\Delta\theta)|^2$ is called the combining loss due to phase error. The numerical result is given in Table 6.3.

In conclusion, the balanced amplifier provides good input (and output) VSWR if the input (and output) reflection coefficients of the individual amplifiers in the configuration are almost identical. Furthermore, the output power is theoretically twice that of the individual amplifiers. These properties make balanced amplifiers attractive in many communication systems. It can be shown that if one amplifier fails completely ($s_{21}=0$), then the output power drops to one-fourth of its original (6 dB loss). The balanced configuration can be used as subsystem in a larger n-way amplifier, but care must be exercised because of the 90° phase shift produced by the directional couplers. Phase shift compensation must be implemented when directional couplers are employed together with split-T power divider/combiners. Note that the output signal of a balanced amplifier is shifted 90° from the input signal as seen from the (2,1) term of $\hat{S}$ in (6.218). A four-way amplifier is shown in Fig. 6.34 using 3 dB 90° hybrids in its combining/dividing networks.

6.10.5 Variable Combiner/Divider

We have seen from the previous analysis that an n-way amplifier can provide a degradation combining efficiency $\eta=(n-m)^2/n^2$, that is, if m amplifiers in the system fail, the output power drops to η times the original level. Theoretically, since there still $n-m$ working amplifiers in the system, we expect to obtain an output power of $(n-m)/n$ times the original level, that is, to have a degradation efficiency $\eta_t = (n-m)/n$. To obtain this theoretical combining efficiency, variable combiner/dividers must be employed instead of the fixed combiner/dividers discussed previously. Consider the network N shown in Fig. 6.35 that has a phase shifter with variable phase shift ϕ cascaded between two identical 3 dB directional couplers with port 4 of the second coupler terminated in a matched load. We will show that this is a variable two-way combiner/divider that can produce a truly graceful degradation ($\eta_t = \frac{1}{2}$) in the case that either input signals a_1 or a_2 are shut off. Let $a_i = |a_i|\exp(j\theta)$, $i=1,2$ be the two input incident power waves at port 1 and 2 of N with the assumption that their in-phase relationship can be adjusted accordingly. Then using (6.157) and (6.161) with $\alpha=\beta=1/\sqrt{2}$, noting that the phase shifter will change the phase of the output reflected wave at port 3 of C_1 by ϕ radians, we have the following output reflected wave b_3 at port 3 of C_2.

$$b_3 = \tfrac{1}{2}\sqrt{|a_1|^2+|a_2|^2}\,\left[\exp(j\theta_1)+\exp(j\theta_2)\right] \tag{6.221a}$$

where

$$\theta_1 = \theta + \tan^{-1}\left|\frac{a_1}{a_2}\right| \tag{6.221b}$$

$$\theta_2 = \theta - \phi + \frac{\pi}{2} + \tan^{-1}\left|\frac{a_2}{a_1}\right| \tag{6.221c}$$

In order for the output power $|b_3|^2$ to be equal to the sum of the input powers $|a_1|^2+|a_2|^2$ in normal operation, it is sufficient that $\theta_1=\theta_2$ as seen from (6.221a), that is,

$$\phi = \frac{\pi}{2} + \tan^{-1}\left|\frac{a_2}{a_1}\right| - \tan^{-1}\left|\frac{a_1}{a_2}\right| \tag{6.222}$$

In this case $b_3 = \sqrt{|a_1|^2+|a_2|^2}\,\exp(j\theta_1)$ and thus $|b_3|^2 = |a_1|^2+|a_2|^2$. In the situation of failure, suppose $a_i = \gamma_i|a_i|\exp(j\theta)$, $i=1,2$, with $0 \leqslant \gamma_i \leqslant 1$ (again we assume the phase shift of a_i can be adjusted at the inputs, by another phase

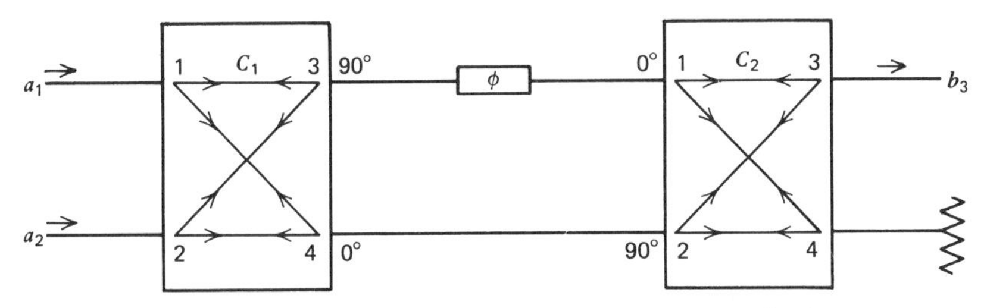

Fig. 6.35. A variable two-way power combiner/divider.

Table 6.4 Power split ratios for a two-way variable divider

ϕ	$\|b_1\|^2/\|a_{in}\|^2$	$\|b_2\|^2/\|a_{in}\|^2$
0°	0	1
70.5°	$\frac{1}{3}$	$\frac{2}{3}$
90°	$\frac{1}{2}$	$\frac{1}{2}$
109.5°	$\frac{2}{3}$	$\frac{1}{3}$
180°	1	0

shifters, for example), then in order for $|b_3|^2=|\gamma_1 a_1|^2+|\gamma_2 a_2|^2$, that is, to obtain the theoretical combining efficiency, ϕ in (6.222) must be

$$\phi=\frac{\pi}{2}+\tan^{-1}\left|\frac{\gamma_2 a_2}{\gamma_1 a_1}\right|-\tan^{-1}\left|\frac{\gamma_1 a_1}{\gamma_2 a_2}\right| \tag{6.223}$$

In particular if $a_1=0$, then $\phi=\pi$ and if $a_2=0$, then $\phi=0$. If the roles of inputs and outputs are interchanged, we have a variable two-way divider, that is, we can obtain any combination of power split at port 1 and 2 of C_1 from an input at port 3 of C_2. The variable two-way combiner/divider can be employed as a building block in the chain or corporate structures to obtain a variable n-way combining/dividing networks.

Example 6.15

Consider the two-way variable power divider in Fig. 6.35, where the input wave a_{in} is fed to port 3 of C_2 and the output waves b_1 and b_2 are extracted from ports 1 and 2 of C_1. Assume that there is no loss in C_1, C_2 and the phase shifter ϕ. The output waves at ports 1 and 2 of C_2 are $(1/\sqrt{2})a_{in}$ and $(j/\sqrt{2})a_{in}$, respectively. The input wave to port 3 of C_1 is thus $(1/\sqrt{2})a_{in}\exp(j\phi)$

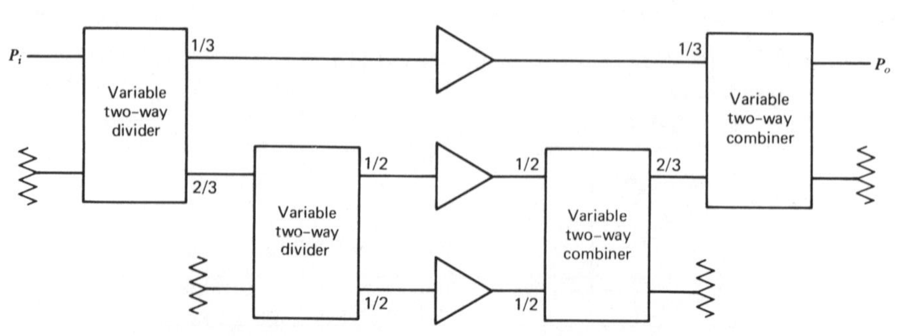

Fig. 6.36. A variable three-way power amplifier.

Table 6.5 Theoretical power loss in fixed and variable n-way amplifiers

n	m	$10\log\frac{(n-m)^2}{n}$ Fixed Combining Loss (dB)	$10\log\frac{n-m}{n}$ Variable Combining Loss (dB)
2	1	−6.00	−3.00
3	1	−3.50	−1.76
3	2	−9.54	−4.77
4	1	−2.50	−1.25
4	2	−6.00	−3.00

and hence the output waves b_1 and b_2 are given by (6.157) as

$$b_1 = \frac{1}{\sqrt{2}}\left[\frac{1}{\sqrt{2}}a_{in}\exp(j\phi)\right] + \frac{j}{\sqrt{2}}\left[\frac{j}{\sqrt{2}}a_{in}\right] = \tfrac{1}{2}a_{in}[\exp(j\phi) - 1]$$

$$b_2 = \frac{j}{\sqrt{2}}\left[\frac{1}{\sqrt{2}}a_{in}\exp(j\phi)\right] + \frac{1}{\sqrt{2}}\left[\frac{j}{\sqrt{2}}a_{in}\right] = \frac{j}{2}a_{in}[\exp(j\phi) + 1]$$

Various power split ratios $|b_1|^2/|a_{in}|^2$ and $|b_2|^2/|a_{in}|^2$ are given in Table 6.4.

A variable three-way power amplifier is shown schematically in Fig. 6.36 with variable power split ratios shown at corresponding ports. In order to control the phase shifters (PIN diode phase shifters can be used, for example), microprocessors can be employed as command networks together with sensors at each output of each individual amplifier in the system. Table 6.5 shows the theoretical net power loss in fixed and variable n-way amplifiers when m amplifiers fail.

6.11 CIRCULATOR-COUPLED AMPLIFIERS

Low noise GaAs FET amplifiers are being used widely in unattended satellite earth stations. Usually several channels share the same low noise amplifier. For example, a light route earth station terminal transmitting at 6 GHz and receiving at 4 GHz with a 4–5 m diameter antenna can accommodate about five channels (single channel per carrier). Therefore reliability of the amplifier is very important for undisrupted transmissions. For low noise amplifiers, the two most serious failures that can happen during operation are the active device failure and the power supply failure. Either failure mechanism can cause a transmission loss of more than 20 dB, which is unacceptable for radio signals. In order to avoid serious degradation because of these failures, redundant amplifiers with active switches or balanced amplifiers can be employed. A low noise amplifier with 50 dB gain or more at 4 GHz would then require four stages with 2 GaAs FETs per stage. This would double the cost without relief in the case of failure of the active switches or the loss of the dc power supply. To meet the low cost requirement and to provide acceptable degradation in the case of

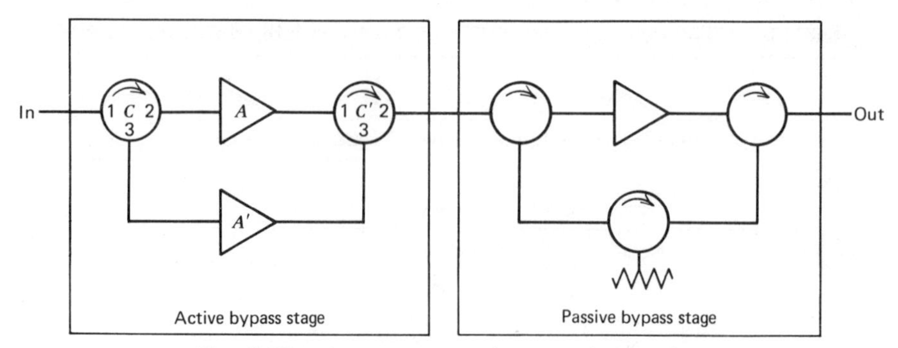

Fig. 6.37. A two-stage circulator-coupled amplifier.

the above two failures, a circulator-coupled amplifier that provides a fail-soft operation is described here. The schematic configuration is shown in Fig. 6.37. In this amplifier, a circulator-coupled active bypass amplifier is used for the first stage and subsequent stages employ passive bypass amplifiers. For high gain amplifiers, this configuration can reduce the cost, since only the first stage is redundant while the subsequent stages have no redundant amplifiers. The failure of the primary amplifier A in the active bypass stage will introduce a loss caused by the return loss at the input and output ports of A (lab test has shown that over the 3–6 GHz band, these return losses combine to about 2–3 dB for GaAs FETs). In practice low loss circulators C and C' are normally used (0.5 dB combined insertion loss is typical). Therefore, if A fails, the signal will be routed through the redundant amplifier A' with a typical loss of about 2.5–3.5 dB, which means that the noise figure of the overall amplifier will increase by the same amount. For C-band earth stations with a 1.5 dB noise figure amplifier (3.7–4.2 GHz), the failure of A will typically increase the figure to 4–5 dB. The failure of any passive bypass stage in the chain will reduce the gain of the amplifier by the amount equal to the gain of the stage in addition to the return losses of the failed amplifiers and the insertion losses of the circulators in that stage.

Because of transmission requirements, it is necessary to have flat gain in microwave amplifiers. The configuration in Fig. 6.37, especially the active bypass stage, may introduce large fluctuations in the amplifier gain over a frequency band, even though both amplifiers A and A' possess flat gain characteristics over the same band. The following discussion is centered on the flat gain criteria for circulator-coupled amplifiers for broadband applications. Let $\mathbf{S}=[S_{ij}]$ and $\mathbf{S}'=[S'_{ij}]$ be the scattering matrices of A and A', respectively. We assume the scattering matrices of the circulators C and C' are $\mathbf{\Sigma}$ and $\mathbf{\Sigma}'$, respectively, as shown in (6.224).

$$\mathbf{\Sigma}=\begin{bmatrix} 0 & 0 & \sigma \\ \sigma & 0 & 0 \\ 0 & \sigma & 0 \end{bmatrix}, \qquad \mathbf{\Sigma}'=\begin{bmatrix} 0 & 0 & \sigma' \\ \sigma' & 0 & 0 \\ 0 & \sigma' & 0 \end{bmatrix} \tag{6.224}$$

In practice the circulators have high isolation from port to port (>30 dB) and very low VSWR at each port (<1.2). Therefore it is sufficient to represent $\mathbf{\Sigma}$ and $\mathbf{\Sigma}'$ by (6.224) where $|\sigma|\approx 1$, $|\sigma'|\approx 1$ are the attenuation factors per pass. According to Fig. 6.37 the incident waves at ports 2 and 3 of C are equal to the reflected waves at the input ports of A and A' respectively, and vice versa. Also the incident waves at port 1 and 3 of C' are equal to the reflected waves at the output ports of A and A', respectively, and vice versa. Therefore, after some manipulations, the scattering matrix $\mathbf{S}$ of the active bypass stage can be obtained as follows:

$$\hat{\mathbf{S}} = \begin{bmatrix} \dfrac{\sigma^3 S_{11} S'_{11}}{1-\sigma\sigma' S_{12} S'_{21}} & \sigma\sigma' S'_{12} + \dfrac{\sigma^2\sigma'^2 S_{12} S'_{11} S'_{22}}{1-\sigma\sigma' S_{12} S'_{21}} \\ \sigma\sigma' S_{21} + \dfrac{\sigma^2\sigma'^2 S_{22} S_{11} S'_{21}}{1-\sigma\sigma' S_{12} S'_{21}} & \dfrac{\sigma^3 S_{22} S'_{22}}{1-\sigma\sigma' S_{12} S'_{21}} \end{bmatrix} \tag{6.225}$$

From the (2,1) term of $\hat{\mathbf{S}}$ it is seen that $|\sigma\sigma' S_{12} S'_{21}| \ll 1$ since $|\sigma|\approx 1$, $|\sigma'|\approx 1$ and $|S_{12} S'_{21}| \ll 1$ for transistors. In low noise amplifiers, S_{11} and S'_{11} are usually fixed once the noise matching networks have been designed; therefore the contribution of the second term in $\hat{S}_{21}$ is mostly determined by $S_{22} S'_{21}$. In order to minimize this term so that the gain of the active bypass configuration can be approximately equal to $|\sigma\sigma' S_{21}|^2$, we must minimize S_{22}, which means that the output port of A must be matched to the impedance at port 1 of C' as close as possible over the operating bandwidth (in practice all the port impedances of circulators are 50 Ω). Furthermore the (2,2) term of $\mathbf{S}$ must be minimized in order to have low VSWR at the output port; therefore it is necessary that S'_{22} also be minimized, that is, the output port of A' must also be matched as close as possible to the impedance at port 3 of C' over the working bandwidth. Since all microwave transistors exhibit a roll-off of about 6 dB/octave, matching the output of A and A' to 50 Ω for small VSWR will also cause the gain of the active bypass stage to roll off 6 dB/octave. This can be compensated by a simple equalizer at the output of the chain to flatten the gain. In the case of the failure of A, that is, $S_{21}=0$, the gain of the active bypass stage is simply determined by the second term of the (2,1) term of $\hat{\mathbf{S}}$, which is approximately $|\sigma^2\sigma'^2 S_{22} S_{11} S'_{21}|^2 = \sigma^4\sigma'^4 |S_{22}|^2 |S_{11}|^2 |S'_{21}|^2$ where S_{11} and S_{22} are now the reflection coefficients of the failed amplifier A. Experiments have shown that $20\log|S_{11}||S_{22}|$ are fairly constant for failed GaAs FETs over 3–6 GHz (2–3 dB typically). Thus in the case of failure of the primary amplifier A the gain of the active bypass stage still exhibits smooth roll-off, which implies that the overall gain of the whole amplifier can be kept fairly flat. The scattering matrix of the passive bypass stage can be obtained from (6.225) by letting $S'_{21}=\sigma'$, $S'_{12}=0$ and $S'_{11}=S'_{22}=0$.

$$\hat{\mathbf{S}}' = \begin{bmatrix} 0 & 0 \\ \sigma\sigma' S_{21} + \dfrac{\sigma^2\sigma'^3 S_{22} S_{11}}{1-\sigma\sigma'^2 S_{12}} & 0 \end{bmatrix} \tag{6.226}$$

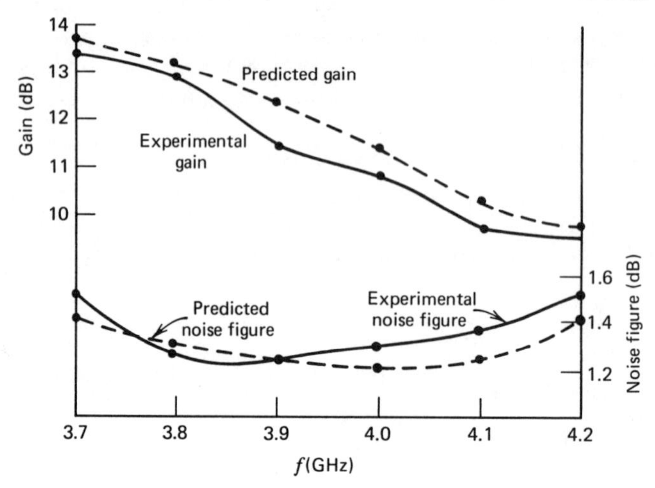

Fig. 6.38. Predicted and experimental gains and noise figures of an active bypass GaAs FET amplifier without gain roll-off compensation.

In the passive bypass stage, the contribution of the second term in the (2, 1) term of $\hat{S}'$ in (6.226) is insignificant compared to the first term; hence the output of the amplifier in the passive bypass stage can be matched to 50 Ω for flat gain instead of matching for minimum VSWR.

Since both A and A' in the active bypass stage operate under the same input noise source, the noise figure F of the active bypass stage can be simply derived

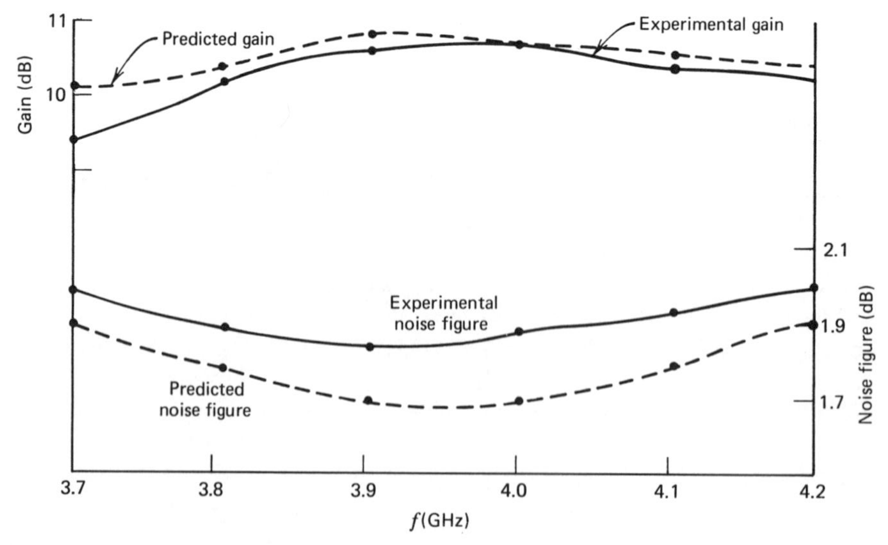

Fig. 6.39. Predicted and experimental gains and noise figures of a passive bypass GaAs FET amplifier.

from the noise figures F and F' of A and A' respectively as follows:

$$\hat{F}=\frac{\dfrac{\sigma'^2}{\sigma^2}\left[F|S_{21}|^2+\dfrac{\sigma'^2}{\sigma^2}F'|S_{11}|^2|S_{22}|^2|S'_{21}|^2\right]}{\left|\sigma\sigma' S_{21}+\dfrac{\sigma^2\sigma'^2 S_{21}S_{11}S'_{21}}{1-\sigma\sigma' S_{12}S'_{21}}\right|^2} \tag{6.227}$$

In the case of the failure of A, the noise figure $\hat{F}$ in dB will be increased by an amount equal to the return loss at the input and output of A plus twice the loss per pass of C and C' (all in dB).

Figs. 6.38 and Fig. 6.39 show the computer-predicted and experimental gains of an active bypass stage (without slope roll-off compensation) using 0.5 μ gate length GaAs FET NE-38806s and those of a passive bypass stage using 1 μ gate length GaAs FET NE-24406s. The input matching networks of A and A' in the active bypass stage are designed for minimum noise figure whereas their output matching networks are designed for minimum VSWR. The input matching network of the passive bypass amplifier is also designed for minimum noise figure whereas its output matching network is designed for flat gain. The amplifier is designed over 3.7–4.2 GHz.

6.12 CONCLUDING REMARKS

Troublesome distortions often occur in communication systems because of their nonlinearities, which can be either instantaneous (memoryless) or with memory. For instantaneous nonlinearities, power series expansion of the output in terms of the input is sufficient to characterize the nonlinearities. When the system possesses memory due to the effects of reactive elements at high frequencies (as in amplifiers, for example), power series representation is not adequate. Indeed, if the distortion is mild, such as that in communication receivers, the system can be characterized by its Volterra functional series, which can be called "power series with memory." Using these mathematical representations, the nonlinearities can be characterized by using the single-frequency input test for gain compression and the AM-to-PM conversion coefficient, and the two-frequency input test for second- and third-order intermodulation products and corresponding intercept points, which are the theoretical output power levels at which the fundamental frequency output power equals the intermodulation power. The intercept point is a convenient measure of the degree of nonlinearity of a system and it can be related to other types of distortion measurement such as cross modulation and dynamic range. In microwave systems such as GaAs FET and IMPATT amplifiers operated in small-signal high gain modes, the distortion characterization can be conveniently predicted by the use of the Volterra functional series with the aid of a digital computer once the device nonlinearities have been determined by measurement.

As solid-state device powers are limited at higher frequency bands, there is a need to obtain more power than the device capability by using power combining techniques employing power combining/dividing networks that can be built from two-way combiner/dividers (couplers). Two popular combing/dividing networks are the chain and corporate structures. The chain structure offers the advantage of ease of changing the number of ports. To add a port to an existing structure, a coupler with predetermined coupling coefficient is added to the chain and the other couplers remain unchanged. The disadvantage of this type of structure is the difficulty of building couplers with high coupling coefficients when large numbers of sources are combined; some couplings are too weak for branch-line couplers or split-T couplers and some are too strong for parallel-line couplers. The corporate structure is not as flexible as the chain structure but usually offers less coupling loss and hence higher efficiency as the number of combining sources increases. A combination of these two structures in many applications can offer a compromise between flexibility and loss. Using combining/dividing networks, an n-way amplifier can be built from n individual (elemental) amplifiers to obtain a theoretical output power of n times the output power of the individual amplifiers in the system. The balanced amplifier is a special case of the n-way amplifier and can offer many advantages over the single-ended amplifier, for example, good input and output VSWR and twice the output power at the expense of using twice the number of devices. Besides the power combining capability, the n-way amplifier also offers a "graceful degradation" property (also called "fail-soft"). In satellite earth terminals, a fail-soft high power amplifier in the transmitter is often more desirable than a switched redundant configuration. The primary advantage of using n-way amplifiers is that the failure of either amplifier in the system does not disrupt transmission even instantaneously. A fixed four-way power amplifier can offer a degradation of 2.5 dB loss in power in the case of failure of either one of the identical amplifiers in the chain. But in many cases when truly graceful degradation is required, variable n-way amplifiers must be used. The time required for the full output power to be achieved depends upon the switching speed of the phase shifters. For earth terminal transmitters at the Ku-band, where the power required is about 500–1000 W, the use of variable combiners can save cost and increase reliability. Furthermore, the loss in a variable three-way waveguide combiner at Ku-band is under 0.4 dB.

PROBLEMS

6.1. Consider the nonlinear circuit shown in Fig. P6.1 where the current through the nonlinear element is given by $i_N = G_3 v^3$ ($G_3 > 0$). Let $i(t)$ be the input and $v(t)$ be the output. Show that the first three nonlinear transfer functions of the circuit

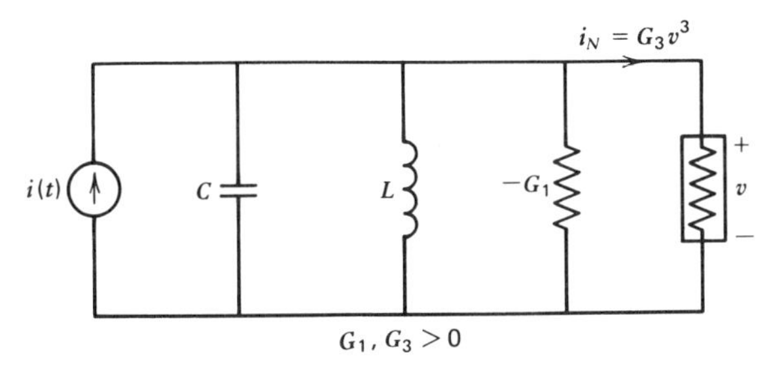

Fig. P6.1. A nonlinear oscillatory circuit.

are given by

$$H_1(j\omega) = \frac{j\omega L}{1-\omega^2 LC - j\omega G_1 L}$$

$$H_2(j\omega_1, j\omega_2) = 0$$

$$H_3(j\omega_1, j\omega_2, j\omega_3) = \frac{-j(\omega_1+\omega_2+\omega_3)H_1(j\omega_1)H_2(j\omega_2)H_3(j\omega_3)G_3 L}{1-(\omega_1+\omega_2+\omega_3)^2 LC - j(\omega_1+\omega_2+\omega_3)G_1 L}$$

6.2. Derive the gain of the hybrid combining negative resistance amplifier shown in Fig. P6.2 assuming that the 3 dB 90° hybrid is ideal when $Z_d \neq Z'_d$ and when $Z_d = Z'_d$. Note that each diode receives only half the input power.

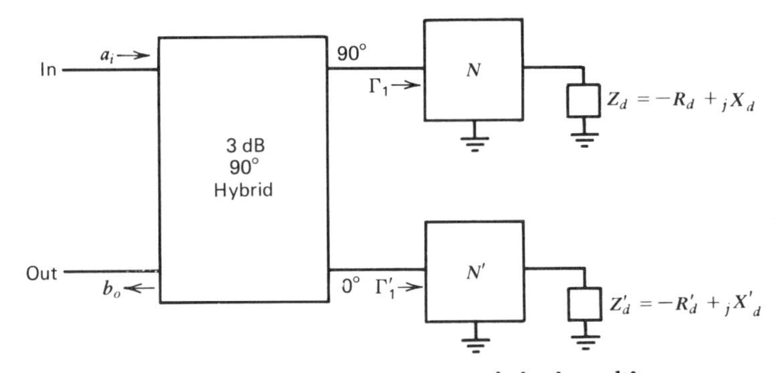

Fig. P6.2. A negative resistance hybrid amplifier.

6.3. In Problem 6.2, derive the output power $|b_o|^2$ in terms of the input power $|a_i|^2$ assuming the coupling loss of the hybrid is 0.3 dB.

6.4. Show that the third-order nonlinear transfer function of the circuit in Fig. 6.9 is given by

$$H_3(j\omega_1, j\omega_2, j\omega_3) = -\tfrac{1}{3}G_2^2 H_1(j\omega_1)H_2(j\omega_2)H_3(j\omega_3)[H_1(j\omega_1+j\omega_2) + H_1(j\omega_2+j\omega_3) + H_1(j\omega_1+j\omega_3)]\, H_1(j\omega_1+j\omega_2+j\omega_3)$$

6.5. Show that a linear system with transfer function $L(j\omega)$ followed by an nth-order nonlinear system with the nth nonlinear transfer function $H_n(j\omega_1, j\omega_2, \ldots, j\omega_n)$ has a cascade nonlinear transfer function K_n given by

$$K_n(j\omega_1, j\omega_2, \ldots, j\omega_n) = H_n(j\omega_1, j\omega_2, \ldots, j\omega_n) L(j\omega_1) L(j\omega_2) \ldots L(j\omega_n)$$

6.6. Show that an nth-order nonlinear system with transform H_n followed by a linear system with transform L has the following cascade transform K_n:

$$K_n(j\omega_1, j\omega_2, \ldots, j\omega_n) = L(j\omega_1 + j\omega_2 + \cdots + j\omega_n) H_n(j\omega_1, j\omega_2, \ldots, j\omega_n)$$

6.7. Consider the nonlinear system represented by the differential equation

$$\frac{d^2y}{dt^2} + a\frac{dy}{dt} + by + cy^3 = x(t)$$

Show that the fifth-order nonlinear transfer function is given by

$$H_5(j\omega_1, j\omega_2, \ldots, j\omega_5) = \frac{-3cH_3(j\omega_1, j\omega_2, j\omega_3)}{b - (\omega_1 + \omega_2 + \cdots + \omega_5)^2 + ja(\omega_1 + \omega_2 + \cdots + \omega_5)} \cdot \frac{1}{b - \omega_4^2 + ja\omega_4} \cdot \frac{1}{b - \omega_5^2 + ja\omega_5}$$

6.8. Derive the 5×5 scattering matrix of the equal phase, equal amplitude four-way corporate combiner shown in Fig. 6.27 assuming a nonideal identical two-way Wilkinson combiner with scattering matrix $\mathbf{S} = [S_{ij}]$. Obtain the result for the corresponding divider by inspection.

6.9. Derive the 5×5 scattering matrix of the four-way chain combiner/divider in Fig. 6.28 assuming the 3 dB, 4.77 dB, and 6 dB two-way combiner/divider scattering matrices are $\mathbf{S} = [S_{ij}]$, $\mathbf{S}' = [S'_{ij}]$ and $\mathbf{S}'' = [S''_{ij}]$, respectively.

6.10. Consider the GaAs FET of Fig. 6.16 where $R_i = 1.15\ \Omega$, $C_{F1} = 0.05$ pF, $C_{F2} = 0$, $C_{F3} = -0.002$ pF/V^2, $C_{G1} = 0.53$ pF, $C_{G2} = 0$, $C_{G3} = 0.014$ pF/V^2, $G_{m1} = g_{m0} = 0.0342$ mhos, $G_{m2} = 0$, $G_{m3} = -0.0013$ mhos/V^2, $C_D = 0.436$ pF, $g_o = 1/466$ mhos, $\tau = 13$ psec. Compute the small-signal scattering parameters of the GaAs FET from 5.5 to 6.5 GHz. Design matching circuits to obtain maximum gain. Write a computer program to compute the intercept point P_I for this amplifier.

6.11. Consider the circuit in Fig. 6.18 where $-G_{d1} = -0.00103$ mhos, $G_{d2} = 0$, $G_{d3} = 0.00086$ mhos/V^2, $C_{d1} = 0.164$ pF, $C_{d2} = 0$, $C_{d3} = 0.000063$ pF/V^2. Design the matching network N to obtain small-signal maximum gain from 5.86 to 5.94 GHz. Write a computer program to compute the intercept point P_I for this amplifier.

6.12. Consider the variable three-way power amplifier shown in Fig. 6.36 assuming that all the individual amplifiers have ideal input and output isolators and all the variable two-way combiner/dividers coupling losses are 0.3 dB. What would be the power split ratios if

(a) Amplifier 1 fails?
(b) Amplifier 3 fails?
(c) Amplifiers 1 and 2 fail, each with half of its output power lost?

6.13. Verify Equations (6.225) and (6.227).

6.14. Assume that all individual amplifiers in an n-way amplifier are perfectly matched and that its transfer scattering parameters s_{21} have the same phase shift. Show that the output power P_o of the n-way amplifier is given by

$$P_o = \frac{1}{n}\left(\sum_{p=1}^{n} \sqrt{P_p}\right)^2$$

where P_p, $p=1,2,\ldots,n$ is the output power of the pth individual amplifier in the system (assuming the combiner/divider is ideal, equal phase, and equal amplitude).

6.15. Design an equal phase, equal amplitude six-way chain power combiner to combine six power sources with the following powers: $P_1=3$ W, $P_2=5$ W, $P_3=3.8$ W, $P_4=6$ W, $P_5=5.4$ W, and $P_6=6.5$ W.

6.16. Design an equal phase, equal amplitude four-way corporate combiner to combine four power sources with the following powers: $P_1=23$ dBm, $P_2=20$ dBm, $P_3=27$ dBm and $P_4=25$ dBm assuming that the coupling loss for each two-way combiner building block is 0.2 dB.

6.17. Evaluate the phase error combining loss of the four-way amplifier in Fig. 6.34 assuming that $s_{21,1}=|s_{21}|\exp(j\theta)$, $s_{21,2}=|s_{21}|\exp(j\theta+j\pi/8)$, $s_{21,3}=|s_{21}|\exp(j\theta+j\pi/16)$, $s_{21,4}=|s_{21}|\exp(j\theta-j\pi/32)$.

6.18. Compute the input VSWR of the balanced amplifier in Fig. 6.33 assuming that $s_{11,1}=|s_{11}|\exp(j\theta)$ and $s_{11,2}=|s_{11}|\exp j(\theta+\Delta\theta)$ for $\Delta\theta=0$, $\pi/32$, $\pi/16$, $\pi/8$, $\pi/4$, and $\pi/2$.

6.19. Derive the noise figure F of the balanced amplifier in terms of the individual amplifier's noise figures F_1 and F_2 assuming the 3 dB 90° combiner/divider is ideal.

6.20. Derive the scattering matrix of the network shown in Fig. P6.20 assuming the diodes are identical with admittance $-jB_d$. Can this network be used as a variable combiner/divider as B_d varies from zero to infinity? How?

6.21. Design a variable four-way power amplifier using variable two-way combiner/dividers. Indicate the power split ratios at outputs and inputs of each two-way divider and combiner.

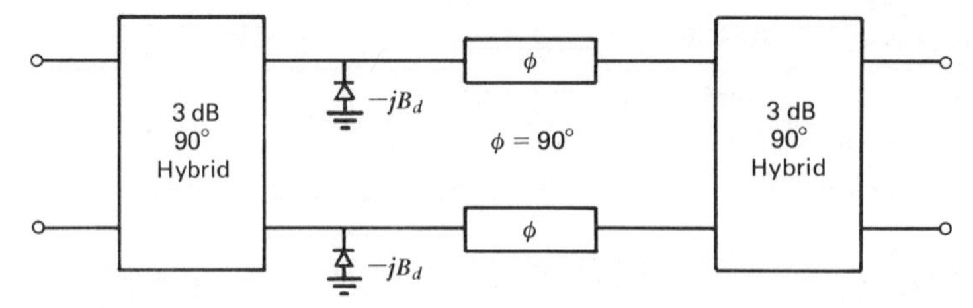

Fig. P6.20. A variable power combiner/divider.

REFERENCES

1. E. J. Wilkinson, "An *N*-way hybrid power divider," *IEEE Trans. Microwave Theory Tech.*, **MTT-8**, no. 1, 116–118, 1960.
2. L. I. Parad and R. L. Moynihan, "Split-tee power divider," *IEEE Trans. Microwave Theory Tech.*, **MTT-13**, no. 1, 91–95, 1965.
3. G. L. Matthaei, L. Young, and E. M. T. Jones, *Microwave Filters, Impedance Matching Networks and Coupling Structures*. New York: McGraw-Hill, 1964.
4. R. Levy, "Directional couplers," in *Advances in Microwaves*, edited by L. Young. New York: Academic Press, 1966.
5. V. Volterra, *Theory of Functionals and of Integrals and Integro-Differential Equations*. New York: Dover, 1959.
6. J. J. Bussgang, L. Ehrman, and J. W. Graham, "Analysis of nonlinear systems with multiple inputs," *Proc. IEEE*, **62**, no. 8, 1088–1119, 1974.
7. B. Bedrosian and S. O. Rice, "The output properties of Volterra systems (nonlinear systems with memory) driven by harmonic and Gaussian inputs," *Proc. IEEE*, **59**, no. 12, 1688–1707, 1971.
8. R. Minassian, "Large-signal GaAs MESFET model and distortion analysis," *Electron. Letters*, **14**, 183–185, 1978.
9. R. Minassian, "Simplified GaAs MESFET model to 10 GHz," *Electron. Letters*, **13**, 549–551, 1977.
10. A. M. Khadr and R. H. Johnston, "Distortion in high-frequency FET amplifiers," *IEEE J. Solid-State Circuits*, **SC-9**, no. 4, 180–189, 1974.
11. R. G. Meyers, M. J. Shensa, and R. Eschenbach, "Cross modulation and intermodulation in amplifiers at high frequencies," *IEEE J. Solid-State Circuits*, **SC-7**, no. 1, 16–23, 1972.
12. L. O. Chua and C. Y. Ng, "Frequency domain analysis of nonlinear systems: general theory," *IEE J. Electron. Circuits Syst.*, **3**, no. 4, 165–185, 1979.
13. Y. L. Kuo, "Distortion analysis of bipolar transistor," *IEEE Trans. Circuit Theory*, **CT-20**, no. 6, 709–716, 1973.
14. S. Narayanan, "Transistor distortion analysis using Volterra series representations," *Bell Syst. Tech. J.*, **46**, no. 5, 991–1024, 1967.
15. M. Schetzen, "Measurement of the kernels of a nonlinear system of finite order," *Internat. J. Control*, **1**, 251–263, 1965.
16. Y. W. Lee and M. Schetzen, "Measurement of the Wiener kernels of a nonlinear system by cross-correlation," *Internat. J. Control*, **2**, 237–254, 1965.

17. K. J. Russell, "Microwave power combining techniques," *IEEE Trans. Microwave Theory Tech.*, **MTT-27**, no. 5, 472–478, 1979.
18. A. W. Morse, "Modify combiner design to team high power amps," *Microwaves*, **17**, no. 1, 70–79, 1978.
19. E. J. Wilkinson and D. J. Sommers, "Variable multiport power combiners," *Microwave J.*, **21**, no. 2, 59–63, 1978.
20. K. Kurokawa, "Design theory of balanced transistor amplifiers," *Bell Syst. Tech. J.*, **44**, no. 10, 1675–1698, 1965.
21. R. E. Collins, *Foundations for Microwave Engineering*. New York: McGraw-Hill, 1966.
22. H. Komizo, Y. Daido, H. Ashida, Y. Ito, and M. Honma, "Improvement of nonlinear distortion in an IMPATT stable amplifier," *IEEE Trans. Microwave Theory Tech.*, **MTT-21**, no. 11, 721–728, 1973.
23. F. T. El-Mokadem, and T. T. Ha, "Low-noise circulator-coupled amplifier for unattended small earth terminals," *IEE Proc.*, **127**, Pt. H, 4, 228–230, 1980.
24. T. T. Ha, "Microwave power combining and graceful degradation," *IEE Proc.*, **127**, Pt. G, 3, 148–152, 1980.

Appendix A1

Unconditional Stability

A two-port network N with scattering matrix $\mathbf{S}$ and source and load reflection coefficients Γ_s and Γ_L, all normalized to the characteristic impedance Z_0, is unconditionally stable if and only if

$$\left| S_{11} + \frac{\Gamma_L S_{12} S_{21}}{1 - \Gamma_L S_{22}} \right| < 1 \tag{A1.1}$$

$$\left| S_{22} + \frac{\Gamma_s S_{12} S_{21}}{1 - \Gamma_s S_{11}} \right| < 1 \tag{A1.2}$$

for all $|\Gamma_s| < 1$ and all $|\Gamma_L| < 1$.

Equation (A1.1) can be rewritten as

$$\left| S_{22}^{-1} \left(\Delta + \frac{S_{12} S_{21}}{1 - S_{22}\Gamma_L} \right) \right| < 1 \tag{A1.3}$$

for all $|\Gamma_L| < 1$, where $\Delta = S_{11}S_{22} - S_{12}S_{21}$. We see that the bilinear transformation $S_{22}^{-1}[\Delta + S_{12}S_{21}(1 - S_{22}\Gamma_L)^{-1}]$ transforms the unit circle $|\Gamma_L| = 1$ into the circle ψ. To find the representation of ψ we note that $1 - S_{22}\Gamma_L$ transforms $|\Gamma_L| = 1$ into a circle with center at 1 and radius $|S_{22}|$, and hence $(1 - S_{22}\Gamma_L)^{-1}$ transforms $|\Gamma_L| = 1$ into a circle with center at $\frac{1}{2}[(1 + |S_{22}|)^{-1} + (1 - |S_{22}|)^{-1}] = (1 - |S_{22}|^2)^{-1}$ and radius $\frac{1}{2}[(1 - |S_{22}|)^{-1} - (1 + |S_{22}|)^{-1}] = |S_{22}|(1 - |S_{22}|^2)^{-1}$. The circle ψ is then represented by

$$\psi = S_{22}^{-1}\Delta + \frac{S_{12}S_{21}S_{22}^{-1}}{1 - |S_{22}|^2} + \frac{|S_{12}S_{21}|}{1 - |S_{22}|^2} e^{j\varphi} \tag{A1.4}$$

where φ varies from 0 to 2π. For (A1.1) to hold, the magnitude of ψ has to be smaller than unity regardless of the value of φ, that is,

$$\frac{1}{|S_{22}|} \left| \Delta + \frac{S_{12}S_{21}}{1 - |S_{22}|^2} \right| + \frac{|S_{12}S_{21}|}{1 - |S_{22}|^2} < 1 \tag{A1.5}$$

or

$$0 \leq \frac{1}{|S_{22}|}\left|\Delta + \frac{S_{12}S_{21}}{1-|S_{22}|^2}\right| < 1 - \frac{|S_{12}S_{21}|}{1-|S_{22}|^2} \tag{A1.6}$$

Equation (A1.6) indicates that

$$1-|S_{22}|^2 > |S_{12}S_{21}| \tag{A1.7}$$

Squaring both sides of (A1.6) and rearranging the terms we have

$$\frac{1-|S_{22}|^2-|S_{11}|^2+|\Delta|^2}{2|S_{12}S_{21}|} > 1 \tag{A1.8}$$

Similarly from (A1.2) we obtain (A1.8) and

$$1-|S_{11}|^2 > |S_{12}S_{21}| \tag{A1.9}$$

By adding (A1.7) and (A1.9) we get

$$2-|S_{11}|^2-|S_{22}|^2 > 2|S_{12}S_{21}| \tag{A1.10}$$

Furthermore,

$$|\Delta| = |S_{11}S_{22}-S_{12}S_{21}| \leq |S_{11}||S_{22}|+|S_{12}S_{21}|.$$

Using this inequality in (A1.10) we get

$$|\Delta| \leq |S_{11}||S_{22}| - \tfrac{1}{2}|S_{11}|^2 - \tfrac{1}{2}|S_{22}|^2 + 1 \tag{A1.11a}$$

$$|\Delta| \leq 1 - \tfrac{1}{2}(|S_{11}|-|S_{22}|)^2 < 1 \tag{A1.11b}$$

Appendix A2

Potential Unstability

If $K<1$, then the two-port is said to be potentially unstable. When considering an amplifier design it is possible to determine the region of Γ_L and Γ_s that make the amplifier stable. From (2.46a) we get

$$\Gamma_L = \frac{S_{11}-\Gamma_i}{\Delta - S_{22}\Gamma_i} = \Delta^{-1}S_{22}^{-1}\left(\Delta + \frac{S_{12}S_{21}}{1-\Delta^{-1}S_{22}\Gamma_i}\right) \tag{A2.1}$$

From (A2.1) we see that the circle $|\Gamma_i|=1$ is bilinearly transformed into the circle ψ, the locus of all Γ_L such that $|\Gamma_i|=1$. Since $1-\Delta^{-1}S_{22}\Gamma_i$ transforms $|\Gamma_i|=1$ into a circle centered at 1 and radius $|\Delta^{-1}S_{22}|$, hence $(1-\Delta^{-1}S_{22}\Gamma_i)^{-1}$ transforms $|\Gamma_i|=1$ into a circle centered at $\frac{1}{2}[(1+|\Delta^{-1}S_{22}|)^{-1}+(1-|\Delta^{-1}S_{22}|)^{-1}]=(1-|\Delta^{-1}S_{22}|^2)^{-1}$ and radius $\frac{1}{2}|[(1-|\Delta^{-1}S_{22}|)^{-1}-(1-|\Delta^{-1}S_{22}|)^{-1}]|=|\Delta^{-1}S_{22}||(1-|\Delta^{-1}S_{22}|^2)^{-1}|$. The circle ψ then has center C_L and radius r_L given as

$$C_L = \Delta^{-1}S_{22}^{-1}\left(\Delta + \frac{S_{12}S_{21}}{1-|\Delta^{-1}S_{22}|^2}\right) = \frac{S_{11}\Delta^* - S_{22}^*}{|\Delta|^2-|S_{22}|^2} \tag{A2.2}$$

$$r_L = |\Delta^{-1}S_{22}^{-1}|\left|\frac{|S_{12}S_{21}||\Delta^{-1}S_{22}|}{1-|\Delta^{-1}S_{22}|^2}\right| = \left|\frac{S_{12}S_{21}}{|\Delta|^2-|S_{22}|^2}\right| \tag{A2.3}$$

The locus of Γ_s can be derived in a similar way from (2.46b).

Appendix A3

Microstrip Line Design Tables

ϵ_r = 2.22

W/H	ϵ_r'	Z_0	W/H	ϵ_r'	Z_0
0.0500	1.6530	236.6581	2.5500	1.8850	56.5952
0.1000	1.6707	203.2641	2.6000	1.8871	55.9037
0.1500	1.6842	183.7372	2.6500	1.8892	55.2291
0.2000	1.6954	169.9053	2.7000	1.8913	54.5709
0.2500	1.7053	159.2016	2.7500	1.8933	53.9284
0.3000	1.7141	150.4811	2.8000	1.8953	53.3011
0.3500	1.7222	143.1322	2.8500	1.8973	52.6884
0.4000	1.7296	136.7894	2.9000	1.8992	52.0899
0.4500	1.7366	131.2169	2.9500	1.9011	51.5051
0.5000	1.7431	126.2536	3.0000	1.9030	50.9336
0.5500	1.7493	121.7843	3.0500	1.9049	50.3748
0.6000	1.7551	117.7241	3.1000	1.9067	49.8284
0.6500	1.7607	114.0085	3.1500	1.9086	49.2941
0.7000	1.7660	110.5871	3.2000	1.9103	48.7713
0.7500	1.7711	107.4201	3.2500	1.9121	48.2597
0.8000	1.7760	104.4755	3.3000	1.9139	47.7591
0.8500	1.7807	101.7268	3.3500	1.9156	47.2690
0.9000	1.7853	99.1524	3.4000	1.9173	46.7891
0.9500	1.7897	96.7338	3.4500	1.9189	46.3192
1.0000	1.7939	94.4557	3.5000	1.9206	45.8589
1.0500	1.7980	92.0765	3.5500	1.9222	45.4079
1.1000	1.8020	89.9390	3.6000	1.9238	44.9660
1.1500	1.8059	87.9428	3.6500	1.9254	44.5328
1.2000	1.8097	86.0701	3.7000	1.9270	44.1082
1.2500	1.8133	84.3063	3.7500	1.9286	43.6919
1.3000	1.8169	82.6388	3.8000	1.9301	43.2837
1.3500	1.8204	81.0573	3.8500	1.9316	42.8833
1.4000	1.8238	79.5527	3.9000	1.9331	42.4904
1.4500	1.8271	78.1174	3.9500	1.9346	42.1050
1.5000	1.8303	76.7449	4.0000	1.9361	41.7268
1.5500	1.8335	75.4296	4.0500	1.9375	41.3555
1.6000	1.8365	74.1666	4.1000	1.9389	40.9911
1.6500	1.8396	72.9516	4.1500	1.9404	40.6332
1.7000	1.8425	71.7811	4.2000	1.9417	40.2818
1.7500	1.8454	70.6517	4.2500	1.9431	39.9367
1.8000	1.8482	69.5608	4.3000	1.9445	39.5976
1.8500	1.8510	68.5058	4.3500	1.9459	39.2645
1.9000	1.8537	67.4846	4.4000	1.9472	38.9372
1.9500	1.8564	66.4952	4.4500	1.9485	38.6155
2.0000	1.8590	65.5357	4.5000	1.9498	38.2993
2.0500	1.8616	64.6047	4.5500	1.9511	37.9885
2.1000	1.8641	63.7008	4.6000	1.9524	37.6829
2.1500	1.8666	62.8225	4.6500	1.9537	37.3823
2.2000	1.8690	61.9687	4.7000	1.9549	37.0867
2.2500	1.8714	61.1383	4.7500	1.9562	36.7960
2.3000	1.8738	60.3303	4.8000	1.9574	36.5099
2.3500	1.8761	59.5437	4.8500	1.9586	36.2285
2.4000	1.8784	58.7777	4.9000	1.9598	35.9516
2.4500	1.8806	58.0315	4.9500	1.9610	35.6790
2.5000	1.8828	57.3042	5.0000	1.9622	35.4107

ϵ_r = 2.55

W/H	ϵ_r'	Z_0	W/H	ϵ_r'	Z_0
0.0500	1.8297	224.9446	2.5500	2.1243	53.3112
0.1000	1.8521	193.0526	2.6000	2.1270	52.6561
0.1500	1.8692	174.4049	2.6500	2.1297	52.0171
0.2000	1.8835	161.1981	2.7000	2.1323	51.3936
0.2500	1.8960	150.9804	2.7500	2.1349	50.7850
0.3000	1.9073	142.6577	2.8000	2.1375	50.1909
0.3500	1.9175	135.6458	2.8500	2.1400	49.6107
0.4000	1.9270	129.5953	2.9000	2.1425	49.0440
0.4500	1.9358	124.2810	2.9500	2.1449	48.4903
0.5000	1.9441	119.5487	3.0000	2.1473	47.9492
0.5500	1.9520	115.2885	3.0500	2.1497	47.4203
0.6000	1.9594	111.4192	3.1000	2.1520	46.9031
0.6500	1.9665	107.8790	3.1500	2.1543	46.3973
0.7000	1.9732	104.6199	3.2000	2.1566	45.9025
0.7500	1.9797	101.6037	3.2500	2.1588	45.4185
0.8000	1.9859	98.7999	3.3000	2.1610	44.9447
0.8500	1.9919	96.1833	3.3500	2.1632	44.4810
0.9000	1.9977	93.7329	3.4000	2.1654	44.0270
0.9500	2.0033	91.4313	3.4500	2.1675	43.5824
1.0000	2.0087	89.2638	3.5000	2.1696	43.1470
1.0500	2.0139	87.0020	3.5500	2.1717	42.7204
1.1000	2.0190	84.9697	3.6000	2.1737	42.3024
1.1500	2.0239	83.0718	3.6500	2.1758	41.8928
1.2000	2.0287	81.2916	3.7000	2.1778	41.4913
1.2500	2.0333	79.6149	3.7500	2.1797	41.0976
1.3000	2.0379	78.0301	3.8000	2.1817	40.7116
1.3500	2.0423	76.5270	3.8500	2.1836	40.3330
1.4000	2.0466	75.0973	3.9000	2.1855	39.9616
1.4500	2.0508	73.7336	3.9500	2.1874	39.5973
1.5000	2.0549	72.4297	4.0000	2.1893	39.2397
1.5500	2.0589	71.1802	4.0500	2.1911	38.8888
1.6000	2.0628	69.9806	4.1000	2.1929	38.5443
1.6500	2.0667	68.8268	4.1500	2.1947	38.2061
1.7000	2.0704	67.7154	4.2000	2.1965	37.8740
1.7500	2.0741	66.6432	4.2500	2.1982	37.5479
1.8000	2.0777	65.6076	4.3000	2.2000	37.2275
1.8500	2.0812	64.6063	4.3500	2.2017	36.9128
1.9000	2.0847	63.6371	4.4000	2.2034	36.6035
1.9500	2.0881	62.6982	4.4500	2.2051	36.2996
2.0000	2.0914	61.7880	4.5000	2.2067	36.0009
2.0500	2.0947	60.9048	4.5500	2.2084	35.7072
2.1000	2.0979	60.0473	4.6000	2.2100	35.4185
2.1500	2.1010	59.2143	4.6500	2.2116	35.1347
2.2000	2.1041	58.4047	4.7000	2.2132	34.8555
2.2500	2.1071	57.6173	4.7500	2.2148	34.5809
2.3000	2.1101	56.8513	4.8000	2.2164	34.3107
2.3500	2.1131	56.1056	4.8500	2.2179	34.0450
2.4000	2.1160	55.3796	4.9000	2.2194	33.7834
2.4500	2.1188	54.6723	4.9500	2.2209	33.5261
2.5000	2.1216	53.9831	5.0000	2.2224	33.2728

$\epsilon_r = 3.80$

W/H	ϵ_r'	Z_0	W/H	ϵ_r'	Z_0
0.0500	2.4987	192.4864	2.5500	3.0311	44.6306
0.1000	2.5393	164.8740	2.6000	3.0360	44.0747
0.1500	2.5702	148.7323	2.6500	3.0408	43.5326
0.2000	2.5960	137.3060	2.7000	3.0455	43.0037
0.2500	2.6186	128.4711	2.7500	3.0502	42.4877
0.3000	2.6389	121.2793	2.8000	3.0548	41.9841
0.3500	2.6574	115.2241	2.8500	3.0593	41.4923
0.4000	2.6746	110.0025	2.9000	3.0638	41.0121
0.4500	2.6905	105.4192	2.9500	3.0682	40.5431
0.5000	2.7055	101.3403	3.0000	3.0725	40.0847
0.5500	2.7197	97.6706	3.0500	3.0768	39.6368
0.6000	2.7331	94.3395	3.1000	3.0810	39.1989
0.6500	2.7459	91.2935	3.1500	3.0852	38.7708
0.7000	2.7581	88.4909	3.2000	3.0893	38.3521
0.7500	2.7698	85.8987	3.2500	3.0934	37.9425
0.8000	2.7810	83.4902	3.3000	3.0974	37.5417
0.8500	2.7919	81.2436	3.3500	3.1013	37.1495
0.9000	2.8023	79.1408	3.4000	3.1052	36.7656
0.9500	2.8124	77.1666	3.4500	3.1090	36.3897
1.0000	2.8221	75.3083	3.5000	3.1128	36.0216
1.0500	2.8316	73.3728	3.5500	3.1166	35.6610
1.1000	2.8407	71.6332	3.6000	3.1203	35.3078
1.1500	2.8496	70.0091	3.6500	3.1239	34.9617
1.2000	2.8583	68.4859	3.7000	3.1276	34.6225
1.2500	2.8667	67.0518	3.7500	3.1311	34.2900
1.3000	2.8749	65.6964	3.8000	3.1346	33.9640
1.3500	2.8828	64.4114	3.8500	3.1381	33.6444
1.4000	2.8906	63.1895	3.9000	3.1416	33.3309
1.4500	2.8982	62.0243	3.9500	3.1450	33.0233
1.5000	2.9056	60.9105	4.0000	3.1483	32.7215
1.5500	2.9129	59.8436	4.0500	3.1517	32.4254
1.6000	2.9199	58.8196	4.1000	3.1549	32.1348
1.6500	2.9269	57.8350	4.1500	3.1582	31.8495
1.7000	2.9337	56.8868	4.2000	3.1614	31.5694
1.7500	2.9403	55.9725	4.2500	3.1646	31.2944
1.8000	2.9468	55.0896	4.3000	3.1677	31.0242
1.8500	2.9532	54.2362	4.3500	3.1708	30.7589
1.9000	2.9594	53.4105	4.4000	3.1739	30.4982
1.9500	2.9655	52.6109	4.4500	3.1769	30.2420
2.0000	2.9715	51.8358	4.5000	3.1799	29.9903
2.0500	2.9774	51.0841	4.5500	3.1829	29.7429
2.1000	2.9832	50.3545	4.6000	3.1858	29.4996
2.1500	2.9889	49.6459	4.6500	3.1887	29.2605
2.2000	2.9945	48.9574	4.7000	3.1916	29.0253
2.2500	3.0000	48.2880	4.7500	3.1945	28.7941
2.3000	3.0054	47.6369	4.8000	3.1973	28.5666
2.3500	3.0107	47.0034	4.8500	3.2001	28.3429
2.4000	3.0159	46.3867	4.9000	3.2028	28.1227
2.4500	3.0210	45.7860	4.9500	3.2056	27.9061
2.5000	3.0261	45.2009	5.0000	3.2083	27.6929

ϵ_r = 4.40

W/H	ϵ_r'	Z_0	W/H	ϵ_r'	Z_0
0.0500	2.8199	181.1940	2.5500	3.4663	41.7348
0.1000	2.8692	155.1074	2.6000	3.4722	41.2129
0.1500	2.9067	139.8591	2.6500	3.4781	40.7040
0.2000	2.9380	129.0672	2.7000	3.4838	40.2075
0.2500	2.9655	120.7244	2.7500	3.4895	39.7232
0.3000	2.9901	113.9346	2.8000	3.4951	39.2505
0.3500	3.0126	108.2191	2.8500	3.5006	38.7890
0.4000	3.0334	103.2915	2.9000	3.5060	38.3384
0.4500	3.0528	98.9670	2.9500	3.5114	37.8982
0.5000	3.0710	95.1193	3.0000	3.5167	37.4682
0.5500	3.0882	91.6582	3.0500	3.5219	37.0479
0.6000	3.1045	88.5171	3.1000	3.5270	36.6371
0.6500	3.1200	85.6453	3.1500	3.5320	36.2355
0.7000	3.1348	83.0035	3.2000	3.5370	35.8427
0.7500	3.1490	80.5604	3.2500	3.5419	35.4585
0.8000	3.1627	78.2908	3.3000	3.5468	35.0826
0.8500	3.1758	76.1741	3.3500	3.5516	34.7148
0.9000	3.1885	74.1932	3.4000	3.5563	34.3547
0.9500	3.2007	72.3337	3.4500	3.5610	34.0022
1.0000	3.2126	70.5836	3.5000	3.5656	33.6570
1.0500	3.2240	68.7619	3.5500	3.5701	33.3189
1.1000	3.2352	67.1245	3.6000	3.5746	32.9877
1.1500	3.2460	65.5958	3.6500	3.5791	32.6632
1.2000	3.2565	64.1622	3.7000	3.5835	32.3452
1.2500	3.2667	62.8125	3.7500	3.5878	32.0335
1.3000	3.2766	61.5372	3.8000	3.5921	31.7279
1.3500	3.2863	60.3281	3.8500	3.5963	31.4282
1.4000	3.2957	59.1783	3.9000	3.6005	31.1343
1.4500	3.3050	58.0822	3.9500	3.6046	30.8461
1.5000	3.3140	57.0345	4.0000	3.6087	30.5632
1.5500	3.3228	56.0310	4.0500	3.6127	30.2857
1.6000	3.3314	55.0679	4.1000	3.6167	30.0133
1.6500	3.3398	54.1420	4.1500	3.6207	29.7460
1.7000	3.3480	53.2504	4.2000	3.6245	29.4835
1.7500	3.3561	52.3907	4.2500	3.6284	29.2257
1.8000	3.3640	51.5607	4.3000	3.6322	28.9726
1.8500	3.3717	50.7585	4.3500	3.6360	28.7240
1.9000	3.3793	49.9824	4.4000	3.6397	28.4797
1.9500	3.3867	49.2309	4.4500	3.6434	28.2397
2.0000	3.3940	48.5025	4.5000	3.6470	28.0039
2.0500	3.4012	47.7961	4.5500	3.6507	27.7721
2.1000	3.4082	47.1106	4.6000	3.6542	27.5442
2.1500	3.4151	46.4449	4.6500	3.6578	27.3202
2.2000	3.4219	45.7981	4.7000	3.6613	27.0999
2.2500	3.4286	45.1693	4.7500	3.6647	26.8833
2.3000	3.4351	44.5578	4.8000	3.6681	26.6702
2.3500	3.4416	43.9628	4.8500	3.6715	26.4607
2.4000	3.4479	43.3836	4.9000	3.6749	26.2545
2.4500	3.4541	42.8196	4.9500	3.6782	26.0516
2.5000	3.4603	42.2702	5.0000	3.6815	25.8519

$\epsilon_r = 9.50$

W/H	ϵ_r'	Z_0	W/H	ϵ_r'	Z_0
0.0500	5.5498	129.1587	2.5500	7.1657	29.0269
0.1000	5.6729	110.3081	2.6000	7.1806	28.6588
0.1500	5.7667	99.2947	2.6500	7.1952	28.2999
0.2000	5.8451	91.5058	2.7000	7.2096	27.9500
0.2500	5.9137	85.4895	2.7500	7.2238	27.6086
0.3000	5.9753	80.5972	2.8000	7.2378	27.2755
0.3500	6.0315	76.4824	2.8500	7.2515	26.9505
0.4000	6.0835	72.9378	2.9000	7.2651	26.6331
0.4500	6.1319	69.8295	2.9500	7.2785	26.3232
0.5000	6.1774	67.0660	3.0000	7.2916	26.0205
0.5500	6.2204	64.5821	3.0500	7.3046	25.7247
0.6000	6.2611	62.3294	3.1000	7.3174	25.4357
0.6500	6.3000	60.2714	3.1500	7.3301	25.1531
0.7000	6.3370	58.3793	3.2000	7.3426	24.8769
0.7500	6.3726	56.6308	3.2500	7.3549	24.6067
0.8000	6.4067	55.0074	3.3000	7.3670	24.3425
0.8500	6.4396	53.4943	3.3500	7.3790	24.0839
0.9000	6.4712	52.0791	3.4000	7.3908	23.8309
0.9500	6.5018	50.7514	3.4500	7.4025	23.5833
1.0000	6.5314	49.5024	3.5000	7.4140	23.3408
1.0500	6.5601	48.2051	3.5500	7.4254	23.1033
1.1000	6.5879	47.0387	3.6000	7.4366	22.8708
1.1500	6.6149	45.9500	3.6500	7.4477	22.6430
1.2000	6.6411	44.9294	3.7000	7.4587	22.4198
1.2500	6.6667	43.9688	3.7500	7.4695	22.2010
1.3000	6.6915	43.0613	3.8000	7.4802	21.9866
1.3500	6.7157	42.2013	3.8500	7.4908	21.7764
1.4000	6.7394	41.3838	3.9000	7.5012	21.5702
1.4500	6.7624	40.6046	3.9500	7.5115	21.3681
1.5000	6.7849	39.8602	4.0000	7.5217	21.1697
1.5500	6.8069	39.1474	4.0500	7.5318	20.9752
1.6000	6.8284	38.4636	4.1000	7.5418	20.7842
1.6500	6.8494	37.8064	4.1500	7.5516	20.5969
1.7000	6.8700	37.1738	4.2000	7.5614	20.4129
1.7500	6.8902	36.5641	4.2500	7.5710	20.2324
1.8000	6.9099	35.9756	4.3000	7.5805	20.0550
1.8500	6.9293	35.4071	4.3500	7.5900	19.8809
1.9000	6.9482	34.8572	4.4000	7.5993	19.7098
1.9500	6.9668	34.3250	4.4500	7.6085	19.5418
2.0000	6.9851	33.8093	4.5000	7.6176	19.3767
2.0500	7.0030	33.3094	4.5500	7.6266	19.2144
2.1000	7.0205	32.8244	4.6000	7.6356	19.0549
2.1500	7.0378	32.3536	4.6500	7.6444	18.8982
2.2000	7.0548	31.8963	4.7000	7.6531	18.7440
2.2500	7.0714	31.4519	4.7500	7.6618	18.5925
2.3000	7.0878	31.0198	4.8000	7.6704	18.4435
2.3500	7.1039	30.5995	4.8500	7.6788	18.2969
2.4000	7.1198	30.1905	4.9000	7.6872	18.1527
2.4500	7.1353	29.7924	4.9500	7.6955	18.0108
2.5000	7.1507	29.4046	5.0000	7.7037	17.8712

ϵ_r = 10.50

W/H	ϵ'_r	Z_0	W/H	ϵ'_r	Z_0
0.0500	5.8174	126.1527	2.5500	7.5284	28.3190
0.1000	5.9478	107.7290	2.6000	7.5442	27.9597
0.1500	6.0470	96.9653	2.6500	7.5596	27.6094
0.2000	6.1301	89.3533	2.7000	7.5749	27.2677
0.2500	6.2028	83.4738	2.7500	7.5899	26.9345
0.3000	6.2680	78.6931	2.8000	7.6047	26.6094
0.3500	6.3275	74.6722	2.8500	7.6193	26.2920
0.4000	6.3825	71.2087	2.9000	7.6336	25.9823
0.4500	6.4338	68.1716	2.9500	7.6478	25.6797
0.5000	6.4820	65.4715	3.0000	7.6617	25.3842
0.5500	6.5275	63.0447	3.0500	7.6755	25.0955
0.6000	6.5706	60.8439	3.1000	7.6891	24.8134
0.6500	6.6117	58.8332	3.1500	7.7024	24.5376
0.7000	6.6510	56.9849	3.2000	7.7156	24.2680
0.7500	6.6886	55.2767	3.2500	7.7287	24.0043
0.8000	6.7247	53.6909	3.3000	7.7415	23.7464
0.8500	6.7595	52.2128	3.3500	7.7542	23.4940
0.9000	6.7931	50.8304	3.4000	7.7667	23.2470
0.9500	6.8255	49.5335	3.4500	7.7791	23.0053
1.0000	6.8568	48.3136	3.5000	7.7913	22.7686
1.0500	6.8872	47.0466	3.5500	7.8033	22.5369
1.1000	6.9166	45.9074	3.6000	7.8152	22.3099
1.1500	6.9452	44.8441	3.6500	7.8270	22.0876
1.2000	6.9730	43.8473	3.7000	7.8386	21.8697
1.2500	7.0000	42.9091	3.7500	7.8500	21.6562
1.3000	7.0263	42.0229	3.8000	7.8614	21.4469
1.3500	7.0520	41.1830	3.8500	7.8726	21.2417
1.4000	7.0770	40.3846	3.9000	7.8836	21.0405
1.4500	7.1014	39.6237	3.9500	7.8946	20.8432
1.5000	7.1252	38.8967	4.0000	7.9054	20.6497
1.5500	7.1485	38.2006	4.0500	7.9160	20.4598
1.6000	7.1713	37.5329	4.1000	7.9266	20.2735
1.6500	7.1935	36.8911	4.1500	7.9370	20.0906
1.7000	7.2153	36.2734	4.2000	7.9473	19.9111
1.7500	7.2367	35.6780	4.2500	7.9575	19.7349
1.8000	7.2576	35.1034	4.3000	7.9676	19.5618
1.8500	7.2780	34.5483	4.3500	7.9776	19.3919
1.9000	7.2981	34.0114	4.4000	7.9875	19.2249
1.9500	7.3178	33.4917	4.4500	7.9972	19.0609
2.0000	7.3371	32.9882	4.5000	8.0069	18.8998
2.0500	7.3561	32.5001	4.5500	8.0164	18.7414
2.1000	7.3747	32.0266	4.6000	8.0259	18.5858
2.1500	7.3930	31.5669	4.6500	8.0352	18.4328
2.2000	7.4109	31.1204	4.7000	8.0445	18.2824
2.2500	7.4286	30.6865	4.7500	8.0537	18.1345
2.3000	7.4459	30.2647	4.8000	8.0627	17.9891
2.3500	7.4630	29.8543	4.8500	8.0717	17.8460
2.4000	7.4797	29.4550	4.9000	8.0806	17.7053
2.4500	7.4962	29.0663	4.9500	8.0894	17.5669
2.5000	7.5125	28.6878	5.0000	8.0981	17.4307

$\epsilon_r = 10$

W/H	ϵ'_r	Z_0	W/H	ϵ'_r	Z_0
0.0500	6.0850	123.3473	2.5500	7.8911	27.6606
0.1000	6.2226	105.3228	2.6000	7.9077	27.3094
0.1500	6.3274	94.7925	2.6500	7.9241	26.9670
0.2000	6.4151	87.3459	2.7000	7.9401	26.6332
0.2500	6.4918	81.5944	2.7500	7.9560	26.3075
0.3000	6.5607	76.9178	2.8000	7.9716	25.9898
0.3500	6.6235	72.9848	2.8500	7.9870	25.6796
0.4000	6.6816	69.5970	2.9000	8.0022	25.3769
0.4500	6.7357	66.6264	2.9500	8.0171	25.0813
0.5000	6.7865	63.9856	3.0000	8.0318	24.7925
0.5500	6.8345	61.6121	3.0500	8.0464	24.5104
0.6000	6.8801	59.4597	3.1000	8.0607	24.2347
0.6500	6.9235	57.4933	3.1500	8.0748	23.9652
0.7000	6.9649	55.6858	3.2000	8.0887	23.7017
0.7500	7.0046	54.0153	3.2500	8.1025	23.4440
0.8000	7.0428	52.4646	3.3000	8.1161	23.1920
0.8500	7.0795	51.0192	3.3500	8.1294	22.9454
0.9000	7.1149	49.6675	3.4000	8.1427	22.7040
0.9500	7.1491	48.3994	3.4500	8.1557	22.4678
1.0000	7.1822	47.2065	3.5000	8.1686	22.2366
1.0500	7.2142	45.9678	3.5500	8.1813	22.0101
1.1000	7.2453	44.8539	3.6000	8.1939	21.7884
1.1500	7.2755	43.8144	3.6500	8.2063	21.5711
1.2000	7.3048	42.8398	3.7000	8.2185	21.3582
1.2500	7.3333	41.9226	3.7500	8.2306	21.1496
1.3000	7.3611	41.0561	3.8000	8.2426	20.9451
1.3500	7.3882	40.2350	3.8500	8.2544	20.7446
1.4000	7.4146	39.4545	3.9000	8.2660	20.5481
1.4500	7.4403	38.7106	3.9500	8.2776	20.3553
1.5000	7.4655	37.9999	4.0000	8.2890	20.1662
1.5500	7.4901	37.3194	4.0500	8.3003	19.9806
1.6000	7.5141	36.6666	4.1000	8.3114	19.7986
1.6500	7.5376	36.0393	4.1500	8.3224	19.6199
1.7000	7.5606	35.4354	4.2000	8.3333	19.4445
1.7500	7.5831	34.8534	4.2500	8.3441	19.2723
1.8000	7.6052	34.2917	4.3000	8.3547	19.1033
1.8500	7.6268	33.7491	4.3500	8.3652	18.9372
1.9000	7.6480	33.2243	4.4000	8.3757	18.7741
1.9500	7.6688	32.7163	4.4500	8.3860	18.6139
2.0000	7.6892	32.2242	4.5000	8.3962	18.4565
2.0500	7.7092	31.7470	4.5500	8.4062	18.3018
2.1000	7.7288	31.2842	4.6000	8.4162	18.1497
2.1500	7.7481	30.8349	4.6500	8.4261	18.0002
2.2000	7.7671	30.3985	4.7000	8.4359	17.8533
2.2500	7.7857	29.9744	4.7500	8.4455	17.7088
2.3000	7.8040	29.5621	4.8000	8.4551	17.5667
2.3500	7.8220	29.1611	4.8500	8.4646	17.4270
2.4000	7.8397	28.7708	4.9000	8.4739	17.2895
2.4500	7.8571	28.3909	4.9500	8.4832	17.1542
2.5000	7.8743	28.0210	5.0000	8.4924	17.0212

Appendix A4

Proof of Inequality

If $1-|S_{22}|^2>|S_{12}S_{21}|$ then $B_1=1+|S_{11}|^2-|S_{22}|^2-|\Delta|^2>0$ where $\Delta=S_{11}S_{22}-S_{12}S_{21}$.

Proof.

$$
\begin{aligned}
|\Delta|^2 &= |S_{11}S_{22}-S_{12}S_{21}|^2 \leqslant (|S_{11}S_{22}|+|S_{12}S_{21}|)^2 \\
&\leqslant |S_{11}S_{22}|^2+2|S_{11}S_{22}||S_{12}S_{21}|+|S_{12}S_{21}|^2
\end{aligned}
$$

Thus

$$
\begin{aligned}
B_1 &> |S_{11}|^2+|S_{12}S_{21}|-|S_{11}S_{22}|^2-2|S_{11}S_{22}||S_{12}S_{21}|-|S_{12}S_{21}|^2 \\
&> |S_{11}|^2(1-|S_{22}|^2)+|S_{12}S_{21}|(1-2|S_{11}S_{22}|-|S_{12}S_{21}|) \\
&> |S_{11}|^2|S_{12}S_{21}|+|S_{12}S_{21}|(1-2|S_{11}S_{22}|-1+|S_{22}|^2) \\
&> |S_{11}|^2|S_{12}S_{21}|-|S_{12}S_{21}|(2|S_{11}S_{22}|-|S_{22}|^2) \\
&> |S_{12}S_{21}|(|S_{11}|^2-2|S_{11}S_{22}|+|S_{22}|^2) \\
&> |S_{12}S_{21}|(|S_{11}|-|S_{22}|)^2>0.
\end{aligned}
$$

The proof of $B_2>0$ is similar to the above.

Appendix A5

Decibel Units for Gain and Power

The gain of an amplifier is expressed in decibels and its power is measured in dBm or dBW.

THE DECIBEL (dB)

The decibel is a logarithmic unit of power ratio, although it is commonly also used for voltage and current ratio. If the input power P_i and the output power P_o of a network are expressed in the same units, then the network power gain (or loss) is

$$G = 10 \log \frac{P_o}{P_i} \quad \text{dB} \tag{A5.1}$$

For example, if $P_o = 4$ W and $P_1 = 2$ W, then $G = 10 \log 2 = 3$ dB. If $P_o = 1$ W and $P_i = 2$ W, then $G = -10 \log 2 = -3$ dB and the network is said to have a power loss of 3 dB.

THE dBm

The unit most commonly used to describe the input and output power of an amplifier is the dBm, which is defined as the power level P in reference to 1 mW, that is,

$$P\,(\text{dBm}) = 10 \log \frac{P\,(\text{mW})}{1\ \text{mW}} \tag{A5.2}$$

Thus $P = 1$ mW $= 0$ dBm and $P = 5$ W $= +37$ dBm and $P = 0.5$ mW $= -3$ dBm. The output power of an amplifier that has 10 dB gain is $+20$ dBm if its input power is $+10$ dBm.

THE dBW

Another unit that finds common use in microwave systems is the dBW, defined as the power level P in reference to 1 W, that is,

$$P\,(\mathrm{dBW}) = 10 \log \frac{P\,(\mathrm{W})}{1\ \mathrm{W}} \tag{A5.3}$$

Thus $P = 1$ W $= 0$ dBW and $P = 5$ W $= 7$ dBW. Furthermore we see that

$$P\,(\mathrm{dBW}) = P\,(\mathrm{dBm}) - 30 \tag{A5.4}$$

Appendix A6

Chebyshev and Butterworth Responses

This appendix provides information about the two types of familiar responses used in matching networks and filter design, namely, the Chebyshev and the Butterworth responses. We will focus our discussion on the lowpass type and obtain the corresponding bandpass type by appropriate transformation. Both responses can be expressed mathematically as

$$G(\omega^2)=\frac{K_n}{1+P_n^2(\omega)}, \qquad 0\leqslant K_n \leqslant 1 \tag{A6.1}$$

where $P_n(\omega)=\epsilon C_n(\omega)$ is used for the Chebyshev type response and $P_n(\omega)=\omega^n$ is used for the Butterworth response. We note that $G(\omega^2)$ in (A6.1) has the cutoff frequency $\omega_c=1$ rad/sec.

CHEBYSHEV RESPONSE

The polynomial $C_n(\omega)$ is the Chebyshev polynomial of order n defined as

$$C_n(\omega)=\cos(n\cos^{-1}\omega) \tag{A6.2a}$$

$$=\cosh(n\cosh^{-1}\omega) \tag{A6.2b}$$

Using the relation

$$\cos(n+1)y=2\cos ny\cos y-\cos(n-1)y$$

in (A6.2a) we obtain

$$C_{n+1}(\omega)=2\omega C_n(\omega)-C_{n-1}(\omega) \tag{A6.3}$$

and with $C_0(\omega)=1$ and $C_1(\omega)=\omega$, the higher orders can be evaluated as

$$
\begin{aligned}
C_2(\omega)&=2\omega^2-1\\
C_3(\omega)&=4\omega^3-3\omega\\
C_4(\omega)&=8\omega^4-8\omega^2+1\\
C_5(\omega)&=16\omega^5-20\omega^3+5\omega\\
C_6(\omega)&=32\omega^6-48\omega^4+18\omega^2-1
\end{aligned}
\tag{A6.4}
$$

From (A6.2a) we note that $-1\leq C_n(\omega)\leq 1$. Furthermore,

$$
\begin{aligned}
C_n(0)&=(-1)^{n/2}, && n \text{ even}\\
&=0, && n \text{ odd}\\
C_n(\pm 1)&=1, && n \text{ even}\\
&=\pm 1, && n \text{ odd}
\end{aligned}
$$

For $\omega>1$, $C_n(\omega)$ tends to infinity monotonically. This can be seen as (with $y=\cos^{-1}\omega$)

$$
\begin{aligned}
\cos ny&=\tfrac{1}{2}\left(e^{jny}+e^{-jny}\right)\\
&=\tfrac{1}{2}\left[\left(\cos y+\sqrt{\cos^2 y-1}\right)^n+\left(\cos y+\sqrt{\cos^2 y-1}\right)^{-n}\right]\\
&=\tfrac{1}{2}\left[\left(\omega+\sqrt{\omega^2-1}\right)^n+\left(\omega+\sqrt{\omega^2-1}\right)^{-n}\right]
\end{aligned}
$$

For $\omega\to\infty$, $\cos nx\to 2^{n-1}\omega^n$.

The poles of the Chebyshev response $G(-s^2)$ where ω is replaced by $-js$ are the zeros of the polynomial

$$1+\epsilon^2 C_n^2(-js)=0 \tag{A6.5}$$

Since $C_n(-js)=\cos ny$ where $y=\cos^{-1}(-js)=v+j\mu$, we have

$$
\begin{aligned}
\cos ny&=\cos\left[n(v+j\mu)\right]\\
&=\cos nv\cosh n\mu-j\sin nv\sinh n\mu=\pm\frac{j}{\epsilon}
\end{aligned}
\tag{A6.6}
$$

Equating the real and imaginary parts of both sides of (A6.6), we get

$$\cos nv\cosh n\mu=0 \tag{A6.7a}$$

$$\sin nv\sinh n\mu=\pm\frac{1}{\epsilon} \tag{A6.7b}$$

Since $\cosh n\mu > 0$, we obtain

$$v_k = \pm \frac{2k-1}{2n}\pi, \qquad k = 1, 2, \ldots \tag{A6.8}$$

and since $\sin nv_k = \pm 1$, hence

$$\sinh n\mu_k = \pm \frac{1}{\epsilon}$$

or

$$\mu_k = \pm \frac{1}{n}\sinh^{-1}\frac{1}{\epsilon} = \pm \frac{1}{n}\ln\left(\frac{1}{\epsilon} + \sqrt{\frac{1}{\epsilon^2} + 1}\right) \tag{A6.9}$$

Since $s_k = j\cos y_k = j\cos(v_k + j\mu_k) = \sin v_k \sinh \mu_k + j\cos v_k \cosh \mu_k = \sigma_k + j\omega_k$, hence

$$\sigma_k = \sin v_k \sinh \mu_k$$

$$\sigma_k = \pm \sin\left(\frac{2k-1}{2n}\pi\right)\sinh x,\ x = \frac{1}{n}\sinh^{-1}\frac{1}{\epsilon} \tag{A6.10a}$$

$$\omega_k = \cos v_k \cosh \mu_k$$

$$\omega_k = \cos\left(\frac{2k-1}{2n}\pi\right)\cosh x \tag{A6.10b}$$

for $k = 1, 2, \ldots, n$.

The locus of the $s_k = \sigma_k + j\omega_k$ in the s-plane can be found by solving the two equations in (A6.10) for $\sin[(2k-1)\pi/2n]$ and $\cos[(2k-1)\pi/2n]$ and then squaring and adding the two relations. This gives

$$\frac{\sigma_k^2}{\sinh^2 x} + \frac{\omega_k^2}{\cosh^2 x} = 1 \tag{A6.11}$$

Hence the locus if an ellipse with minor semiaxis $\sinh x$ and major semiaxis $\cosh x$. Using the poles in (A6.10), (A6.5) can be decomposed into a product of Hurwitz (with all roots in $\mathrm{Re}\, s < 0$) and anti-Hurwitz (with all roots in $\mathrm{Re}\, s > 0$) polynomials as

$$1 + \epsilon^2 C_n^2(-js) = 2^{2n-2}\epsilon^2 P(s)P(-s) \tag{A6.12}$$

where $P(s)$ in Hurwitz and $P(-s)$ is anti-Hurwitz, and

$$P(s) = s^n + a_{n-1}s^{n-1} + \cdots + a_1 s + a_0 \tag{A6.13}$$

is formed by the roots in $\operatorname{Re} s<0$, which are

$$s_k = -\sinh\left(\frac{2k-1}{2n}\pi\right)\sinh x + j\cos\left(\frac{2k-1}{2n}\pi\right)\cosh x \qquad \text{for } k=1,2,\ldots,n \tag{A6.14}$$

Multiplying the roots of $P(s)$ together one can prove that

$$\begin{aligned}
a_0 &= 2^{1-n}\sinh nx \qquad n \text{ odd}\\
&= 2^{1-n}\cosh nx \qquad n \text{ even}\\
a_{n-3} &= \frac{\sinh x}{\sin u_1}\left(\frac{n}{4} - \frac{\cos^2 u_1 \sin u_1}{\sin u_3} + \frac{\sinh^2 x \cos u_2}{2\sin u_1 \sin u_3}\right)\\
a_{n-2} &= \frac{n}{4} + \frac{\sinh^2 x \cos u_1}{\sin u_1 \sin 2u_1}\\
a_{n-1} &= \frac{\sinh x}{\sin u_1}
\end{aligned} \tag{A6.15}$$

where $u_m = m\pi/2n$.

BUTTERWORTH RESPONSE

The poles of Butterworth response $G(-s^2)$ are the zero of the polynomial

$$1+(-1)^n s^{2n} = 0 \tag{A6.16}$$

which are given by

$$s_k = \exp\left[j(2k-1+n)\pi/2n\right], \qquad k=1,2,\ldots,2n \tag{A6.17}$$

which are located on the unit circle in the s-plane. The polynomial in (A6.16) can also be decomposed into Hurwitz and anti-Hurwitz polynomials $Q(s)$ and $Q(-s)$ as

$$1+(-1)^n s^{2n} = Q(s)Q(-s) \tag{A6.18a}$$

where

$$Q(s) = s^n + a_{n-1}s^{n-1} + \cdots + a_1 s + a_0 \tag{A6.18b}$$

is formed by the zeros of $1+(-1)^n s^{2n}=0$ in Re $s<0$, which are given by

$$s_k=\exp\left[j(2k-1+n)\pi/2n\right], \qquad k=1,2,\ldots,n \tag{A6.19}$$

Upon using (A6.19) in (A6.18b) yields

$$a_0=1, \qquad a_k=\prod_{m=1}^{k}\frac{\cos\left[(m-1)\pi/2n\right]}{\sin(m\pi/2n)} \tag{A6.20}$$

LOWPASS TO BANDPASS TRANSFORMATION

The lowpass response in (A6.1) can be transformed to a bandpass response by the following transformation

$$s'=\frac{s}{B}+\frac{\omega_0^2}{Bs} \tag{A6.21}$$

where

$$\omega_0^2=\omega_1\omega_2, \qquad B=\omega_2-\omega_1$$

and ω_1 and ω_2 are called the lower and upper passband cut-off frequencies. The transformation (A6.21) maps the interval $-j1$ to $j1$ in the s'-plane to the intervals $j\omega_1$ to $j\omega_2$, and $-j\omega_1$ to $-j\omega_2$ and vice versa. Thus for any inductance L and C in the s'-plane, the transformation requires that

$$s'L=\frac{sL}{B}+\frac{\omega_0^2 L}{Bs}=sL_1+\frac{1}{sC_1} \tag{A6.22a}$$

and

$$\frac{1}{s'C}=\frac{1}{sC/B+\omega_0^2/Bs}=\frac{1}{sC_2+1/sL_2} \tag{A6.22b}$$

where

$$L_1=L/B, \qquad C_1=B/\omega_0^2 L \tag{A6.22c}$$

$$L_2=B/\omega_0^2 C, \qquad C_2=C/B \tag{A6.22d}$$

Thus a bandpass network is obtained from the corresponding lowpass network where each inductance L in the lowpass network is replaced by a series resonant L_1 and C_1 circuit whose resonant frequency is $\omega_0=1/\sqrt{L_1C_1}$, and where each capacitance C in the lowpass network is replaced by a parallel resonant L_2 and

C_2 circuit whose resonant frequency is again $\omega_0 = 1/\sqrt{L_2C_2}$. Using the above result, the spectral factorization of $1-G(-s^2)$ can be written as

$$\rho(s)\rho(-s) = 1 - G(-s^2) = 1 - \frac{K_n}{1+P_n^2(\omega)}$$

$$= \frac{(1-K_n)+P_n^2(\omega)}{1+P_n^2(\omega)}$$

$$= (1-K_n)\frac{1+(1-K_n)^{-1}P_n^2(\omega)}{1+P_n^2(\omega)} \tag{A6.23}$$

For the Chebychev response, we get

$$\rho(s)\rho(-s) = (1-K_n)\frac{1+\epsilon'^2C_n^2(-js)}{1+\epsilon^2C_n^2(-js)}$$

$$= (1-K_n)\frac{2^{2n-2}\epsilon'^2P'(s)P'(-s)}{2^{n-2}\epsilon^2P(s)P(-s)} = \frac{P'(s)P'(-s)}{P(s)P(-s)} \tag{A6.24}$$

where $\epsilon' = \epsilon/\sqrt{1-K_n}$ and $P'(s)$ and $P'(-s)$ are the Hurwitz and anti-Hurwitz polynomials of $1+\epsilon'^2C_n(-js)$. It can be easily seen that all the results derived for $1+\epsilon^2C_n(-js)=0$ hold for $1+\epsilon'^2C_n(-js)=0$ with ϵ replaced by ϵ'. In general $P'(s)$ can be written as

$$P'(s) = s^n + a'_{n-1}s^{n-1} + \cdots + a'_1s + a'_0$$

where a'_i's are obtained from a_i's in (A6.15) by replacing ϵ with ϵ'. Thus $\rho(s)$ can be obtained from (A6.24) by

$$\rho(s) = \frac{P'(s)}{P(s)} = \frac{s^n + a'_{n-1}s^{n-1} + \cdots + a'_1s + a'_0}{s^n + a_{n-1}s^{n-1} + \cdots + a_1s + a_0} \tag{A6.25a}$$

If the bandpass transformation (A6.21) is applied to (A6.25) we obtain the corresponding bandpass $\rho(s')$ as

$$\rho(s') = \frac{P'(s')}{P(s')} = \frac{s'^n + a'_{n-1}s'^{n-1} + \cdots + a'_1s' + a'_0}{s'^n + a_{n-1}s'^{n-1} + \cdots + a_1s' + a_0} \tag{A6.25b}$$

Using the relation

$$(a+b)^n = \sum_{k=0}^{n} \binom{n}{k} a^{n-k}b^k$$

where

$$\binom{n}{k} = \frac{n!}{k!(n-k)!}$$

We get

$$a_{n-i}s'^{n-i} = \frac{a_{n-i}}{B^{n-i}} \sum_{k=0}^{n-i} \binom{n-i}{k} \omega_0^{2k} s'^{n-i-2k} \tag{A6.26}$$

Substituting (A6.26) into (A6.25b) yields

$$\rho(s') = \frac{s^{2n} + \hat{a}'_{2n-1}s^{2n-1} + \cdots + \hat{a}'_1 s + \hat{a}'_0}{s^{2n} + \hat{a}_{2n-1}s^{2n-1} + \cdots + \hat{a}_1 s + \hat{a}_0} \tag{A6.27}$$

where some of the $\hat{a}_i$ are given in terms of the a_i as

$$\begin{aligned}
\hat{a}_{2n-1} &= Ba_{n-1} \\
\hat{a}_{2n-2} &= n\omega_0^2 + B^2 a_{n-2} \\
\hat{a}_{2n-3} &= (n-1)Ba_{n-1}\omega_0^2 + B^3 a_{n-3} \\
\hat{a}_1 &= \omega_0^{2n-2} Ba_{n-1} \\
\hat{a}_0 &= \omega_0^{2n}
\end{aligned} \tag{A6.28}$$

Using (A6.15) and (A6.20) in (A6.28), the relations in (5.42) and (5.44) can be obtained easily.

Appendix A7

Impedance Transformation With Inductive and Capacitive T and pi Networks

In designing the matching networks for microwave amplifiers, the designer usually ends up with an ideal transformer at the source and at the load side. In many cases, these ideal transformers can be replaced by equivalent T or pi networks and this allows the use of distributed elements for approximation.

Consider the inductive L-section in cascade with an ideal transformer of turn ratio $1:t$ and terminated in the load Z_L in Fig. A7.1. We wish to find its equivalent T network shown in Fig. A7.2. Let $\mathbf{Z}=[Z_{ij}]$ and $\mathbf{Z}'=[Z'_{ij}]$ be the impedance matrices of the two-port networks N and N', respectively. It can be easily shown that

$$\mathbf{Z}=\begin{bmatrix} s(L_s+L_p) & stL_p \\ stL_p & st^2L_p \end{bmatrix} \tag{A7.1}$$

and

$$Z_1=Z'_{11}-Z'_{12} \tag{A7.2a}$$

$$Z_2=Z'_{12}=Z'_{21} \tag{A7.2b}$$

$$Z_3=Z'_{22}-Z'_{12} \tag{A7.2c}$$

For N and N' to be equivalent we must have $\mathbf{Z}=\mathbf{Z}'$ and thus, from (A7.2) we obtain

$$Z_1=s(L_s+L_p)-stL_p=s\left[L_s+(1-t)L_p\right] \tag{A7.3a}$$

$$Z_2=stL_p \tag{A7.3b}$$

$$Z_3=st^2L_p-stL_p=s(t^2-t)L_p \tag{A7.3c}$$

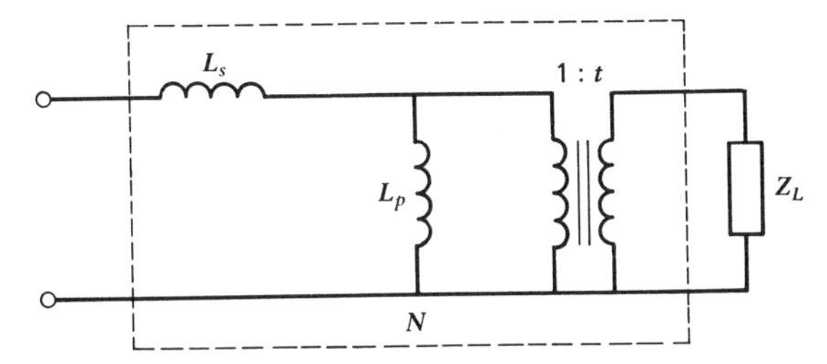

Fig. A7.1. An inductive L-section with an ideal transformer.

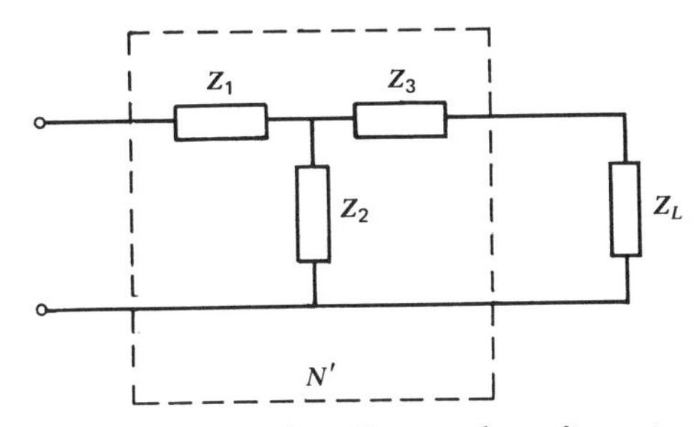

Fig. A7.2. Equivalent T network configuration.

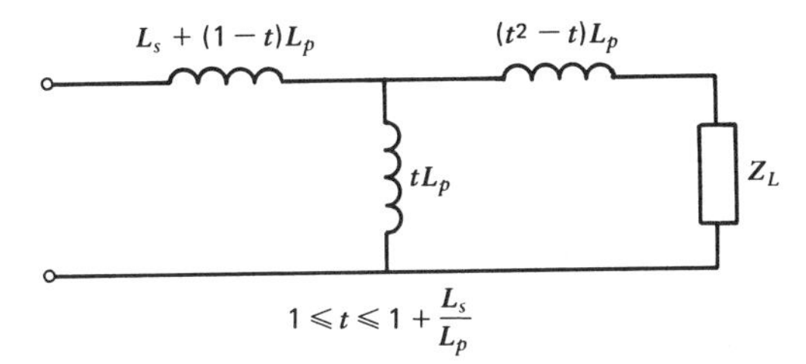

Fig. A7.3. Equivalent inductive T network.

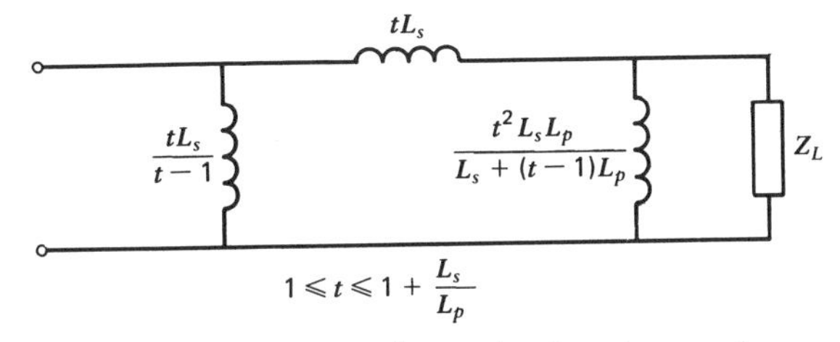

Fig. A7.4. Equivalent inductive pi network.

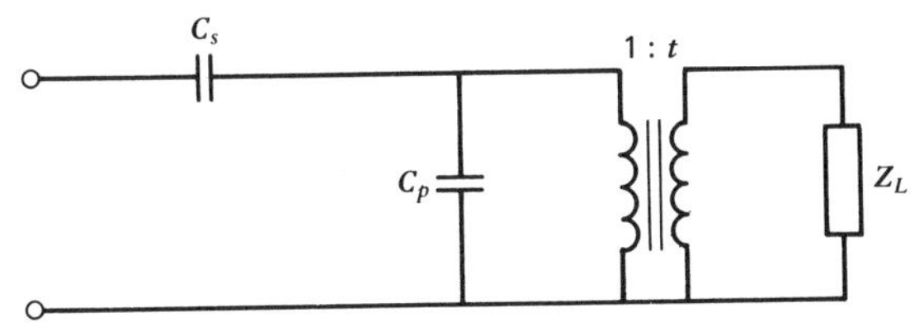

Fig. A7.5. A capacitive L-section with an ideal transformer.

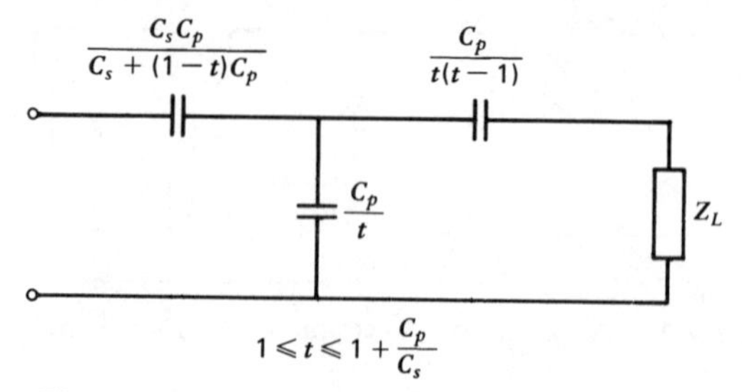

Fig. A7.6. Equivalent capacitive T network.

For Z_1 and Z_3 to be passive we must have

$$L_s + (1-t)L_p \geqslant 0$$

$$(t^2 - t)L_p \geqslant 0$$

which yields

$$1 \leqslant t \leqslant 1 + \frac{L_s}{L_p} \tag{A7.4}$$

The equivalent networks of Fig. A7.1 are shown in Figs. A7.3 and A7.4. Similarly, the equivalent capacitive T and pi networks of a capacitive *L*-section in cascade with an ideal transformer are shown in Figs. A7.5–A7.7.

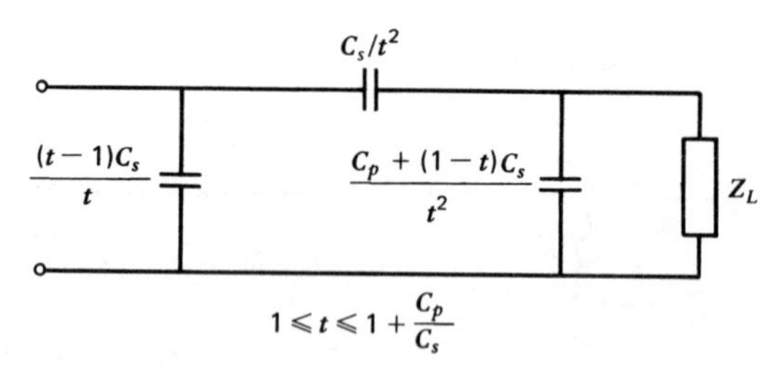

Fig. A7.7. Equivalent capacitive pi network.

Appendix A8

Theory of Noisy Two-Port Networks

In a two-port amplifier there is always a weak output signal called noise even when there is no input signal. Therefore the sensitivity of a detection system is set by the level of the noise presented to the system with the signal, adding to its own noise voltage. There are various noise sources in an amplifier; the two main sources are thermal noise and shot noise.

THERMAL NOISE

Thermal noise, also called Johnson noise, is a phenomenon associated with the motion of the electrons in a conductor. These electrons possess varying amounts of energy due to the temperatures of the conductor, and the random motion gives rise to an ac voltage within the conductor. The available power in a 1 Hz bandwidth from such a thermal noise source is given by

$$p_n = kT \quad \text{W/Hz} \tag{A8.1}$$

where $k = 1.3805(10^{-23})$ J/°K is Boltzmann's constant and T is the absolute temperature in degrees Kelvin of the thermal noise source. At room temperature, that is, $T = 290$°K, $p_n = 4(10^{-21})$ W/Hz or $p_n = -174$ dBm/Hz.

In a detection system with bandwidth B, the available noise power from a thermal noise source is

$$p_a = Bp_n = kTB \quad \text{W} \tag{A8.2}$$

An example of a thermal noise source is a resistor. A noise equivalent circuit for a resistor consists of a Thevenin's noise voltage source e_n in series with a hypothetically noiseless resistor R having the same resistance. The maximum power available from e_n can be obtained by connecting it to a load resistor $R_L = R$, that is, $p_a = e_{n,\text{rms}}^2/4R$. Since the available power from the thermal noise source is kTB, we see that

$$\overline{e_n^2} = e_{n,\text{rms}}^2 = 4kTBR \tag{A8.3}$$

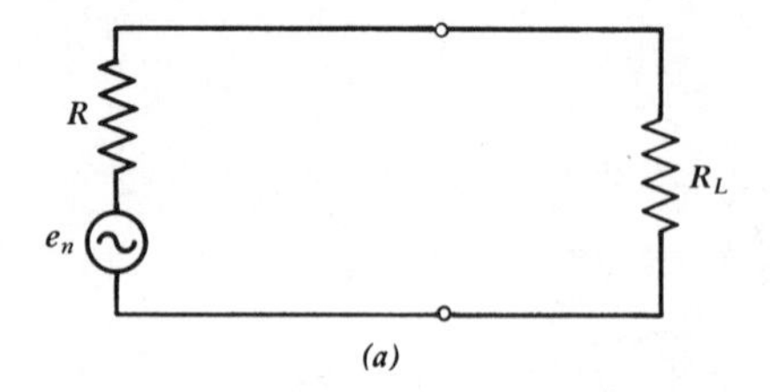

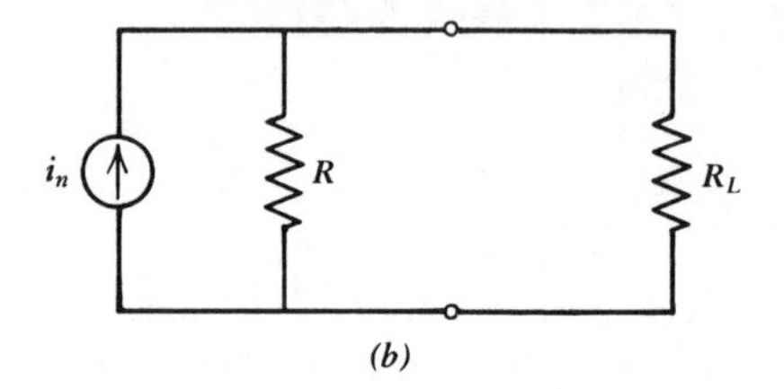

Fig. A8.1. (a)–(b) Equivalent circuits of a noisy resistor.

In a similar way, the rms value of the Norton's noise current source i_n in parallel with $G=1/R$ is given by

$$\overline{i_n^2} = i_{n,\text{rms}}^2 = 4kTBG \tag{A8.4}$$

The concept is shown in Figs. A8.1a and b.

SHOT NOISE

Shot noise, also called Schottky noise, is present in all active devices because of the discrete nature of the electron flow. For example, the drain-to-source current in a GaAs FET is due to the flow of the electrons from the source to the drain electrodes across the active layer, and the fluctuation in the number of carriers is the shot noise. The mean-square shot noise current is given by

$$\overline{i^2} = i_{\text{rms}}^2 = 2qI_{dc}B \tag{A8.5}$$

where $q=1.6(10^{-19})$ coulombs is the electron charge, I_{dc} is the direct current through the device and B is the bandwidth.

In this discussion we will present the theory for noisy two-ports with internal noise sources. Consider the noisy two-port shown in Fig. A8.2a with an admittance $\mathbf{Y}$ (or impedance $\mathbf{Z}$). This two-port can be represented by the equivalent circuits in Figs. A8.2b and c, where the two-port is replaced by a noise-free but unchanged two-port with the noise current sources i_1 and i_2 (or noise voltage sources e_1 and e_2). In the analysis of noisy two-ports, it is convenient to place all the noise sources preceding the noise-free two-ports as

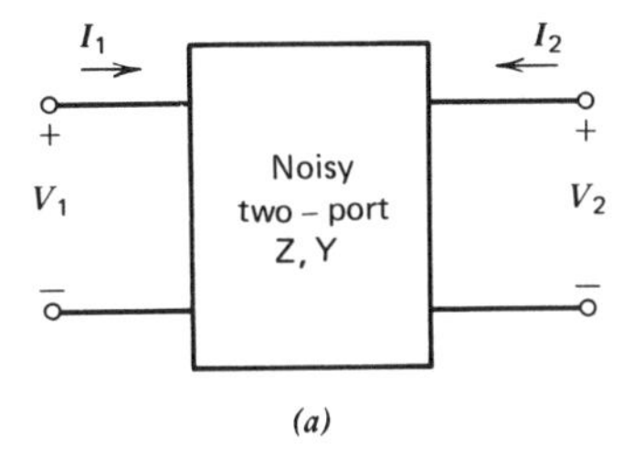

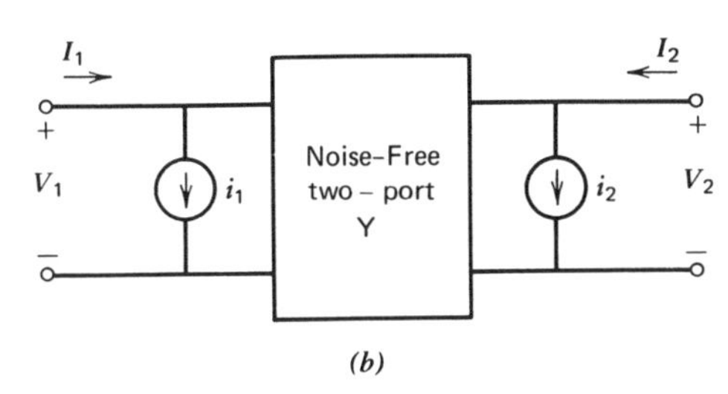

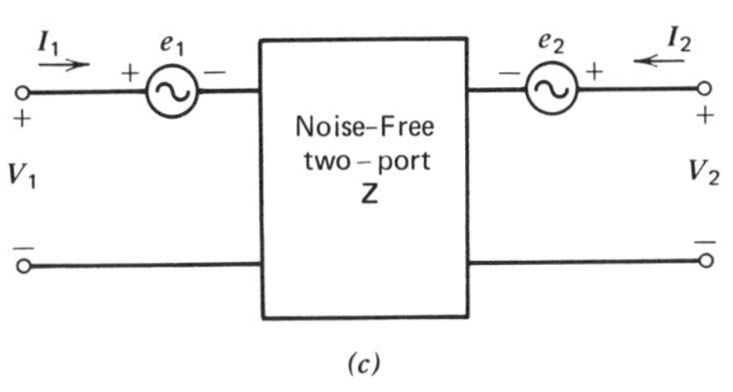

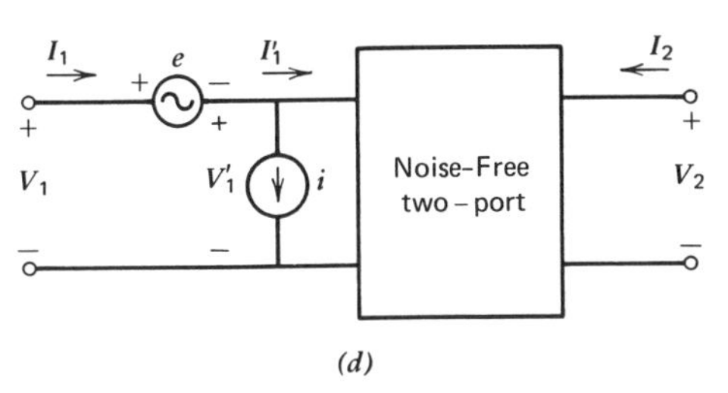

Fig. A8.2. (*a*) A two-port with internal noise sources. (*b*) Equivalent circuit with external noise current sources. (*c*) Equivalent circuit with external noise voltage sources. (*d*) Equivalent circuit with external noise voltage and current source at the input.

shown in Fig. A8.2*d*. The noise voltage and current sources e and i can be obtained from i_1, i_2, or e_1, e_2 as follows. We note that

$$I_1 = I_1' + i \tag{A8.6a}$$

$$V_1 = V_1' + e \tag{A8.6b}$$

But

$$I_1' = Y_{11}V_1' + Y_{12}V_2 \tag{A8.7a}$$

$$I_2 = Y_{21}V_1' + Y_{22}V_2 \tag{A8.7b}$$

Substituting (A8.7) into (A8.6) yields

$$I_1 = Y_{11}(V_1 - e) + Y_{12}V_2 + i \tag{A8.8a}$$

$$I_2 = Y_{21}(V_1 - e) + Y_{22}V_2 \tag{A8.8b}$$

Also for the two-port in Fig. A8.2*b* we have

$$I_1 = Y_{11}V_1 + Y_{12}V_2 + i_1 \tag{A8.9a}$$

$$I_2 = Y_{21}V_1 + Y_{22}V_2 + i_2 \tag{A8.9b}$$

By comparing (A8.8) and (A8.9) we obtain

$$e = -i_2 / Y_{21} \tag{A8.10a}$$

$$i = i_1 + Y_{11} e = i_1 - (Y_{11}/Y_{12}) i_2 \tag{A8.10b}$$

In a similar way, it can be shown that

$$e = e_1 - (Z_{11}/Z_{21}) e_2 \tag{A8.11a}$$

$$i = -e_2 / Z_{21} \tag{A8.11b}$$

Now consider the noisy two-port in Fig. A8.2*d* with a source comprising an internal admittance Y_s and a current source i_s connected to its input as shown in Fig. A8.3. This circuit will be used to derive the noise figure F of the network. By definition, the noise figure F is the ratio of the total output noise power per unit bandwidth available at the output port to that portion of the total output noise power engendered by the input termination at the standard temperature $T_0 = 290°K$. To proceed we assume that the noise of the two-port is not correlated to the noise from the source. From Fig. A8.3, the mean-square of the short circuit current i_{sc} at the terminal $1-1'$ is given as

$$\overline{i_{sc}^2} = \overline{i_s^2} + \overline{|i + Y_s e|^2} = \overline{i_s^2} + \overline{i^2} + |Y_s|^2 \overline{e^2} + Y_s^* \overline{ie^*} + Y_s \overline{i^* e} \tag{A8.12}$$

Note that the total output noise power is proportional to $\overline{i_{sc}^2}$ and the noise power due to the source alone is proportional to $\overline{i_s^2}$. Thus the noise figure F can be expressed as

$$F = \frac{\overline{i_s^2} + \overline{|i + Y_s e|^2}}{\overline{i_s^2}} = 1 + \frac{\overline{|i + Y_s e|^2}}{\overline{i_s^2}} \tag{A8.13}$$

Normally there is a correlation between the two noise sources e and i, and the noise current i can be divided into one part i_u not correlated to e and a second part $i - i_u$ fully correlated to e. It is necessary to introduce the correlation

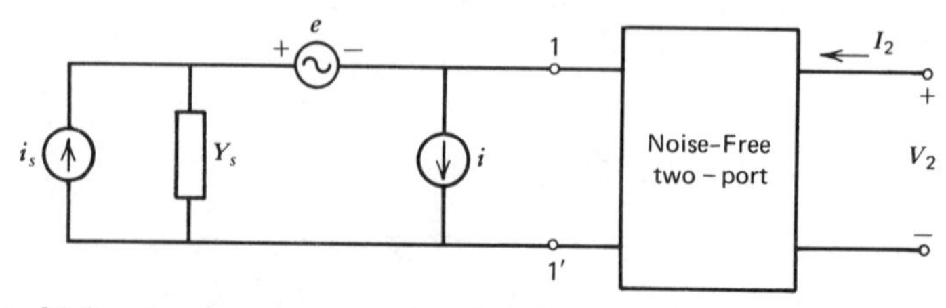

Fig. A8.3. A noisy two-port with signal source for noise figure computation.

admittance $Y_c = G_c + jB_c$ such that

$$i = i_u + Y_c e \tag{A8.14}$$

Thus

$$\overline{e i_u^*} = 0 \tag{A8.15a}$$

$$\overline{(i - i_u) i_u^*} = 0 \tag{A8.15b}$$

$$\overline{e i^*} = \overline{e(i - i_u)^*} = Y_c^* \overline{e^2} \tag{A8.16}$$

From (A8.3)–(A8.4) the noise voltage e, the noise current i_u and the noise current source i_s can be written as

$$\overline{e^2} = 4kT_0 R_n B \tag{A8.17a}$$

$$\overline{i_u^2} = 4kT_0 G_u B \tag{A8.17b}$$

$$\overline{i_s^2} = 4kT_0 G_s B \tag{A8.17c}$$

where R_n and G_u are the equivalent noise resistance of e and the equivalent noise conductance of i_u, respectively, and $G_s = \text{Re}\, Y_s$. The mean-square of the noise current i is then

$$\begin{aligned} \overline{i^2} &= \overline{i_u^2} + \overline{|i - i_u|^2} \\ &= 4kT_0 B\left(G_u + R_n |Y_c|^2\right) \end{aligned} \tag{A8.18}$$

The noise figure F can now be expressed as

$$\begin{aligned} F &= 1 + \frac{\overline{i_u^2} + |Y_s + Y_c|^2 \overline{e^2}}{4kT_0 G_s B} \\ &= 1 + \frac{G_u}{G_s} + \frac{R_n}{G_s}\left[(G_s + G_c)^2 + (B_s + B_c)^2\right] \end{aligned} \tag{A8.19}$$

Since the noise figure F is a function of the source admittance Y_s, it can be shown that F achieves its minimum value F_m at an optimum source admittance $Y_m = G_m + jB_m$ where

$$G_m = \left(G_c^2 + \frac{G_u}{R_n}\right)^{1/2} \tag{A8.20}$$

$$B_m = -B_c \tag{A8.21}$$

and

$$F_m = 1 + 2R_n\left[G_c + \left(G_c^2 + \frac{G_u}{R_n}\right)^{1/2}\right] \tag{A8.22}$$

In terms of G_m, B_m, and F_m, F can be expressed as

$$F = F_m + \frac{R_n}{G_s}\left[(G_s - G_m)^2 + (B_s - B_m)^2\right] \tag{A8.23}$$

Using the above result, the change of the noise parameters can be predicted for any change of the linear two-port due to their interconnection. Consider the series connection of a noisy two-port N with the impedance matrix $\mathbf{Z}$ and a noise-free lossless two-port N' with the impedance matrix $\mathbf{Z}'$ as shown in Fig. A8.4*a*. Our purpose is to evaluate the change in the noise parameters F_m, R_n, G_m, and B_m in (A8.23) of N due to adding of N' into the network. To achieve this goal, it is necessary to transform the two inner-correlated noise sources e and i into two outer-correlated noise sources e' and i'. From Fig. A8.4*a* we get

$$\begin{aligned} V_1 &= e + \hat{V}_1 + V_1' = e + Z_{11}\hat{I}_1 + Z_{12}I_2 + Z_{11}'I_1 + Z_{12}'I_2 \\ &= e + Z_{11}(I_1 - i) + Z_{12}I_2 + Z_{11}'I_1 + Z_{12}'I_2 \\ &= e + (Z_{11} + Z_{11}')I_1 + (Z_{12} + Z_{12}')I_2 - Z_{11}i \end{aligned} \tag{A8.24}$$

$$\begin{aligned} V_2 &= \hat{V}_2 + V_2' = Z_{21}\hat{I}_1 + Z_{22}I_2 + Z_{21}'I_1 + Z_{22}'I_2 \\ &= Z_{21}(I_1 - i) + Z_{22}I_2 + Z_{21}'I_1 + Z_{22}'I_2 \\ &= (Z_{21} + Z_{21}')I_1 + (Z_{22} + Z_{22}')I_2 - Z_{21}i \end{aligned} \tag{A8.25}$$

From Fig. A8.4*b* we get

$$\begin{aligned} V_1 &= e' + \hat{V}_1 + \hat{V}_1' = e' + Z_{11}\hat{I}_1 + Z_{12}I_2 + Z_{11}'\hat{I}_1 + Z_{12}'I_2 \\ &= e' + Z_{11}(I_1 - i') + Z_{12}I_2 + Z_{11}'(I_1 - i') + Z_{12}'I_2 \\ &= e' + (Z_{11} + Z_{11}')I_1 + (Z_{12} + Z_{12}')I_2 - (Z_{11} + Z_{11}')i' \end{aligned} \tag{A8.26}$$

$$\begin{aligned} V_2 &= \hat{V}_2 + \hat{V}_2' = Z_{21}\hat{I}_1 + Z_{22}I_2 + Z_{21}'\hat{I}_1 + Z_{22}'I_2 \\ &= Z_{21}(I_1 - i') + Z_{22}I_2 + Z_{21}'(I_1 - i') + Z_{22}'I_2 \\ &= (Z_{21} + Z_{21}')I_1 + (Z_{22} + Z_{22}')I_2 - (Z_{21} + Z_{21}')i' \end{aligned} \tag{A8.27}$$

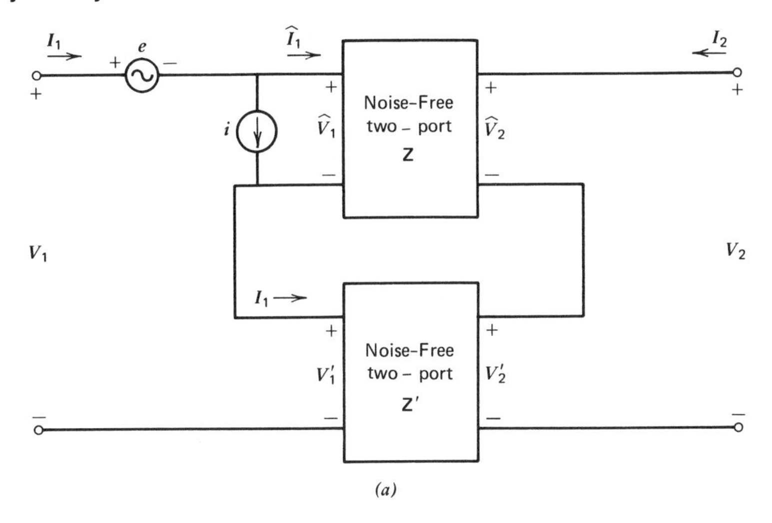

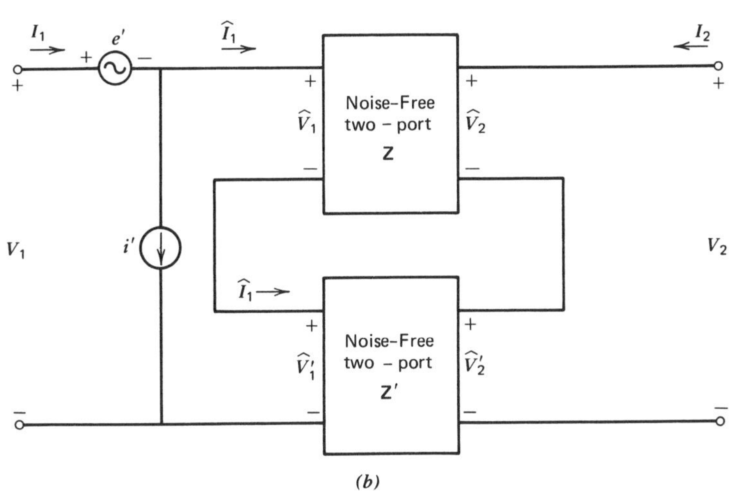

Fig. A8.4. (a)–(b) Transformation of two inner-correlated noise sources e and i into two outer-correlated noise sources e' and i' for series connection.

By comparing (A8.25) with (A8.27) and (A8.24) with (A8.26) we obtain

$$(Z_{21}+Z'_{21})i'=Z_{21}i$$

Hence

$$i'=\frac{Z_{21}}{Z_{21}+Z'_{21}}i \tag{A8.28}$$

and

$$e-Z_{11}i=e'-(Z_{11}+Z'_{11})i'$$

Hence

$$e' = e - Z_{11} i + (Z_{11} + Z'_{11}) i'$$

$$e' = e - Z_{11} i + (Z_{11} + Z'_{11}) \frac{Z_{21}}{Z_{21} + Z'_{21}} i$$

$$e' = e + \frac{Z'_{11} Z_{21} - Z_{11} Z'_{21}}{Z_{21} + Z'_{21}} i \qquad \text{(A8.29)}$$

Let $\mathbf{S}$ and $\mathbf{S}'$ be the corresponding scattering matrix of $\mathbf{Z}$ and $\mathbf{Z}'$, normalized to Z_0, that is,

$$\mathbf{S} = (\mathbf{Z} + Z_0 \mathbf{1}_2)^{-1} (\mathbf{Z} - Z_0 \mathbf{1}_2)$$

$$\mathbf{S}' = (\mathbf{Z}' + Z_0 \mathbf{1}_2)^{-1} (\mathbf{Z}' - Z_0 \mathbf{1}_2)^{-1}$$

Then

$$\mathbf{Z} = Z_0 (\mathbf{1}_2 + \mathbf{S})(\mathbf{1}_2 - \mathbf{S})^{-1}$$

$$\mathbf{Z}' = Z_0 (\mathbf{1}_2 + \mathbf{S}')(\mathbf{1}_2 - \mathbf{S}')^{-1}$$

The parameters Z_{ij} of $\mathbf{Z}$ and Z'_{ij} of $\mathbf{Z}'$ can be expressed as

$$Z_{11} = Z_0 C_2 / C_1, \quad Z_{12} = 2 Z_0 S_{12} / C_1, \quad Z_{21} = 2 Z_0 S_{21} / C_1, \quad Z_{22} = 2 Z_0 C_3 / C_1 \qquad \text{(A8.30)}$$

$$Z'_{11} = 2 Z_0 C'_2 / C_1, \quad Z'_{12} = 2 Z_0 S'_{12} / C'_1, \quad Z'_{21} = 2 Z_0 S'_{21} / C'_1, \quad Z'_{22} = 2 Z_0 C'_3 / C'_1 \qquad \text{(A8.31)}$$

where

$$C_1 = (1 - S_{11})(1 - S_{22}) - S_{12} S_{21}$$

$$C_2 = (1 + S_{11})(1 - S_{22}) + S_{12} S_{21}$$

$$C_3 = (1 + S_{22})(1 - S_{11}) + S_{12} S_{21}$$

and C'_i, $i = 1, 2, 3$ are obtained from C_i by replacing S_{ij} by S'_{ij}. Substituting (A8.30) and (A8.31) into (A8.28) yields

$$\begin{bmatrix} e' \\ i' \end{bmatrix} = \begin{bmatrix} n_{11} & n_{12} \\ n_{21} & n_{22} \end{bmatrix} \begin{bmatrix} e \\ i \end{bmatrix} \qquad \text{(A8.32)}$$

where

$$n_{11}=1, \quad n_{21}=0 \tag{A8.33a}$$

$$n_{12}=Z_0\frac{S_{21}C_2'-S_{21}'C_2}{S_{21}C_1'+S_{21}'C_1} \tag{A8.33b}$$

$$n_{22}=\frac{S_{21}C_1'}{S_{21}C_1'+S_{21}'C_1} \tag{A8.33c}$$

Next consider the parallel connection of the noisy two-port N with admittance matrix $\mathbf{Y}$ and the lossless two-port N' with admittance matrix $\mathbf{Y}'$ in Fig. A8.5a.

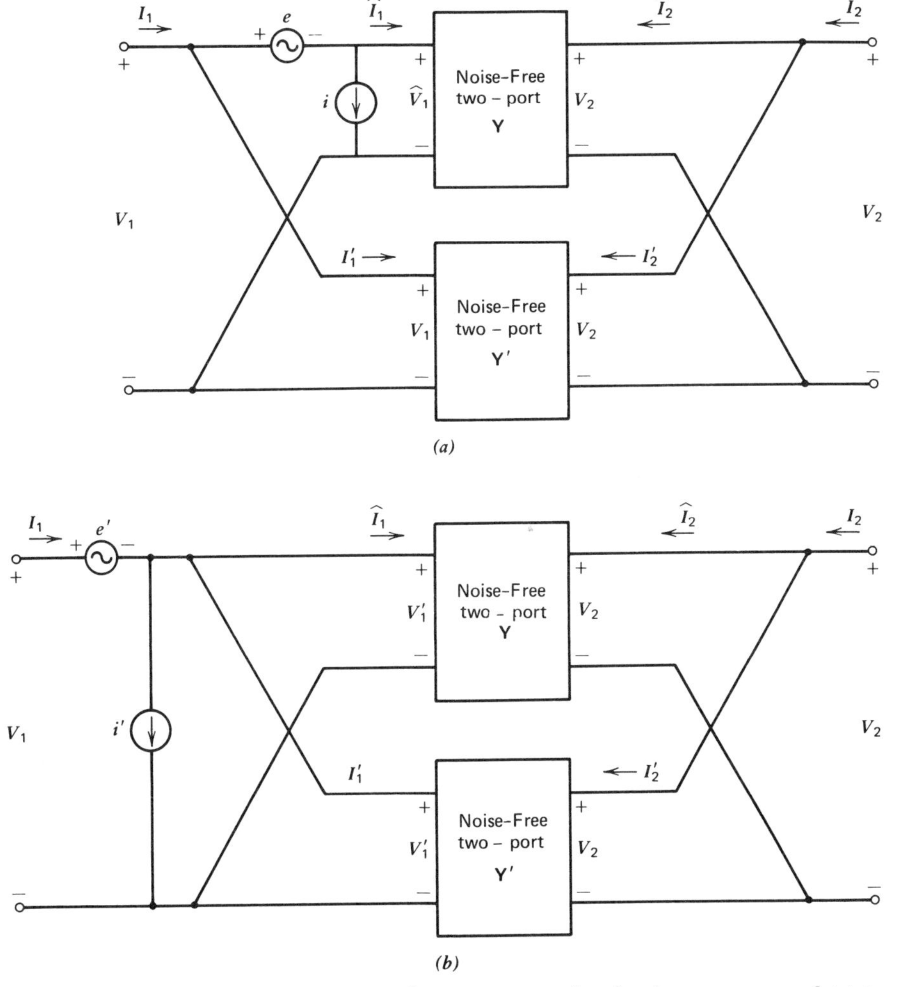

Fig. A8.5. (a)–(b) Transformation of two inner-correlated noise sources e and i into two outer-correlated noise sources e' and i' for parallel connection.

We have

$$I_1 = \hat{I}_1 + I_1' + i = Y_{11}\hat{V}_1 + Y_{12}V_2 + Y_{11}'V_1 + Y_{12}'V_2 + i$$
$$= Y_{11}(V_1 - e) + Y_{12}V_2 + Y_{11}'V_1 + Y_{12}'V_2 + i$$
$$= (Y_{11} + Y_{11}')V_1 + (Y_{12} + Y_{12}')V_2 - Y_{11}e + i \quad \text{(A8.34)}$$

$$I_2 = \hat{I}_2 + I_2' = Y_{21}\hat{V}_1 + Y_{22}V_2 + Y_{21}'V_1 + Y_{22}'V_2$$
$$= Y_{21}(V_1 - e) + Y_{22}V_2 + Y_{21}'V_1 + Y_{22}'V_2$$
$$= (Y_{21} + Y_{21}')V_1 + (Y_{22} + Y_{22}')V_2 - Y_{21}e \quad \text{(A8.35)}$$

From Fig. A8.5*b* we have

$$I_1 = \hat{I}_1 + \hat{I}_1' + i' = Y_{11}V_1' + Y_{12}V_2 + Y_{11}'V_1' + Y_{12}'V_2 + i'$$
$$= Y_{11}(V_1 - e') + Y_{12}V_2 + Y_{11}'(V_1 - e') + Y_{12}'V_2 + i'$$
$$= (Y_{11} + Y_{11}')V_1 + (Y_{12} + Y_{12}')V_2 - (Y_{11} + Y_{11}')e' + i' \quad \text{(A8.36)}$$

$$I_2 = \hat{I}_2 + I_2' = Y_{21}V_1' + Y_{22}V_2 + Y_{21}'V_1' + Y_{22}'V_2$$
$$= Y_{21}(V_1 - e') + Y_{22}V_2 + Y_{21}'(V_1 - e') + Y_{22}'V_2$$
$$= (Y_{21} + Y_{21}')V_1 + (Y_{22} + Y_{22}')V_2 - (Y_{21} + Y_{21}')e' \quad \text{(A8.37)}$$

Comparing (A8.35) to (A8.37) and (A8.34) to (A8.36) yields

$$(Y_{21} + Y_{21}')e' = Y_{21}e$$

hence

$$e' = \frac{Y_{21}}{Y_{21} + Y_{21}'}e \quad \text{(A8.38)}$$

and

$$i - Y_{11}e = i' - (Y_{11} + Y_{11}')e'$$

hence

$$i' = i - Y_{11}e + (Y_{11} + Y_{11}')\frac{Y_{21}}{Y_{21} + Y_{21}'}e$$

$$i' = i - \frac{Y_{11}'Y_{21} - Y_{11}Y_{21}'}{Y_{21} + Y_{21}'}e \quad \text{(A8.39)}$$

Using the relations ($Y_0 = 1/Z_0$)

$$\mathbf{Y} = (Y_0 \mathbf{1}_2 - \mathbf{S})(Y_0 \mathbf{1}_2 + \mathbf{S})^{-1}$$

$$\mathbf{Y}' = (Y_0 \mathbf{1}_2 - \mathbf{S}')(Y_0 \mathbf{1}_2 + \mathbf{S}')^{-1}$$

we obtain, for the parallel connection

$$\begin{bmatrix} e' \\ i' \end{bmatrix} = \begin{bmatrix} n_{11} & n_{12} \\ n_{21} & n_{22} \end{bmatrix} \begin{bmatrix} e \\ i \end{bmatrix}$$

where

$$n_{12} = 0, \quad n_{22} = 1 \tag{A8.40a}$$

$$n_{11} = \frac{S_{21} D_1'}{S_{21} D_1' + S_{21}' D_1} \tag{A8.40b}$$

$$n_{21} = Y_0 \frac{S_{21} D_2' - S_{21}' D_2}{S_{21} D_1' + S_{21}' D_1} \tag{A8.40c}$$

with

$$D_1 = (1 + S_{11})(1 + S_{22}) - S_{12} S_{21}$$

$$D_2 = (1 + S_{22})(1 - S_{11}) + S_{12} S_{21}$$

$$D_1' = (1 + S_{11}')(1 + S_{22}') - S_{12}' S_{21}'$$

$$D_2' = (1 + S_{22}')(1 - S_{11}') + S_{12}' S_{21}'$$

For the cascade connection shown in Figs. A8.6*a* and *b* it can be shown similarly that

$$n_{11} = \frac{(1 + S_{11}')(1 - S_{22}') + S_{12}' S_{21}'}{2 S_{21}'} \tag{A8.41a}$$

$$n_{12} = Z_0 \frac{(1 + S_{11}')(1 + S_{22}') - S_{12}' S_{21}'}{2 S_{21}'} \tag{A8.41b}$$

$$n_{21} = \frac{1}{Z_0} \frac{(1 - S_{11}')(1 - S_{22}') - S_{12}' S_{21}'}{2 S_{21}'} \tag{A8.41c}$$

$$n_{22} = \frac{(1 - S_{11}')(1 + S_{22}') + S_{12}' S_{21}'}{2 S_{21}'} \tag{A8.41d}$$

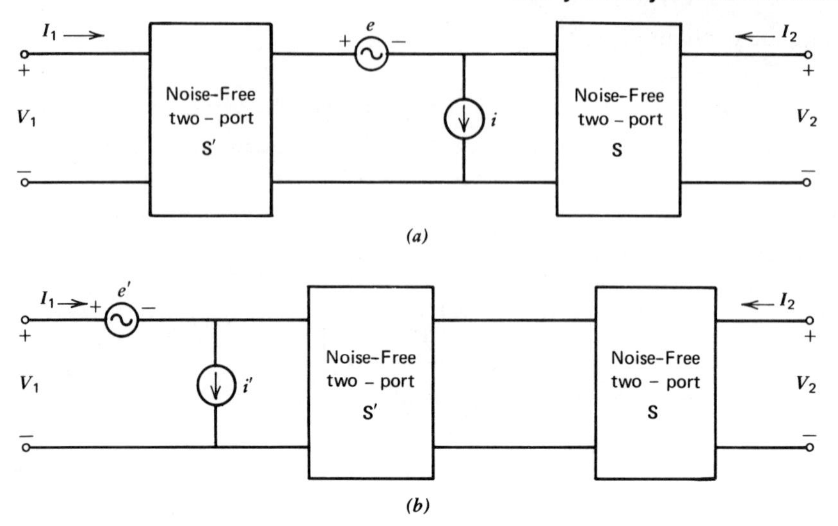

Fig. A8.6. Transformation of two inner-correlated noise sources e and i into two outer-correlated noise sources e' and i' for cascade connection.

In summary, if the two-ports N and N' are linear, the values e' and i' for the transformed noise sources of the interconnected network $\hat{N}$ are given by

$$\begin{bmatrix} e' \\ i' \end{bmatrix} = \begin{bmatrix} n_{11} & n_{12} \\ n_{21} & n_{22} \end{bmatrix} \begin{bmatrix} e \\ i \end{bmatrix} \tag{A8.42}$$

Using the above results, the noise parameters of the interconnected network $\hat{N}$ can be computed from the noise parameters of N. Let $\hat{R}_n$, $\hat{G}_u$, $\hat{Y}_c$, and $\hat{Y}_m$ be the equivalent noise resistance of e', the equivalent noise conductance of i'_u, the correlation admittance of e' and i', and the optimum noise admittance of the interconnected network $\hat{N}$ where

$$\overline{e'^2} = 4kT_0\hat{R}_n B \tag{A8.43}$$

$$\overline{i'^2_u} = 4kT_0\hat{G}_u B \tag{A8.44}$$

$$i' = i'_u + \hat{Y}_c e' \tag{A8.45}$$

Then

$$\hat{R}_n = \frac{\overline{e'^2}}{4kT_0B} = \frac{\overline{(n_{11}e + n_{12}i)^2}}{4kT_0B}$$

$$= \frac{\overline{[n_{11}e + n_{12}(i_u + Y_c e)]^2}}{4kT_0B} = \frac{\overline{[(n_{11} + n_{12}Y_c)e + n_{12}i_u]^2}}{4kT_0B}$$

$$= |n_{11} + n_{12}Y_c|^2 \frac{\overline{e^2}}{4kT_0B} + |n_{12}|^2 \frac{\overline{i_u^2}}{4kT_0B}$$

$$\hat{R}_n = R_n|n_{11} + n_{12}Y_c|^2 + G_u|n_{12}|^2 \tag{A8.46}$$

From (A8.16) and (A8.15a) we get

$$\hat{Y}_c = \frac{\overline{e'^* i'}}{\overline{e'^2}} = \frac{\overline{(n_{11}^* e^* + n_{12}^* i^*)(n_{21} e + n_{22} i)}}{4kT_0 B\hat{R}_n}$$

$$= \frac{\overline{[(n_{11}^* + n_{12}^* Y_c^*)e + n_{12}^* i_u^*][(n_{21} + n_{22} Y_c)e + n_{22} i_u]}}{4kT_0 B\hat{R}_n}$$

$$= \frac{(n_{11}^* + n_{12}^* Y_c^*)(n_{21} + n_{22} Y_c)\,\overline{e^2} + n_{12}^* n_{22}\,\overline{|i_u|^2}}{4kT_0 B\hat{R}_n}$$

$$Y_c = \frac{R_n(n_{11}^* + n_{12}^* Y_c^*)(n_{21} + n_{22} Y_c) + G_u n_{12}^* n_{22}}{R_n|n_{11} + n_{12} Y_c|^2 + G_u|n_{12}|^2} \tag{A8.47}$$

From (A8.18) we have

$$\overline{i'^2} = 4kT_0 B\left(\hat{G}_u + \hat{R}_n|\hat{Y}_c|^2\right)$$

Thus

$$\hat{G}_u = \frac{\overline{i'^2}}{4kT_0 B} - \hat{R}_n|\hat{Y}_c|^2$$

$$= \frac{\overline{(n_{21} e + n_{22} i)^2}}{4kT_0 B} - \hat{R}_n|\hat{Y}_c|^2$$

$$= \frac{\overline{[(n_{21} + n_{22} Y_c)e + n_{22} i_u]^2}}{4kT_0 B} - \hat{R}_n|\hat{Y}_c|^2$$

$$= |n_{21} + n_{22} Y_c|^2 R_n + |n_{22}|^2 G_u - \hat{R}_n|\hat{Y}_c|^2 \tag{A8.48}$$

Substituting (A8.46) and (A8.47) into (A8.48) yields

$$\hat{G}_u = \frac{G_u R_n|n_{11} n_{22} - n_{12} n_{21}|^2}{R_n|n_{11} + n_{12} Y_c|^2 + G_u|n_{12}|^2} \tag{A8.49}$$

The corresponding optimum noise source admittance $\hat{Y}_m = \hat{G}_m + j\hat{B}_m$ and optimum noise figure $\hat{F}_m$ of the interconnected network $\hat{N}$ can be computed from $\hat{R}_n$, $\hat{Y}_c$, and $\hat{G}_u$ by (A8.20)–(A8.22). Equation (A8.22) can also be rewritten as

$$F_m = 1 + 2R_n G_c + 2\sqrt{R_n^2 G_c^2 + R_n G_u} \tag{A8.50}$$

and hence

$$\hat{F}_m = 1 + 2\hat{R}_n \hat{G}_c + 2\sqrt{\hat{R}_n^2 \hat{G}_c^2 + \hat{R}_n \hat{G}_u} \tag{A8.51}$$

We observe that if the products R_nG_c and R_nG_u remain constant, that is,

$$R_nG_c = \hat{R}_n\hat{G}_c \tag{A8.52a}$$

$$R_nG_u = \hat{R}_n\hat{G}_u \tag{A8.52b}$$

then $F_m = \hat{F}_m$, that is, the minimum noise figure of the interconnected network $\hat{N}$ remains constant. Using (A8.47) and (A8.49), conditions (A8.52) can be expressed equivalently as

$$|n_{11}n_{22} - n_{12}n_{21}| = 1 \tag{A8.53}$$

Equations (A8.46), (A8.47), (A8.48), and (A8.51) form the cornerstone for the design of low noise feedback amplifiers, and can be easily programmed on digital computers.

Index